Ancestral Sequence Reconstruction

Ancestral Sequence Reconstruction

EDITED BY

David A. Liberles
University of Wyoming, Laramie, WY, USA

OXFORD
UNIVERSITY PRESS

Great Clarendon Street, Oxford OX2 6DP

Oxford University Press is a department of the University of Oxford.
It furthers the University's objective of excellence in research, scholarship,
and education by publishing worldwide in

Oxford New York

Auckland Cape Town Dar es Salaam Hong Kong Karachi
Kuala Lumpur Madrid Melbourne Mexico City Nairobi
New Delhi Shanghai Taipei Toronto

With offices in

Argentina Austria Brazil Chile Czech Republic France Greece
Guatemala Hungary Italy Japan Poland Portugal Singapore
South Korea Switzerland Thailand Turkey Ukraine Vietnam

Oxford is a registered trade mark of Oxford University Press
in the UK and in certain other countries

Published in the United States
by Oxford University Press Inc., New York

© Oxford University Press 2007

The moral rights of the author have been asserted
Database right Oxford University Press (maker)

First published 2007

All rights reserved. No part of this publication may be reproduced,
stored in a retrieval system, or transmitted, in any form or by any means,
without the prior permission in writing of Oxford University Press,
or as expressly permitted by law, or under terms agreed with the appropriate
reprographics rights organization. Enquiries concerning reproduction
outside the scope of the above should be sent to the Rights Department,
Oxford University Press, at the address above

You must not circulate this book in any other binding or cover
and you must impose the same condition on any acquirer

British Library Cataloguing in Publication Data
Data available

Library of Congress Cataloging in Publication Data
Data available

Typeset by Newgen Imaging Systems (P) Ltd.,
Printed in Great Britain
on acid-free paper by
Antony Rowe, Chippenham, Wiltshire

ISBN 978 0 19 929918 8

10 9 8 7 6 5 4 3 2 1

Contents

Foreword and introduction	vii
Introduction to the meeting in Kristineberg, Sweden	x
Contributors	xii

I Introductory scientific overview

1 The early days of paleogenomics: connecting molecules to the planet   3
Steven A. Benner

2 Ancestral sequence reconstruction as a tool to understand natural history and guide synthetic biology: realizing and extending the vision of Zuckerkandl and Pauling   20
Eric A. Gaucher

3 Linking sequence to function in drug design with ancestral sequence reconstruction   34
Janos T. Kodra, Marie Skovgaard, Dennis Madsen, and David A. Liberles

II Computational methodology and concerns

4 Probabilistic models and their impact on the accuracy of reconstructed ancestral protein sequences   43
Tal Pupko, Adi Doron-Faigenboim, David A. Liberles, and Gina M. Cannarozzi

5 Probabilistic ancestral sequences based on the Markovian model of evolution: algorithms and applications   58
Gina M. Cannarozzi, Adrian Schneider, and Gaston H. Gonnet

6 Estimating the history of mutations on a phylogeny   69
Jonathan P. Bollback, Paul P. Gardner, and Rasmus Nielsen

7 Coarse projections of the protein-mutational fitness landscape   80
F. Nicholas Braun

8 Dealing with uncertainty in ancestral sequence reconstruction: sampling from the posterior distribution   85
David D. Pollock and Belinda S.W. Chang

9 Evolutionary properties of sequences and ancestral state reconstruction   95
Lesley J. Collins and Peter J. Lockhart

10 Reconstructing the ancestral eukaryote: lessons from the past 103
Mary J. O'Connell and James O. McInerney

III Computational applications of ancestral sequence reconstruction

11 Using ancestral sequence inference to determine the trend of functional divergence after gene duplication 117
Xun Gu, Ying Zheng, Yong Huang, and Dongping Xu

12 Reconstruction of ancestral proteomes 128
Toni Gabaldón and Martijn A. Huynen

13 Computational reconstruction of ancestral genomic regions from evolutionarily conserved gene clusters 139
Etienne G.J. Danchin, Eric A. Gaucher, and Pierre Pontarotti

IV Experimental methodology and concerns

14 Experimental resurrection of ancient biomolecules: gene synthesis, heterologous protein expression, and functional assays 153
Eric A. Gaucher

15 Dealing with model uncertainty in reconstructing ancestral proteins in the laboratory: examples from archosaur visual pigments and coral fluorescent proteins 164
Belinda S.W. Chang, Mikhail V. Matz, Steven F. Field, Johannes Müller, and Ilke van Hazel

V Experimental synthesis of ancestral proteins to test biological hypotheses

16 Using ancestral gene resurrection to unravel the evolution of protein function 183
Joseph W. Thornton and Jamie T. Bridgham

17 A thermophilic last universal ancestor inferred from its estimated amino acid composition 200
Dawn J. Brooks and Eric A. Gaucher

18 The resurrection of ribonucleases from mammals: from ecology to medicine 208
Slim O. Sassi and Steven A. Benner

19 Evolution of specificity and diversity 225
Denis C. Shields, Catriona R. Johnston, Iain M. Wallace, and Richard J. Edwards

Conclusions and a way forward 236
David A. Liberles

Index 239

Foreword and introduction

With the realization that the combination of computational reconstruction of ancestral protein sequences and the experimental synthesis of these proteins could be used to test specific molecular, biomedical, ecological, and evolutionary hypotheses, this methodological combination has been used with increasing popularity. Because a number of scientific issues surround the use of ancestral sequence reconstruction that need to be fleshed out, a scientific meeting was organized to discuss the use of ancestral sequence reconstruction. Beyond procedures and pitfalls, a number of new applications of ancestral sequence reconstruction have begun to emerge and a presentation of several of these was deemed valuable.

With funding from the European Science Foundation (ESF), Vetenskapsrådet (the Swedish Research Council), and the Linnaeus Centre for Bioinformatics (Uppsala University, Sweden), David Ardell (Uppsala University, Sweden), Giorgio Matassi (University of Paris VI, France), and I organized a meeting entitled, "Using Ancestral Sequence Reconstruction to Understand Protein Function" in Kristineberg, Sweden, on 30–31 March 2005. The meeting consisted of 38 participants from 12 different countries attending 18 scientific presentations. Following the meeting and the vibrant discussion, it was decided that a book involving chapters by those attending the meeting and others in the field would be worthwhile, which was the origin of this project.

One philosophical discussion that emerged was on the true meaning of homology and what a homologous site is when a sequence slides through a structure generating diverging alignments from sequence and structure-based methods. David Ardell (in his lecture in Sweden) presented examples of cases where the two diverged and recommended sequence analysis using a DNA-based view of homology, where a single position within a gene represented the homologous site, where substitution models should then be applied to characterize its evolution. This is the traditional view of homology as embodied in the vast literatures of molecular evolution and population genetics. However, Richard Goldstein (who has in the past generated substitution matrices that characterize substitution differentially between different structural elements) and David Pollock took a structural perspective on homology, arguing that a homologous position might sometimes be better defined by the structural attributes constraining it in a three-dimensional structure rather than the position within a gene sequence. For example, position 3 in an α-helix could be aligned with position 3 in the homologous α-helix of another protein, even if they represent positions 65 and 68 in the gene sequence and no insertion or deletion events have occurred. This latter view requires the use of different types of substitution models than the former view, so the divergence of opinion has practical as well as philosophical concerns.

Another active area of discussion involved sources of bias and an ongoing discussion of the validity of using maximum-likelihood or -parsimony ancestral sequence reconstructions compared with a sampling from the posterior distribution of a Bayesian ancestral sequence reconstruction. The discussion at the meeting (in addition to Chapter 8 in this volume) has spawned an active discussion in the peer-reviewed scientific literature. The argument is that the maximum-likelihood or maximum-parsimony ancestral sequence is under-represented by rare variants, such as hydrophobic residues on the surface, that ultimately attribute overly stable or overly active properties to the reconstructed ancestor. It is argued that this is avoided by sampling from the posterior distribution, even if sampling from the posterior results in less accurate reconstruction at the sequence level. The experimental implications of this proposal are presented in both Chapters 8 and 15. A brief rebuttal to this view and defense of maximum likelihood is presented by Eric Gaucher in Chapter 2. Further analysis and discussions of this topic are sure to appear in the literature over the coming years.

A third topic raised for discussion at the meeting by Giorgio Matassi was, "Are all proteins reconstructable?" Clearly some proteins, like the green fluorescent protein-like proteins worked on by Mikhail Matz and colleagues (and presented in Chapter 15) are more amenable to experimental study than other proteins. However, functional assays *in vitro* or *in vivo* are indeed available for a great many proteins. Chapter 17 presents a reconstruction back to the last universal ancestor and other chapters deal with various complexities in sequence evolution that will enable more accurate reconstruction.

The first two chapters provide a historical and scientific overview of ancestral sequence reconstruction, and Chapter 3 extends the use of the technique to applications of drug design and mentions the companion technique of substitutional mapping. A discussion of standard approaches for ancestral sequence reconstruction is presented in Chapters 4 and 5, with Chapter 6 presenting a method (with a companion software package) for substitutional mapping.

Chapters 7 and 8 present some of the limitations and considerations that should go into computationally reconstructing ancestors, including methodological sources of bias and biophysical implications. Chapter 9 presents a discussion of covarion or heterotacheous processes, where sites shift rates due to intra- or intermolecular coevolution, and their effects on ancestral sequence reconstruction. Chapter 10 analyzes some controversies in our knowledge of the reference species tree and how different topologies can affect reconstructed ancestral sequences. The covarion processes discussed in Chapter 9, while sometimes neutral, are also sometimes linked to functional shifts. Chapter 11 discusses methodology for linking this process to functional shifts after gene duplication using ancestral sequences. Chapters 12 and 13 present computational strategies and applications of using ancestral sequence data to reconstruct entire proteomes. In work not presented in this book, David Haussler and colleagues have extended this type of approach to reconstructing the entire genome of the last common ancestor of mammals. The thoughtful introduction by Emile Zuckerkandl proposes further extension of the analysis from entire proteomes to interactomes and the field, although not there yet, will surely move in this direction.

Moving to experimental work to test computational hypotheses, Chapters 14 and 15 present strategies for converting computationally reconstructed ancestral sequences to proteins resurrected in the laboratory. Chapter 15 includes an expanded discussion of how to accommodate the controversial computational strategy suggested in Chapter 8. Chapters 16–19 then address various biological questions using ancestral sequence reconstruction and resurrection, across different evolutionary depths and drawing on widely different scientific disciplines.

Rather than presenting a view that is consistent from chapter to chapter, several contradictory views are presented differently by different authors to give readers a chance to appreciate ongoing debates in the field and formulate their own opinions. In the concluding section, I provide a list of several available software packages that are available to perform different analyses described in the book. I also attempt to tie together some of the discussion to present the experimental molecular biologist with a potential way forward in attempting these methods in their own laboratory.

The image of crocodilians on the book cover was generated with the enthusiastic help of John Brueggen at the St. Augustine Alligator Farm Zoological Park (http://www.alligatorfarm.us). The picture shows all 23 extant species of crocodilans and as a Postdoctoral Researcher at University of Florida, I always enjoyed visiting the alligator farm and comparing the species in my mind. While I have never worked with crocodilians, one of the constant battles that my lab has faced is the search for DNA from different closley related species. Ultimately in this process, we are interested in addressing the question, "What were the molecular events that made each species unique from its closest relatives?" So, as you look at the crocodilians on the cover of the book, ask yourself how these species are different, what the molecular underpinnings of this are, what the selective forces that drove this were, and how the techniques described in this book can help us answer these questions.

I am grateful to the external reviewers of chapters for this book, notably Aoife McLysaght (Trinity College, Ireland), Arthur Lesk (Pennsylvania State University, USA), my research group (especially Alexander Churbanov and Steven Massey), and my wife Jessica, as well as to authors who took the time to review other chapters in this effort (a special thanks for extra effort go to Eric Gaucher, Tal Pupko, and Denis Shields). I also need to thank Ian Sherman and Stefanie Gehrig at Oxford University Press for their patience. Thank you for your interest in the growing research field.

David A. Liberles
University of Wyoming,
Laramie, WY, USA

Introduction to the Meeting in Kristineberg, Sweden

This introduction was written by Emile Zuckerkandl and read at the meeting by Giorgio Matassi.

Learning about this meeting was exciting news. It is a great honor for me to be permitted to address to you all a wholehearted welcome. I very much regret not to be able to do so in person and to miss out on two full days of attractively diverse and promising contributions to the meeting's theme: ancestral sequence reconstruction and its use in the study of the evolution of protein function. We must hope that the approach referred to by the meeting's title will emerge strengthened rather than weakened, because it appears to be of irreplaceable value to the study of the evolution of informational macromolecules. This study indeed is obliged to resort largely to deductive methods, since a direct determination of ancestral macromolecules is impossible in most cases. It thus becomes very important maximally to clarify the limitations of the methods of inference of ancestral sequences and their higher-order structures as well as to try to devise ways of overcoming the limitations. This is one of the aims of the present meeting. Likewise, it is very important to devise new approaches to help the reconstruction process.

Beyond the methodologies, however, lies the *most* exciting: the analysis of their results. In fact, the greatest interest of the reconstruction of ancestral informational macromolecules may well lie in the reconstruction of their interactions. Let me refer to informational macromolecules as semantides, as proposed 40 years ago, and thus use three syllables for the concept in lieu of ten. The ambition to reconstruct semantide interactions poses great additional challenges. It can be successful only with the help of differing but convergent approaches. One such approach will probably take a large set of deduced structures of individual ancestral semantides and arrange this set according to an ancestral network of evolutionarily conserved semantide interactions. Such networks of ancestral semantide interactions would be inferred from comparisons of whole genomes over various—at times very wide—spans of the phylogenetic tree, taking into account the interactions as observable in contemporary organisms. New technologies will hopefully be developed to help carry out this gigantic task, and the fruits of the labor should justify the effort. By determining the macromolecular interactions that were conserved at different evolutionary times, one might achieve two remarkable feats. The first would be to establish a sort of skeleton of molecular evolution, represented by the conserved interactions among semantides along various evolutionary lines of descent. One would trace along distinct phylogenetic branches the different degrees of persistence of semantide interactions, taking into account variations in the mutual fit of many semantides that despite structural alterations continue to interact functionally. Such differences in interactions are often attributable to mutational damage followed by functional restoration through a modified fit. The second feat would consist of attempts to discover, on the flip side of conservation, the molecular pathways taken at various times by evolutionary novelties. Analyzing the molecular pathways of past evolutionary novelties in the making may well be the greatest challenge of all, but again contemporary organisms are likely, here, to extend a helping hand.

By applying the various methodologies to be further developed—and, in part, yet to be invented—the most fascinating general discovery to be made, it seems to me, is that of the evolution of gene regulation; namely, in particular, of transcriptional regulation and its evolutionary history. Knowing about the precise pathways of the evolution of the various modes of transcriptional regulation would greatly assist another undertaking that may be considered one of the broadest and deepest aims of biology: understanding more fully the molecular nature and molecular evolution of development. The present meeting may well be considered, albeit indirectly, as a stepping stone in this direction, too.

One may think that all life in the universe has to be built on linear heteropolymers capable of forming higher-order structures and that these structures, possibly with the help of other structures, have to be in turn capable of copying the linear heteropolymers that they contain. Perhaps heteropolymer complementarity—with a sequence strand generating its complementary strand—is likewise an absolutely general aspect of living systems. In light of such a generalization, at any rate, the methods discussed or introduced at the present meeting are likely to be applicable to all forms of life in the universe.

I am glad, right now, that I am not standing on this podium so that I shall not myself receive the tomatoes that may very well be thrown at the speaker at this point. Perhaps, however, there is in this hall at least one scheduled speaker who might agree with the statement just made, since he has put astrobiology in his title. There is little doubt that the potential extensions from the topics of this meeting are vast, whether their vastness approaches infinity or not. Welcome then, once again, to a gathering that I am sure is going to be most substantive, in good measure by its intense and patient scrutiny of detailed mechanics, short of which there can never be a successful completion of any interstellar voyage.

Emile Zuckerkandl
Stanford University and Institute of
Molecular Medical Sciences,
Palo Alto, CA, USA

Contributors

Steven A. Benner, Foundation for Applied Molecular Evolution, 1115 NW 4th Street, Gainesville, FL 32601, USA

Jonathan P. Bollback, Center for Bioinformatics and Institute of Biology, University of Copenhagen, Universitetsparken 15, 2100 Copenhagen Ø, Denmark

F. Nicholas Braun, Institute of Medical Biology, University of Tromsö, N-9037 Tromsö, Norway

Jamie T. Bridgham, Center for Ecology and Evolutionary Biology, University of Oregon, Eugene, OR 97403, USA

Dawn J. Brooks, Foundation for Applied Molecular Evolution, 1115 NW 4th Street, Gainesville, FL 32601, USA

Gina M. Cannarozzi, Institute of Computational Science, ETH Zurich, 8092 Zürich, Switzerland

Belinda S.W. Chang, Departments of Ecology and Evolutionary Biology, and Cell and Systems Biology, University of Toronto, 25 Harbord Street, Toronto, ON M5S 3G5, Canada

Lesley J. Collins, Allan Wilson Centre for Molecular Ecology and Evolution, Massey University, Palmerston North, New Zealand

Etienne G.J. Danchin, Glycogenomics and Biomedical Strucural Biology, AFMB, UMR 6098 CNRS/Université de Provence/Université de la Méditerrannée, Marseilles, France

Adi Doron-Faigenboim, Department of Cell Research and Immunology, George S. Wise Faculty of Life Sciences, Tel Aviv University, Ramat Aviv 69978, Israel

Richard J. Edwards, Bioinformatics, Conway Institute for Biomolecular and Biomedical Research, University College Dublin, Dublin 4, Ireland

Steven F. Field, Whitney Laboratory for Marine Bioscience, University of Florida, 9505 Ocean Shore Blvd, Saint Augustine, FL 32080, USA

Toni Gabaldon, Bioinformatics Department, Centro de Investigación Príncipe Felipe, Autopista del Saler 16, 46013 Valencia, Spain

Paul P. Gardner, Center for Bioinformatics and Institute of Biology, University of Copenhagen, Universitetsparken 15, 2100 Copenhagen Ø, Denmark

Eric A. Gaucher, Foundation for Applied Molecular Evolution, 1115 NW 4th Street, Gainesville, FL 32601, USA

Gaston H. Gonnet, Institute of Computational Science, ETH Zurich, 8092 Zürich, Switzerland

Xun Gu, Department of Genetics, Development and Cell Biology and Center for Bioinformatics and Biological Statistics, Iowa State University, Ames, IA, USA

Yong Huang, Department of Genetics, Development and Cell Biology and Center for Bioinformatics and Biological Statistics, Iowa State University, Ames, IA, USA

Martijn A. Huynen, Center for Molecular and Biomolecular Informatics and Nijmegen Center for Molecular Life Sciences, University Medical Center St. Radboud, Toernoooiveld 1, 6525 ED Nijmegen, The Netherlands

Catriona R. Johnston, Bioinformatics, Conway Institute for Biomolecular and Biomedical Research, University College Dublin, Dublin 4, Ireland

Janos T. Kodra, Novo Nordisk A/S, Novo Alle, 2760 Måløv, Denmark

David A. Liberles, Department of Molecular Biology, University of Wyoming, Laramie, WY 82071, USA

Peter J. Lockhart, Allan Wilson Centre for Molecular Ecology and Evolution, Massey University, Palmerston North, New Zealand

Dennis Madsen, Novo Nordisk A/S, Novo Alle, 2760 Måløv, Denmark

Mikhail V. Matz, Whitney Section of Integrative Biology, University of Texas at Austin, 1 University Station C0930, Austin, TX 78712, USA

James O. McInerney, Department of Biology, Callan Building, National University of Ireland Maynooth, Maynooth, Co. Kildare, Ireland

Johannes Müller, Humboldt-Universität zu Berlin, Museum für Naturkunde, D-10099 Berlin, Germany

Rasmus Nielsen, Center for Bioinformatics and Institute of Biology, University of Copenhagen, Universitetsparken 15, 2100 Copenhagen Ø, Denmark

Mary J. O'Connell, School of Biotechnology, Dublin City University, Glasnevin, Dublin 9, Ireland

David D. Pollock, Department of Biochemistry and Molecular Genetics, University of Colorado Health Sciences Center, Aurora, CO 80045, USA

Pierre Pontarotti, Phylogenomics Laboratory, EA 3781 Evolution Biologique, Université de Provence, Marseilles, France.

Tal Pupko, Department of Cell Research and Immunology, George S. Wise Faculty of Life Sciences, Tel Aviv University, Ramat Aviv 69978, Israel

Slim O. Sassi, Foundation for Applied Molecular Evolution, 1115 NW 4th Street, Gainesville, FL 32601, USA

Adrian Schneider, Institute of Computational Science, ETH Zurich, 8092 Zürich, Switzerland

Denis C. Shields, Bioinformatics, Conway Institute for Biomolecular and Biomedical Research, University College Dublin, Dublin 4, Ireland

Marie Skovgaard, Novo Nordisk A/S, Novo Alle, 2760 Måløv, Denmark

Joseph W. Thornton, Center for Ecology and Evolutionary Biology, University of Oregon, Eugene, OR 97403, USA

Ilke van Hazel, Department of Ecology and Evolutionary Biology, University of Toronto, 25 Harbord Street, Toronto, ON M5S 3G5, Canada

Iain M. Wallace, Bioinformatics, Conway Institute for Biomolecular and Biomedical Research, University College Dublin, Dublin 4, Ireland

Dongping Xu, Department of Genetics, Development and Cell Biology and Center for Bioinformatics and Biological Statistics, Iowa State University, Ames, IA, USA

Ying Zheng, Department of Genetics, Development and Cell Biology and Center for Bioinformatics and Biological Statistics, Iowa State University, Ames, IA, USA

Emile Zuckerkandl, Department of Biological Sciences, Stanford University and Institute of Molecular Medical Sciences, Palo Alto, CA, USA

I
Introductory scientific overview

CHAPTER 1

The early days of paleogenetics: connecting molecules to the planet

Steven A. Benner

1.1 Introduction

Anyone asked to write about the early days feels elderly. Fortunately, Emile Zuckerkandl's introduction shows that the ideas that led to this volume have been around for some time, at least in their basic form, and are rooted in ideas of many heroes of modern molecular biology, including Pauling, Anfinsen, and Zuckerkandl himself.

In 1980, my laboratory was unaware of the Pauling–Zuckerkandl paper (Pauling and Zuckerkandl, 1963; see Chapter 2 in this volume for a fuller discussion of the implications of this paper) when we set out to resurrect ancient proteins from extinct organisms. My group, then consisting of only Krishnan Nambiar and Joseph Stackhouse, was trained to describe the chemical structures and behaviors of enzymes. In those days technology was allowing molecular scientists to extend these descriptions to atomic resolution, the picosecond time scale, and the microscopic rate constant.

But what good were clever experiments to determine, for example, which of two hydrogens was removed by a dehydrogenase (Allemann *et al.*, 1988), or whether the replacement of carbon dioxide by a proton on acetoacetate proceeded with retention or inversion of stereochemical configuration (Benner *et al.*, 1981)? It occurred to us that we might be doing the biochemical equivalent of studying a Picasso with an electron microscope. Were we not describing biomolecular systems to resolutions far greater than they were designed? Biomolecules are not designed, however. They are the products of natural selection imposed upon random variation in their chemical structures. As the result of a combination of historical accident, selective pressures, and vestigiality, all constrained by physical and chemical law, different behaviors must be interesting at different levels. Biomolecular behaviors that influenced the ability of a host organism to survive, mate, and reproduce were especially interesting, as these had been fashioned by natural selection. Behaviors that did not, were not, because they had not. As a criterion for selecting interesting chemical features of a biomolecule to study in detail, an understanding of the relation between biomolecular structure and behavior and fitness was important.

It did not take long at Harvard to realize that this relation was going to be difficult to understand. There, Martin Kreitman, Robert Dorit, and others, including some very dialectical biologists (Levins and Lewontin, 1985), were struggling to make this connection starting from the side of biology (Kreitman and Akashi, 1995). Despite this interest, it was proving difficult to connect *any* biomolecular structure or behavior with the survival of an organism, at least in a way that would be compelling to those who chose to deny it (Lewontin, 1974; Clarke, 1975; Gillespie, 1984, 1991; Somero, 1995; Powers and Schulte, 1998). In fact, the discussion was central to the most hotly disputed dispute in molecular evolution, between neutralists and selectionists, where both sides of the dialectic were populated by individuals who were professionally intent on showing how any data interpretable in favor of one side could equally well support the other.

As chemists, we had no part in this fight. However, a review of the contending sides of these disputes (Benner and Ellington, 1988) reminded us of analogous disputes in organic chemistry.

These were often Seinfeld arguments about nothing. For example, chemists had for years discussed the non-classical carbocation problem (Brown, 1977). This was a disagreement about whether the structures of positively charged organic molecules, in general, were better modeled by a formula with dotted lines, or by two formulas without dotted lines. Rational observers realized that one model was undoubtedly better for some molecules, whereas the other was better for others. After all, similar issues had been addressed and resolved in many molecular systems. For example, the structures of benzene and many boron-containing compounds both contained dotted lines. Which model was best undoubtedly depended on the exact structure of the molecule being discussed. By 1980, this dispute had forced chemists to appreciate a certain truism about molecules: organic molecules are never productively discussed in terms of a general molecular structure; they must always be considered individually. This truism, of course, recognized that the discussion of models for the structure of *individual* molecules could nevertheless be interesting.

To chemists, the neutralist/selectionist dispute was directly analogous. This was essentially a disagreement about whether changes in the chemical structure of the generic protein would, in general, change its behavior enough to change its contribution to the fitness of the generic organism. Again, the rational answer was in some cases yes, and in other cases no, depending on the exact structure of the system. Proteins are, after all, organic molecules, suggesting that they must be considered individually. As expected by those who understood this truism, the neutralist/selectionist dispute, in its general form, melted away as soon as our ability to analyze the behavior of individual proteins improved (Hey, 1999).

Even in 1980, however, it was clear that connecting fitness to the behavior of *individual* biomolecules would always remain interesting, for many reasons. First, that understanding would certainly help us select behaviors of those biomolecules to study in detail. If a behavior was important to fitness, it might be highly optimized. Detailed study might therefore instruct us about the interaction between chemical structure and biomolecular behavior, instruction worthy of the growing armamentarium of biophysics and molecular biology.

1.2 History as an essential tool to understand chemistry

It was clear, however, that Structure Theory in chemistry would not support a deep understanding of biological molecules. With simple molecules, like methane, one does not ask about its purpose. This is not true about complex systems, or living systems, where it is appropriate to ask: *why* does it exist? History can be key to any answer to why? questions. Any system, natural or human-made, can be understood better if we understand *both* its structure *and* its history. We would not understand the QWERTY computer keyboard, the Microsoft Windows operating system, or the US Federal Reserve Bank (for example) if we simply deconstructed each into its parts. An understanding of the history of each is essential to an understanding of the systems themselves.

Structure Theory from chemistry had absolutely no historical component. Methane is how it is because of its structure. It always has been this way, and always will be. Where the methane came from and how it got to us was fully irrelevant to our understanding of this molecule. This raised the next in a series of questions leading to experimental paleogenetics: how were Structure Theory and natural history to be combined to better understand biomolecules? Fragments of the history of life on Earth are found in the geological strata, of course. But the fossil record is notoriously incomplete, and would not provide information about proteins even were it not. Molecular fossils (such as those found in petroleum) can be informative, but generally not about individual protein function. Further, any analysis of molecular function must recognize that the behaviors that confer fitness are determined by the system, including other organisms (ecology), the physical environment (planetary biology), and even the cosmos (astrobiology). This level of complexity defeats most theoretical contexts.

It was clear, however, that the chemical structures of proteins themselves contain historical information. The historical relationships between proteins related by common ancestry can be inferred by comparing their amino acid sequences, a theme that was already well developed by 1980 (Dayhoff et al., 1978). Analysis of protein sequences could generate the basic elements of an evolutionary model: a multiple sequence alignment, a tree, and sequences of ancestral protein sequences inferred from these. From these, it might be possible to construct narratives connecting biomolecular structure to fitness.

This process was analogous to processes well known in the field of historical linguistics (Lehman, 1973), which Robert Breedlove had described to me when I was an undergraduate. This field infers the features in ancestral languages by analyzing the features of their descendent languages. For example, the Proto-Indoeuropean word for snow (*$sneig^{w}h$-) can be reconstructed from the descendant words for snow in the descendant Indoeuropean languages (German *schnee*, French *neige*, Irish *sneachta*, Russian *sneg*, Sanskrit *snihyati*, and so on). Other features of the histories of these languages, such as the universal replacement of *sn*- by *n*- in the Romance languages, can also be inferred from this analysis. The analogous inferences about ancestral structures could also be done for proteins.

The reconstruction of ancestral languages provides paleoanthropological information as well. From the ancestral features of reconstructed ancestral languages, one can extract information about the people who spoke them. For example, the ease with which we reconstructed the Proto-Indoeuropean word for snow (with some concessions; the Sanskrit word cited above actually means "he gets wet") tells the story that the Proto-Indoeuropeans themselves lived in a locale where it snowed. In 1980, we hoped to tell analogous stories using proteins inferred to have been present in ancestral forms of life on Earth.

1.3 Swapping places: biologists become chemists and statisticians, just as chemists become natural historians

But would these be only just-so stories? The just-so story is one of the worst insults that a biologist can direct at another. This epithet accuses a professional adversary of building *ad hoc* explanations for specific facts (how the zebra got his stripes). The events behind a just-so story (an ancestral zebra took a nap under a ladder) cannot be independently verified, and are not mathematically modelable. Further, the story could easily be replaced by a different story, just as compelling, had the observations been the opposite. It was clear in 1980 that once the insult stuck, papers would be rejected, grant applications would be turned down, and tenure would be denied.

Curiously, this issue also had a parallel in organic chemistry. Chemists are well known for their ability to use Structure Theory to explain a set of facts, only to be told that the facts are opposite, and then to explain the counter-facts using the same theory. Chemists are rarely defensive about this. In part, this is because Structure Theory as a heuristic has been so successful. If one can make petrochemicals and pharmaceuticals (and much in between) using a theory based on plastic tinker toy models, who can argue?

The success of non-mathematical Structure Theory from chemistry makes a larger point about human knowledge; that it is intrinsically heuristic and intuitive. This is true even for knowledge that is cast in the language of mathematics. This conclusion had been reached by the last century of epistemology as well (Suppe, 1977). Nothing is "proven" (Galison, 1987); the perception of proof is only a function of the number of logical steps that must be taken to premises that are intuitive and heuristic. Experiments end when a burden of proof is met, where that burden is defined by the culture, not by logic.

This point is not fully appreciated by many modern biologists. Many modern biologists seek to avoid the just-so story epithet, and the perception of theirs being a heuristic and/or intuitive theory, by placing a mathematical formalism on top of their models. This drives them towards statistics, which analyzes collections of things. Statistics, in turn, nearly always requires the statistician to deny the truism in chemistry that there is no such thing as general molecular behavior. This, in turn, means that statisticians, in their pursuit of general models framed in mathematical language, are not able to exploit the only research paradigm that has shown itself to be successful in understanding molecules.

In fact, the barrier between mathematics and molecular science is still higher. Statisticians are taught that a model is not scientific if it is *not* formulated in the language of mathematics. Therefore, statisticians are perplexed that a field like chemistry can be successful. And even as they acknowledge that proteins are chemicals, statisticians insist that unless protein sequences are studied as collections, the studies are "unscientific" (Robson and Garnier, 1993). Thus, statisticians actively work to deny to all other scientists the one research paradigm that has been successful to understand molecules.

This cultural phenomenology has set up a role reversal of a sort. With their training in heuristic science, chemists may have been better prepared to make the connection between chemical behavior and biological fitness than biologists. As physical scientists, chemists were trained in mathematics and statistics. Because they understood heuristic models, however, they used statistics and mathematics as tools, not as the way to respond to the complaint, "You are not doing real science".

Further, chemical theory grows by accretion, rather than revolution; it adds theories, ideas, and perspectives to its heuristic theory. This is exactly what is needed to understand the broader picture in contemporary biology. Here, by the end of the current century, we expect to see a global view of reality that combines chemical models, systems models, physiological models, paleontological models, and geological models. If the output is still dissatisfying, then the global view will add still more models. We expect (or, perhaps better, hope) that over time, an increasingly dense set of models, all interconnecting, would eventually converge upon a global picture for biology, just as it has for chemistry over the past century.

1.4 Managing heuristic science

This discussion should not be viewed as a defense of so-called soft science. Rather, it is simply an observation of how human knowledge really works. The observation need not be viewed pejoratively. Human scientists can be creative and productive *because* human understanding is intuitive and heuristic. Thus, while the scientific method taught in middle school emphasizes the importance of making unfiltered observations, analyzing data without prejudice, and doing value-neutral experiments, the productivity of scientists does not depend on the extent to which they meet this largely fictitious ideal, but rather how they manage the closedness of mind, the values, and the filters that come naturally with human cognition.

This concept of management is important. Chemistry does not ignore the natural tendency of humans to convince themselves that data contain patterns that they do not, or that patterns compel models when they need not, or that models are reality, which they are not. Rather, chemistry establishes processes that manage this tendency.

Key to this management is the use of experiment on systems that have been synthesized (Benner and Sismour, 2005). The use of synthesis to create new forms of matter, whose behavior is expected from a heuristic theory, provides an opportunity for an independent test of the heuristic. *De novo* synthesis in not available in many other disciplines. For example, planetary scientists cannot synthesize a new planet to test their theories on how planets work. If they could, the field would be dramatically transformed.

How could we use synthesis and experiment to manage the development of our historical view of biomolecules? In 1980, the answer was materializing before our eyes. Jeffrey Miller, Michael Brown, Alan Fersht, and others were developing the technology to create a protein having any sequence that might be desired. Most protein engineering was targeted to replace single amino acids in extant proteins for the purpose of understanding their role in a protein's catalytic behavior. But it was clear that protein engineering technology could also be used to synthesize ancestral proteins whose sequences had been inferred using ideas outlined by Pauling and Zuckerkandl, where the resurrected proteins could then be experimentally studied in the laboratory.

This is how the idea for experimental paleogenetics (which we originally called paleobiochemistry) began in our laboratories in 1980. We wanted to bring ancient proteins back to life to examine their behaviors. This would use synthesis to add an experimental component to our understanding of the history of biomolecules. The historical component was necessary to understand biomolecules, just as it was to understand the US Federal Reserve banking system. This experimental component would also manage the problems intrinsic to a heuristic science. Through this combination, we hoped to understand how an interaction between chance, history, vestigiality, selection, and physical law determined the structures and behaviors of individual protein families. From there, we could perhaps make inferences about how these were related to fitness and physiological function. Then, perhaps, we could select interesting biomolecular behaviors to study.

1.5 Selecting proteins to begin experimental paleoscience

But what individual protein should we look at? While the Maxam–Gilbert and Sanger papers on DNA sequencing made clear that the sequencing of the human genome was only a matter of time, databases in 1980 contained very few protein sequences. The only families of proteins that were sufficiently well represented to support experimental paleogenetics were the cytochromes, the hemoglobins, and the ribonucleases (RNases). Cytochromes were, of course, substrates for cytochrome oxidases. With no funding for this project (the National Institutes of Health routinely disapproved our proposals in this area) we could not possibly resurrect ancestral cytochromes, only to then need to resurrect their ancient oxidases. Hemoglobins were complicated to express, a problem solved only later. This left the RNases.

Fortunately, Jaap Beintema and his colleagues in Groningen had done the yeoman's job of sequencing RNases (at the level of the protein) from a wide range of ruminants and closely related non-ruminant mammals (Cho *et al.*, 2005). They had, Dayhoff-style (Dayhoff *et al.*, 1978), inferred the sequences of the ancestral proteins throughout the recent history of the digestive enzymes. Barnard had raised an interesting hypothesis suggesting that digestive RNases might be unique to ruminants, and be an adaptation to their unique ruminant digestive physiology (Barnard, 1969). And so, we had a place to start.

The story of RNase resurrections is told in a separate chapter in this volume (see Chapter 18). This story illustrates well the value of paleomolecular resurrections for creating an understanding of the relation between organismic and molecular biology on one hand, and the changing ecosystems wrought by a changing planet on the other. It also, in the process, showed how we might use paleogenetics to select *in vitro* biomolecular behaviors to study in a way that considers physiological relevance (Nambiar *et al.*, 1984, 1987; McGeehan and Benner, 1989; Benner and Allemann, 1989; Stackhouse *et al.*, 1990; Allemann *et al.*, 1991; Jermann *et al.*, 1995).

But our work with RNases, and work in other laboratories in other systems, also showed that experimental paleogenetics could create contentious disputes of its own. Many of these relate to the reliability of statistically grounded tools to infer the structures of ancestral proteins, and how the outcome of paleogenetics experiments should be interpreted. These issues will be addressed in this chapter, and as they are in other chapters in this volume. I will describe the use of paleogenetic experiments to manage them in one system, the alcohol dehydrogenases.

1.6 Mathematical models are nevertheless important

Mathematical formalism is useful in the inference of ancestral sequences from the sequences of their descendants. Protein sequences lend themselves to representations as linear strings of letters. As organic molecules, such linear representations do not capture much of their organic chemical behavior, of course. Nor do such linear representations capture the behavior of protein sequences during divergent evolution. Homoplasy, correlated change, and a host of other features reveal protein sequences for what they really are: poor models of the structures of real organic molecules.

Nevertheless, mathematical formalisms that treat proteins as linear strings of letters turn out to be useful (Benner *et al.*, 1997). Any model that treats protein sequences as linear strings diverging via a Markovian process provides a null hypothesis, a description of protein evolution that *would have happened* if proteins *were* formless, functionless linear strings. By observing how proteins divergently evolve, and comparing this reality to the null hypothesis, one extracts a signal about form and function (Benner *et al.*, 1997).

The null hypothesis provides a serviceable starting point for ancestral reconstruction as well. The underlying Darwinian process is, of course, semi-random. Its departures from randomness, arising from biases in the DNA-polymerization or error-repair mechanisms, nucleosome structure, or other features of the DNA molecule itself, are not likely to be strongly correlated with protein structure and behavior (again analogous to language; the conversion of *sn-* to *n-* is largely unrelated to the dictional meaning of the word snow). Hence, it is not surprising that respectable inferences of ancestral states can be made using the linear string model.

There is nevertheless an ongoing dispute over which methods are precisely best for inferring ancestral sequences (Yang *et al.*, 1995; Zhang and Nei, 1997; Pagel, 1999; Nielsen, 2002; this volume, see Chapters 4 and 8 for example). From a practical perspective, these disputes do not have a large impact on the practice of experimental paleoscience. In practice, the principal ambiguities do not generally arise from subtleties in models for inferring ancestral sequences. Rather, they arise from incomplete sequence data-sets, uncertain gap placement in multiple sequence alignments, uncertain tree topology, and too much sequence divergence relative to tree articulation. This creates uncertainties in inferred ancestral character states long before the choice of the model becomes determinative.

Thus, if an evolutionary tree is highly articulated, the branching topology is secure, and the overall extent of sequence divergence is small, then different mathematical models infer more or less the same ancestral sequences. When the tree is not highly articulated, the branching topology is not secure, and the overall extent of sequence divergence is large, even the most mathematically sophisticated analysis cannot help much.

Today, a practicing paleogeneticist is advised to apply mathematical models at all levels of sophistication to build many different candidate multiple sequence alignments, candidate evolutionary trees, and candidate ancestral sequences. Additional information, such as crystallographic and paleontological data, should be both used and not used. From this will come a view of the ambiguity in the ancestral sequences that arises from the ambiguity and bias in the input.

Four strategies can then be considered to manage this ambiguity. The first relies on improving the statistical models of sequence divergence, in the hope that an increase in the sophistication of the mathematical formalism will resolve ambiguity. The second involves collecting more sequences in the hope of eliminating the ambiguity. The third ignores the ambiguity, in the hope that the ambiguity occurs only at sites that are not critical for the biological interpretation.

The fourth involves synthesizing and studying many candidate ancestral sequences to cover all plausible alternative reconstructions, or to sample among the plausible alternative reconstructions. The experimentalist then asks whether the behavior that supports a biological interpretation (and therefore the interpretation itself) is robust with respect to the ambiguity arising from uncertainties in the models, insufficient data, poorly articulated trees, or other issues in practice. This is our preferred method. The preference reflects a belief (perhaps better described as a faith) about a feature of protein chemistry that is presently unknown, but is not unknowable in principle. If the hypersurfaces relating protein behavior to protein sequence were extremely rugged, and if every amino acid replacement caused a significant change in behavior, then ambiguity would defeat the paleogenetic research approach in all but the most ideal cases. Fortunately, biochemical reality appears to be different. For nearly all proteins, some amino acid replacements at some sites have large impacts on functional behaviors. Replacements at other sites have only a modest impact on those behaviors, and replacements at still other sites have even less impact on most behaviors.

These facts would tend to ameliorate the extent to which ambiguity compromises the interpretation of data extracted from paleoscience experiments. Ambiguities in inferred ancestral characters generally are found at sites that have suffered many amino acid replacements. Multiple replacements often (but not always) reflect the possibility of neutral drift at a site. Neutral drift implies that the choice of a residue at the site does not have a significant impact on fitness. This generally (but not always) means that replacement of an amino acid at that site does not have any impact on the behavior of a protein.

Stringing these together, we might expect that biologically interpretable behavior will not differ greatly between ancestral sequences that differ only at ambiguous sites. To the extent that the premises are true, ambiguity in general will not limit our ability to draw inferences about the behavior of ancestral proteins by experimental analysis of ancestral sequences, even if our analysis does not capture all of the ambiguity in those sequences. This, in turn, means that we will generally be able to use those behaviors to generate interesting biological interpretations. In fact, this is the case with the dozen or so examples of experimental paleogenetics where the issue has been examined over the past two decades.

1.7 Alcohol dehydrogenase

The ultimate goal of molecular paleoscience is to connect the molecular records for all proteins from all organisms in the modern biosphere with the geological, paleontological, and cosmological records to create a broadly based, coherent narrative for life on Earth (Benner et al., 2002). Because much of natural selection is driven by species–species interactions, developing this narrative will require tools that broadly connect genomes from different species, as well as interconnect events within a single species. It remains an open question, of course, how much of the record has been lost through extinction, erosion, and poor fossil preservation.

The literature so far contains only the very first case studies where such a broad interconnection is conceivable. For example, modern yeast living in modern fleshy fruits rapidly convert sugars into bulk ethanol via pyruvate (Figure 1.1). Pyruvate

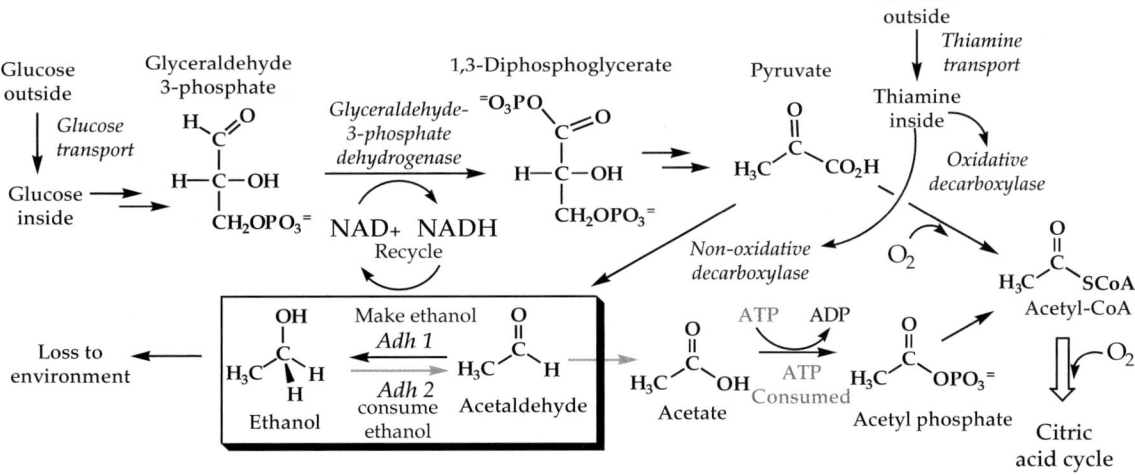

Figure 1.1 The formation of ethanol from glucose by the yeast *Saccharomyces cerevisiae* is an energetically expensive diversion of carbon in the overall degradation of glucose to give acetyl-CoA for the citric acid cycle. The yeast genome has two genes that catalyze the ethanol–acetaldehyde interconversion. One (Adh 1) is used to make ethanol; the other (Adh 2) is used to consume ethanol. Why does the yeast genome have these two in the genome, as either can catalyze this reaction in both directions? Enzymes in italics are associated with gene duplications that, according to the transition redundant exchange (TREx) clock (Benner et al., 2002), arose nearly contemporaneously. The make–accumulate–consume pathway is boxed. Note that the shunting of the carbon atoms from pyruvate into (and then out of, open arrows) ethanol is energy-expensive, consuming a molecule of ATP for every molecule of ethanol generated. This ATP is not consumed if pyruvate is oxidatively decarboxylated directly to give acetyl-CoA to enter the citric acid cycle directly (open arrow to the right). If dioxygen is available, the recycling of NADH does not need the acetaldehyde-to-ethanol reduction. Reprinted from Benner, S.A. and Sismour, A.M. (2005) Synthetic biology. *Nat. Rev. Genet.* **6**: 533–543.

then loses carbon dioxide to give acetaldehyde, which is reduced by alcohol dehydrogenase 1 (Adh 1) to give ethanol, which accumulates. Yeast later consumes the accumulated ethanol, exploiting Adh 2 and Adh 1 homologs differing by 24 (of 348) amino acids.

Generating ethanol from glucose in the presence of dioxygen, only to then re-oxidize the ethanol, is energetically expensive (Figure 1.1). For each molecule of ethanol converted to acetyl-CoA, a molecule of ATP is used. This ATP would not be wasted if the pyruvate that is made initially from glucose were delivered directly to the citric acid cycle.

This implies that yeast has a reason, transcending simple energetic efficiency, for rapidly converting available sugar in fruit to give bulk ethanol in the presence of dioxygen. One just-so story to explain this inefficiency holds that yeast, which is relatively resistant to ethanol toxicity, may accumulate ethanol to defend resources in the fruit from competing microorganisms (Boulton *et al.*, 1996). While the ecology of wine yeasts is certainly more complex than this simple hypothesis implies (Fleet and Heard, 1993), fleshy fruits do offer a large reservoir of carbohydrate. This resource must have value to competing organisms as well as to yeast. For example, humans have exploited the preservative value of ethanol since prehistory (McGovern, 2004).

The timing of Adh expression in *Saccharomyces cerevisiae* and the properties of the expressed proteins are both consistent with this story. The yeast genome encodes two major Adhs that interconvert ethanol and acetaldehyde (Figure 1.1; Wills, 1976). The first (Adh 1) is expressed at high levels constitutively. Its kinetic properties optimize it as a catalyst to make ethanol from acetaldehyde (Fersht, 1977; Ellington and Benner, 1987). In particular, the Michaelis constant (K_m) for ethanol in Adh 1 is high (17 000–20 000 µM), consistent with ethanol being a product of the reaction. After the sugar concentration drops, the second dehydrogenase (Adh 2) is derepressed. This paralog oxidizes ethanol to acetaldehyde with kinetic parameters suited for this role. The K_m for ethanol for Adh 2 is low (600–800 µM), consistent with ethanol at low concentrations becoming its substrate.

Adh 1 and Adh 2 are homologs differing by 24 of 348 amino acids. Their common ancestor, termed ADH$_A$, had an unknown role. If ADH$_A$ existed in a yeast that made, but did not accumulate, ethanol, its physiological role would presumably have been the same as the role of lactate dehydrogenase in mammals during anaerobic glycolysis: to recycle NADH generated by the oxidation of glyceraldehyde 3-phosphate (Figure 1.1; Stryer, 1995). Lactate in human muscle is removed by the bloodstream; ethanol would be lost by the yeast to the environment. If so, ADH$_A$ should have been optimized for ethanol synthesis, as is modern Adh 1. The kinetic behaviors of

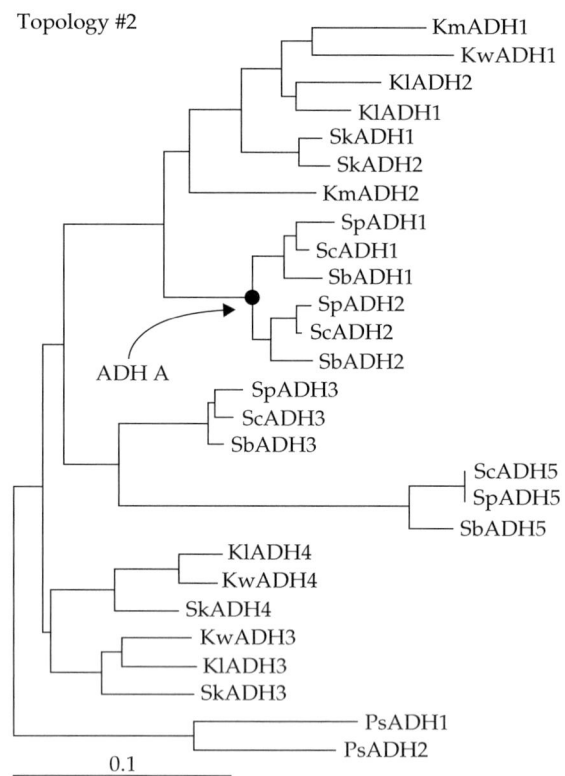

Figure 1.2 Maximum-likelihood trees interrelating sequences determined in this work with sequences in the publicly available database. Shown are the two trees with the best (and nearly equal) maximum-likelihood scores using the following parameters estimated from the data. Substitutions A–C, A–T, C–G, and G–T have a score of 1.00, A–G has a score of 2.92, and C–T has a score of 5.89; empirical base frequencies, and proportion of invariable sites and the shape parameter of the gamma distribution are set to 0.33 and 1.31, respectively. The scale bar represents the number of substitutions/codon per unit of evolutionary time. Reprinted from Benner, S.A. and Sismour, A.M. (2005) Synthetic biology. *Nat. Rev. Genet.* **6**: 533–543.

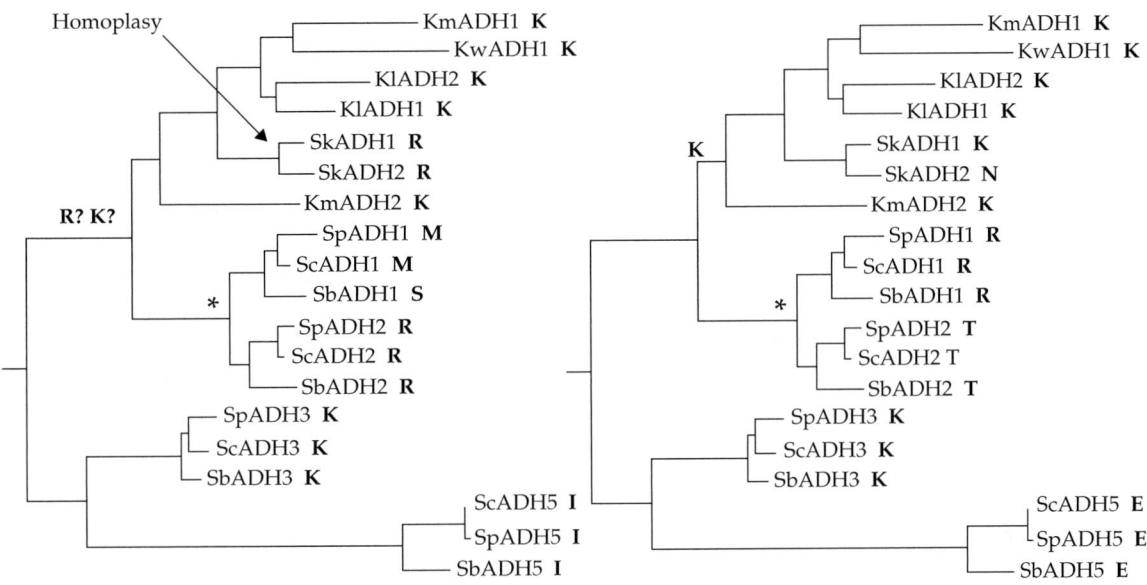

Figure 1.3 The distribution of amino acids at site 168 (left) and site 211 (right) in a set of 19 fungal alcohol dehydrogenases. The node of interest is at the right end of the branch marked by *. Note the difficulty in reconstructing the amino acids at these sites at the node at the right end of the red branch.

ADH$_A$ should resemble those of modern Adh 1 more than Adh 2, with a high K_m for ethanol.

To add paleobiochemical data to convert this just-so story into a more compelling scientific narrative, a collection of Adhs from yeasts related to *S. cerevisiae* was cloned, sequenced, and added to the existing sequences in the database (Thomson et al., 2005). A maximum-likelihood evolutionary tree was constructed using PAUP*4.0 (Figure 1.2; Swofford, 1998). Maximum-likelihood sequences for ADH$_A$ were then reconstructed using both codon and amino acid models in PAML (Yang, 1997). When the posterior probability that a particular amino acid occupied a particular site was >80%, that amino acid was assigned at that site in ADH$_A$.

When the posterior probability was <80% and/or the most probabilistic ancestral state estimated using the codon and amino acid models were not in agreement, the site was considered ambiguous, and alternative ancestral genes were considered. For example, the posterior probabilities of two amino acids (methionine and arginine) were nearly equal at site 168 in ADH$_A$, three amino acids (lysine, arginine, and threonine) were plausibly present at site 211, and two (aspartic acid and asparagine) were plausible for site 236.

Figure 1.3 shows some of the reason for the ambiguities at sites 168 and 211. As can be seen by the placement of characters (here, amino acids) on the leaves of the trees, it is difficult to infer the ancestral character for the node at the right end of the branch in the tree marked by * in Figure 1.3, representing the last common ancestor of Adh 1 and Adh 2. In part, the difficulties arise because of homoplasy, a historical phenomenon where the same amino acid replacement occurred more than once at different times in the family's history. This suggested that selective constraints were influencing the selection of amino acids at those sites. This ambiguity could therefore not be ignored.

To handle these ambiguities, all 12 (all $2 \times 2 \times 3$ combinations) candidate ADH$_A$s were resurrected by constructing genes that encoded them, transforming these genes into a strain of *S. cerevisiae* from which both Adh 1 and Adh 2 had been deleted, and expressing them from the Adh1 promoter. All of the ancestral sequences could rescue the double-deletion phenotype in the expression yeast.

Table 1.1 Kinetic properties of Adh 1, Adh 2, and candidate ancestral ADH$_A$s. Reprinted from Benner, S.A. and Sismour, A.M. (2005) Synthetic biology. *Nat. Rev. Genet.* **6**: 533–543.

Sample[a]	K_m (μM) Ethanol	NAD$^+$	Acetaldehyde	NADH
Adh1	20 060	218	1492	164
MKD	17 280	511	1019	144
MKN	13 750	814	1067	1106
MRD	11 590	734	1265	287
MRN	10 960	554	1163	894
MTD	10 740	467	959	190
MTN	N/A	N/A	N/A	N/A
RKD	8497	449	1066	142
RKN	7238	407	1085	735
RRD	7784	400	1074	203
RRN	8403	172	1156	1142
RTD	6639	254	1083	316
RTN	7757	564	1158	477
Adh1[b]	24 000	240	3400	140
Adh1[c]	17 000	170	1100	110
Adh2[b]	2700	140	45	28
Adh2[c]	810	110	90	50
Adh3[c]	12 000	240	440	70
Adh1[c] (S. pombe)	14 000	160	1600	100
Adh1(M270L)[c]	19 000	630	1000	80
KlP20369[d]	27 000	2800	1200	110
KlX64397[d]	23 000	2200	1700	180
KlX62766[d]	2570	310	100	20
KlX62767[d]	1560	200	3100	30

[a]The three letters in the sample names designate the amino acids at positions 168, 211, and 236; thus MKD is Met-168, Lys-211, Asp-236. The remaining residues were the same as in Adh 1, except for the following changes (using sequence numbering of Adh1 from *S. cerevisiae*): Asn-15, Pro-30, Thr-58, Ala-74, Glu-147, Leu-213, Ile-232, Cys-259, Val-265, Leu-270, Ser-277, and Asn-324. Kl, *Kluyveromyces lactis*; N/A, not applicable; *S. pombe*, *Schizosaccharomyces pombe*.
[b]From Thomson (2002).
[c]From Ganzhorn *et al.* (1987).
[d]From Bozzi *et al.* (1997).

Table 1.1 lists kinetic data from the candidate ancestral ADH$_A$s (Thomson *et al.*, 2005). Simple kinetic metrics were then used to assess the quality of the data. In particular, the Haldane equation relates the equilibrium constant for the Adh reaction with various of the measured kinetic parameters according to the equation (Segal, 1975).

$$K_{eq} = V_f K_{iq} K_p / V_r K_{ia} K_b$$

where V_f and V_r are forward and reverse maximal velocities, K_{ia} and K_{iq} are disassociation constants for NAD$^+$ and NADH, and K_b and K_p are Michaelis constants for ethanol and acetaldehyde, respectively. These parameters were calculated from the experimental data. The Haldane equation reproduced the literature equilibrium constant for the reaction to within a factor of two. One variant, termed MTN (for the amino acids at sites 168, 211, and 236), had very low catalytic activity in both directions. This suggested that this particular candidate ancestor was not the true ancestor.

Significant to the hypothesis, the kinetic properties of the candidate ancestral ADH$_A$s resembled those of Adh 1 more than Adh 2 (Table 1.1). This included the high K_m for ethanol, a sign of an ancestor that did not have ethanol at low concentrations as its physiological substrate. From this observation, Thomson *et al.* (2005) inferred that the ancestral yeast did not have an Adh specialized for the consumption of ethanol, like modern Adh 2, but rather had an Adh specialized for making ethanol, like modern Adh 1. This, in turn, suggested that the ancestral yeast prior to the time of the duplication did not consume ethanol. This implied that the ancestral yeast also did not make and accumulate ethanol under aerobic conditions for future consumption, and that the make–accumulate–consume strategy emerged after Adh 1 and Adh 2 diverged. These interpretations were robust with respect to the ambiguities in the reconstructions.

Several details are worthy of further comment. For modern Adh 1, the ranges of literature K_m values were 17 000–24 000 μM for ethanol, 170–240 μM for NAD$^+$, 1100–3400 μM for acetaldehyde, and 110–140 μM for NADH (Ganzhorn *et al.*, 1987). These comparisons, together with the Haldane analysis, provide a view of the experimental error in the kinetic parameters reported in the paleoreconstruction. The interpretations about the kinetic behavior of the ancient ADH$_A$ are based on differences well outside of experimental error.

Further, when paralogs are generated by duplication, many believe that the duplicate that then evolves more rapidly is the one that acquires the new functional role (Kellis *et al.*, 2004). If this were generally true, one might identify the functionally innovative duplicate by a simple bioinformatics analysis. Whereas this may be true for many genes, chemical principles do not obligate this outcome, and it is not true with these Adh paralogs. Here, the rate of evolution is not markedly faster in the lineage leading to Adh 2 (having the derived behavior) than in the lineage leading to Adh 1 (having the primitive behavior). The paleobiochemistry experiment was necessary to assign the primitive behavior.

Further, the Haldane ratio relates various kinetic parameters (K_{cat}, K_m, K_{diss}) that can change via a changing amino acid sequence to the overall equilibrium constant, which the enzyme (being a catalyst) cannot change. Thus, if a lower K_m for ethanol is selected, other terms in the Haldane must change to keep the ratio the same. This is observed in data for the ancestral proteins prepared here and the natural enzymes.

1.8 Interconnecting models

By the end of the current century, we can expect that the divisions between branches of biology (molecular, cell, systems, organismic, environmental, geo-, and astro-biology) will be subsumed within a broad model of the phenomenon we know as life. This will include, of course, the products of the reductionist paradigm that has placed chemical structures upon many of the phenomena unique to biology, including genetics, emergent behavior, Darwinian evolution, and functional complexity. It will also incorporate the products of the reductionist paradigm that uses mathematical models to describe the interaction between individuals in a population and different organisms within an ecosystem.

But it will also include a history of the biosphere based on the geological, paleontological, and genomic records. This history will be needed to address the questions of why and how in biology. Here, the answers will come in the form of narratives that describe historical events that fashioned the molecules, cells, systems, organisms, and environments for individual biomolecules. There will be little room for statistics in this model. Rather, the individual traits of individual systems will be understood as the products of chance, necessity, and vestigiality interacting under constraints from physical law.

Further, this model will be heuristic. It will avoid the epithet of being a just-so story by having multiple lines about many systems on Earth that interconnect and intercorrelate in a comprehensive model for the history of the planet, the life that it holds, and the chemistry behind that life. Further, it will depend on the synthesis of ancestral forms to test its heuristics, where experimental paleoscience will repeatedly require the revision of the heuristics.

It is possible to combine the data that were available before the experimental paleoscience done with the Adh system, the data on ADH_A from the experiments described here, and subsequently emerging information, to set us on this path to this complex, intercorrelated, and interconnected future for this individual system. We might begin by asking whether the Adh 1/Adh 2 duplication and the accumulate–consume strategy that it presumably enabled became fixed in the yeast population in response to a particular selective pressure?

Hypothetically, the emergence of a make–accumulate–consume strategy may have been driven by the domestication of yeast by humans selecting for yeast that accumulates ethanol. Alternatively, the strategy might have been driven by the emergence of fleshy fruits that offered a resource worth defending using ethanol accumulation. We might distinguish between the two by estimating the date when the Adh 1/2 duplication occurred. Even with large errors in the estimate, a distinction should be possible, as human domestication occurred in the past million years, while fleshy fruits arose in the Cretaceous, after the first angiosperms appeared in the fossil record 125 million years ago (Sun, 2002), but before the extinction of the dinosaurs 65 million years ago (Collinson and Hooker, 1991; Fernandez-Espinar et al., 2003).

The topology of the evolutionary tree in Figure 1.2 suggests that the Adh 1/Adh 2 duplication occurred before the divergence of the *sensu strictu* species of *Saccharomyces* (Fernandez-Espinar et al., 2003), but after the divergence of *Saccharomyces* and *Kluyveromyces*. The date of divergence of *Saccharomyces* and *Kluyveromyces* is unknown, but might be estimated to have occurred

80±15 million years ago (Berbee and Taylor, 1993). This date is consistent with a transition-redundant exchange (TREx) clock (Benner, 2003), which exploits the fractional identity (f_2) of silent sites in conserved 2-fold-redundant codon systems to estimate the time since the divergence of two genes. Between pairs of presumed orthologs from *Saccharomyces* and *Kluyveromyces*, f_2 is typically 0.82, not much lower than the f_2 value (0.85) separating Adh 1 and Adh 2 (Benner *et al.*, 2002), but much lower than paralog pairs within the *Saccharomyces* genome that appear to have arisen by more recent duplication (approx. 0.98; Lynch and Conery, 2000).

Interestingly, Adh 1 and Adh 2 are not the only pair of paralogs where $0.80 < f_2 < 0.86$ (Benner *et al.*, 2002). Analysis of approximately 350 pairs of paralogs contained in the yeast genome (considering pairs that shared at least 100 silent sites, and diverged by less than 120 point-accepted replacements per 100 sites) identified 15 pairs having $0.80 < f_2 < 0.86$ (Figure 1.4). These represent eight duplications that occurred near the time of the Adh 1 and Adh 2 duplication, if f_2 values are assumed to support a clock.

These duplications are not randomly distributed within the yeast genome. Rather, six of the eight duplications involve proteins that participate in the conversion of glucose to ethanol (Table 1.2). Further, the enzymes arising from the duplicates are those that appear, from expression analysis, to control flux from hexose to ethanol (Schaaff *et al.*, 1989; Pretorius, 2000). These include proteins that import glucose, pyruvate decarboxylases that generate the acetaldehyde from pyruvate, the transporter that imports thiamine for these decarboxylases, and the Adhs (the italicized proteins in Figure 1.1). If the f_2 clock (within its expected variance) is assumed to date paralogs in yeast, this cluster suggests that several genes other than Adh duplicated as part of the emergence of the new make–accumulate–consume strategy, near the time when fleshy fruit arose.

The six gene duplications proposed to be part of the emergence of the make–accumulate–consume strategy (in the $0.80 < f_2 < 0.86$ window) are *not* associated with one of the documented blocks of genes were duplicated in ancient fungi, possibly as part of a whole-genome duplication (Wolfe and Shields, 2001). Two duplications in genes that are

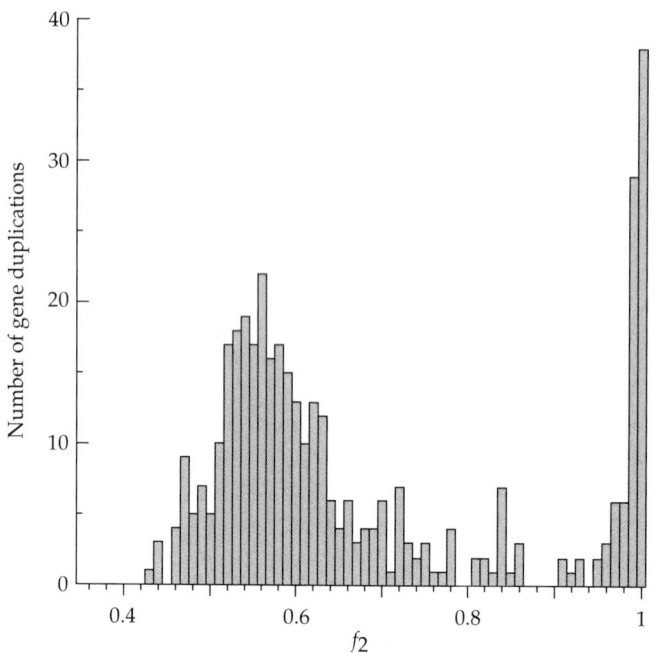

Figure 1.4 A histogram showing all of the pairs of paralogs in the *S. cerevisiae* genome, dated using the transition redundant exchange (TREx) tool (Benner, 2003). The episode of gene duplication where $0.80 < f_2 < 0.86$ is isolated from more ancient duplications (the mode of the distribution at the left) and more recent duplications (represented by the bars at the very right of the plot). Paralog pairs are considered only if they have with at least 100 aligned silent sites, and are not separated by more than 120 point-accepted mutations per 100 aligned amino acid sites (PAM units). Reprinted with permission from Benner *et al.* (2002) Planetary biology: paleontological, geological, and molecular histories of life. *Science* **296**: 864–868, © 2002 AAAS.

Table 1.2 Duplication in the S. cerevisiae genome (SGD), where $0.80 < f_2 < 0.86$. Reprinted from Benner, S.A. and Sismour, A.M. (2005) Synthetic biology. *Nat. Rev. Genet.* **6**: 533–543.

SGD name	gi number	Trivial name	Annotation and comments
Inosine-5′-monophosphate dehydrogenase family (3 paralogs, 3 pairs, 2 duplications)[a]			
$f_2 = 0.803$[c]; pair associated with Wolfe duplication blocks 1 and 44			
YAR073W	gi|456156	IMD1	Nonfunctional homolog, near telomere, not expressed
YLR432W	gi|665971	IMD3	Inosine-5′-monophosphate dehydrogenase
$f_2 = 0.825$[c]			
YLR432W	gi|665971	IMD3	Inosine-5′-monophosphate dehydrogenase
YHR216W	gi|458916	IMD2	Inosine-5′-monophosphate dehydrogenase
Subfamily pair: YHR216W/YAR073W, $f_2 = 0.93$ (proposed recent duplication creating a pseudogene)			
Sugar transporter family A (4 paralogs, 4 pairs, 3 duplications)[b]			
$f_2 = 0.805$[c]; pair not associated with any duplication block			
YJR158W	gi|1015917	HXT16	Sugar transporter repressed by high glucose levels
YNR072W	gi|1302608	HXT17	Sugar transporter repressed by high glucose levels
$f_2 = 0.806$[c]; pair not associated with any duplication block			
YDL245C	gi|1431418	HXT15	Sugar transporter induced by low glucose, repressed by high glucose
YNR072W	gi|1302608	HXT17	Sugar transporter repressed by high glucose levels
$f_2 = 0.809$[c]; pair not associated with any duplication block			
YJR158W	gi|1015917	HXT16	Sugar transporter repressed by high glucose levels
YEL069C	gi|603249	HXT13	Sugar transporter induced by low glucose, repressed by high glucose
$f_2 = 0.810$[c]; pair not associated with any duplication block			
YEL069C	gi|603249	HXT13	Sugar transporter induced by low glucose, repressed by high glucose
YDL245C	gi|1431418	HXT15	Sugar transporter
Subfamily pair: YEL069C/YNR072W, $f_2 = 0.932$ (proposed recent duplication)			
Subfamily pair: YJR158W/YDL245C, $f_2 = 1.000$ (proposed very recent duplication)			
Chaperone family A (2 paralogs, 1 pair, 1 duplication)[a]			
$f_2 = 0.81$; pair associated with Wolfe duplication block 48			
YMR186W	gi|854456	HSC82	Cytoplasmic chaperone, induced 2–3-fold by heat shock
YPL240C	gi|1370495	HSP82	Cytoplasmic chaperone, pheromone signaling, Hsf1p regulation
Phosphatase/thiamine transport family A (2 paralogs, 1 pair, 1 duplication)[b]			
$f_2 = 0.818$; pair not associated with any duplication block			
YBR092C	gi|536363	PHO3	Acid phosphatase implicated in thiamine transport
YBR093C	gi|536365	PHO5	Acid phosphatase, one of three repressible phosphatases
Pyruvate decarboxylase family A (2 paralogs, 1 pair, 1 duplication)[b]			
$f_2 = 0.835$; pair not associated with any duplication block			
YLR044C	gi|1360375		PDC1 Pyruvate decarboxylase, major isoform
YLR134W	gi|1360549		PDC5 Pyruvate decarboxylase, minor isoform

Table 1.2 (Continued)

SGD name	gi number	Trivial name	Annotation and comments
By ortholog analysis, Saccharomyces bayanus (gi|515236) diverged from S. cerevisiae after the $f_2 = 0.835$ duplication, and Kluyveromyces diverged before			
Glyceraldehyde-3-phosphate dehydrogenase family (3 paralogs, 3 pairs, 2 duplications)[b]			
$f_2 = 0.845$[c]; pair not associated with any duplication block			
YJL052W	gi|1008189	TDH1	Glyceraldehyde-3-phosphate dehydrogenase
YGR192C	gi|1323341	TDH3	Glyceraldehyde-3-phosphate dehydrogenase
$f_2 = 0.845$[c]; pair not associated with any duplication block			
YJL052W	gi|1008189	TDH1	Glyceraldehyde-3-phosphate dehydrogenase
YJR009C	gi|1015636	TDH2	Glyceraldehyde-3-phosphate dehydrogenase
Subfamily pair: YJR009C/YGR192C, $f_2 = 0.991$, proposed very recent duplication			
Alcohol dehydrogenase family (2 paralogs, 1 pair, 1 duplication)[b]			
$f_2 = 0.848$; pair not associated with any duplication block			
YMR303C	gi|798945	ADH2	Alcohol dehydrogenase, glucose-repressible
YOL086C	gi|1419926	ADH1	Alcohol dehydrogenase, constitutive
Spermine transporter family (2 paralogs, 1 pair, 1 duplication)[a]			
$f_2 = 0.86$; pair associated with Wolfe duplication block 34			
YGR138C	gi|1323230	TPO2	Spermine-transporter activity
YPR156C	gi|849164	TPO3	Spermine-transporter activity
Sugar transporter family B (3 paralogs, 3 pairs, 2 duplications)[b]			
$f_2 = 0.847$[c]; pair not associated with any duplication block			
YDR343C	gi|1230670	HXT6	Sugar transporter, high affinity, high basal levels
YDR345C	gi|1230672	HXT3	Sugar transporter, low-affinity glucose transporter
$f_2 = 0.854$[c]; pair not associated with any duplication block			
YDR342C	gi|1230669	HXT7	Sugar transporter, high affinity, high basal levels
YDR345C	gi|1230672	HXT3	Sugar transporter, low affinity
Subfamily pair: YDR342C/YDR343C, $f_2 = 0.994$, proposed very recent duplication			

[a] Not associated with fermentation. These are associated with duplication blocks within the yeast genome (Kellis et al., 2004), where the high value of f_2 (typically equilibrated in block paralog pairs) may reflect either variance, or selective pressure to conserve silent sites in individual codons.

[b] Associated with the pathway to make, accumulate and consume ethanol. Genes involved in the fermentation pathway that are not rate-limiting (Schaaff et al., 1989; Pretorius, 2000), generally do not have duplicates in the yeast genome by (e.g. hexokinase, glucose-6-phosphate isomerase, phosphofructokinase, aldolase, triose phosphate isomerase, and phosphoglycerate kinase are all present in one isoform). Enolase has two paralogs (ENO1 and ENO2), where $f_2 = 0.946$. These are distantly related to a homolog known as ERR1, with the silent sites equilibrated. Phosphoglycerate mutase has three paralogs, GM1, GM2 and GM3, with silent sites that are essentially equilibrated.

[c] These pairs represent a family generated with a single duplication with $0.80 < f_2 < 0.86$, and subsequent duplication(s) in the derived lineages. Paralog pairs are considered only if they have with at least 100 aligned silent sites, and are not separated by more than 120 point-accepted mutations per 100 aligned amino acid sites (PAM units). f_2 = fraction of nucleotides conserved at 2-fold-redundant codon sites only, and only at sites where the amino acid is identical.

not associated with fermentation that fall in the $0.80 < f_2 < 0.86$ window *are* part of a duplication block (see Table 1.2). The silent sites for most gene pairs associated with blocks are nearly equilibrated (with the prominent exception of ribosomal proteins), and therefore suggest that most blocks arose by duplications more ancient than duplications in the $0.80 < f_2 < 0.86$ window. Therefore, the hypothesis that a set of six time-correlated duplications (Table 1.2) generated the make–accumulate–consume strategy in yeast near the time when fermentable fruit emerged is not inconsistent with the whole-genome-duplication hypothesis.

This bioinformatics extends the paleoscience experiments across the yeast genome. As of today, this may not convert the story that the paleoscience experiments told into an acceptable narrative. But having a dozen gene duplications correlating with the result from the paleoexperiment helps. The next step is to extend this narrative. If the yeast genome shows evidence of this ecosystem innovation, then so should the genomes of the plants making the fruit, the fruit flies laying eggs in the fermentable fruit, moving down and up in the ecosystem. The end of the Cretaceous saw, in addition to the emergence of fruits, the extinction of the dinosaurs and the emergence of mammals and fruit flies (Baudin *et al.*, 1993; Ashburner, 1998; Barrett and Willis, 2001). For example, females of different fruit fly species position their eggs in fruits with different levels of fermentation (Hougouto *et al.*, 1982). Further, the impact of the introduction of alcohol into the ecosystem should have had impact on microorganisms other than *S. cerevisiae* that had contact with alcohol-rich media.

Likewise, many organisms other than *S. cerevisiae* participate in alcoholic fermentation before yeast takes over. In rotting fruits, *S. cerevisiae* becomes dominant after fermentation begins, while osmotic stress and pH, as well as ethanol, appear to inhibit the growth of competing organisms (Pretorius, 2000). Whereas the genomes of organisms that participate in the initiation of fermentation are not yet available, should they become so, they too can be examined for evidence of adaptive change.

Adding more information will not provide proof. Again, proof is not accessible in the real world. This means that at no point will the narrative evolve to the point where someone who is committed to disagreeing with the narrative not have the option to find a reason not to believe it. This was shown by the experiences with the non-classical carbocation and neutralist/selectionist dispute. But there is only one history of life on Earth. As enough lines of evidence converge on the model, interconnecting enough threads from chemistry, systems biology, ecology, and planetary science, the model will converge. And eventually the model building will end (Galison, 1987).

References

Allemann, R.K., Hung, R., and Benner, S.A. (1988) A stereochemical profile of the dehydrogenases of *Drosophila melanogaster*. *J. Am. Chem. Soc.* **110**: 5555–5560.

Allemann, R.K., Presnell, S.R., and Benner, S.A. (1991) A hybrid of bovine pancreatic ribonuclease and angiogenin. An external loop as a module controlling substrate specificity? *Prot. Eng.* **4**: 831–835.

Ashburner, M. (1998) Speculations on the subject of alcohol dehydrogenase and its properties in *Drosophila* and other flies. *Bioessays* **20**: 949–954.

Barnard, E.A. (1969) Biological function of pancreatic ribonuclease. *Nature* **221**: 340–344.

Barrett, P.M. and Willis, K.J. (2001) Did dinosaurs invent flowers? Dinosaur angiosperm coevolution revisited. *Biol. Rev.* **76**: 411–447.

Baudin, A., Ozier-Kalogeropoulos, O., Denouel, A., Lacroute, F., and Cullin, C. (1993) A simple and efficient method for direct gene deletion in *Saccharomyces cerevisiae*. *Nucleic Acids Res.* **21**: 3329–3330.

Benner, S.A. (2003) Interpretive proteomics: finding biological meaning in genome and proteome databases. *Adv. Enzyme Regul.* **43**: 271–359.

Benner, S.A. and Ellington, A.D. (1988) Interpreting the behavior of enzymes. Purpose or pedigree? *CRC Crit. Rev. Biochem.* **23**: 369–426.

Benner, S.A. and Allemann, R.K. (1989) The return of pancreatic ribonucleases. *Trends Biochem. Sci.* **14**: 396–397.

Benner, S.A. and Sismour, A.M. (2005) Synthetic biology. *Nat. Rev. Genet.* **6**: 533–543.

Benner, S.A., Rozzell, J.D., and Morton, T.H. (1981) Stereospecificity and stereochemical infidelity of acetoacetate decarboxylase (AAD). *J. Am. Chem. Soc.* **103**: 993–994.

Benner, S.A., Cannarozzi, G., Chelvanayagam, G., and Turcotte, M. (1997) *Bona fide* predictions of protein secondary structure using transparent analyses of multiple sequence alignments. *Chem. Rev.* **97**: 2725–2843.

Benner, S.A., Caraco, M.D., Thomson, J.M., and Gaucher, E.A. (2002) Planetary biology: paleontological, geological, and molecular histories of life. *Science* **296**: 864–868.

Berbee, M.L. and Taylor, J.W. (1993) Dating the evolutionary radiations of the true fungi. *Can. J. Bot.* **71**: 1114–1127.

Boulton, B., Singleton, V.L., Bisson, L.F., and Kunkee, R. E. (1996) Yeast and biochemistry of ethanol fermentation. In *Principles and Practices of Winemaking*, pp 139–172. Chapman and Hall, New York.

Bozzi, A., Saliola, M., Falcone, C., Bossa, F., and Martini, F. (1997) Structural and biochemical studies of alcohol dehydrogenase isozymes from *Kluyveromyces lactis*. *Biochim. Biophys. Acta* **1339**: 133–142.

Brown, H.C. (1977) *The Nonclassical Ion Problem*. Plenum Press, New York.

Cho, S., Beintema, J.J., and Zhang, J.Z. (2005) The ribonuclease A superfamily of mammals and birds: identifying new members and tracing evolutionary histories. *Genomics* **85**: 208–220.

Clarke, B. (1975) The contribution of ecological genetics to evolutionary theory: detecting the direct effects of natural selection on particular polymorphic loci. Genetics **79**: 101–108.

Collinson, M.E. and Hooker, J.J. (1991) Fossil evidence of interactions between plants and plant-eating mammals. *Philos. Trans. R. Soc. Lond. Ser. B Biological Sciences* **333**: 197–208.

Dayhoff, M.O., Schwartz, R.M., and Orcutt, B.C. (1978) A model of evolutionary change in proteins. In *Atlas of Protein Sequence and Structure* (Dayhoff, M.O., ed.), pp. 345–352. National Biomedical Research Foundation, Washington DC.

Ellington, A.D. and Benner, S.A. (1987) Free energy differences between enzyme bound states. *J. Theor. Biol.* **127**: 491–506.

Fernandez-Espinar, M.T., Barrio, E., and Querol, A. (2003) Analysis of the genetic variability in the species of the *Saccharomyces* sensu stricto complex. *Yeast* **20**: 1213–1226.

Fersht, A.R. (1977) *Enzyme Structure and Mechanism*. W.H. Freeman, New York.

Fleet, G.H. and Heard, G.M. (1993) Yeast growth during fermentation. In *Wine Microbiology and Biotechnology* (Heard, G.M., ed.), pp 27–54, Harwood Academic Publishers, Chur, Switzerland.

*Galison, P.L. (1987) *How Experiments End*. University of Chicago Press: Chicago.

Ganzhorn, A.J., Green, D.W., Hershey, A.D., Gould, R.M., and Plapp, B.V. (1987) Kinetic characterization of yeast alcohol dehydrogenases. Amino acid residue 294 and substrate specificity. *J. Biol. Chem.* **262**: 3754–3761.

*Gillespie, J.H. (1984) Molecular evolution over the mutational landscape. *Evolution* **38**: 1116–1129.

Gillespie, J.H. (1991) *The Causes of Molecular Evolution*, p. 336. Oxford University Press, Oxford.

Hey, J. (1999) The neutralist, the fly and the selectionist. *Trends Ecol. Evol.* **14**: 35–38.

Hougouto, N., Lietaert, M.C., Libion-Mannaert, M., Feytmans, E., and Elens, A. (1982) Oviposition-site preference and ADH activity in *Drosophila melanogaster*. *Genetica* **58**: 121–128.

Jermann, T.M., Opitz, J.G., Stackhouse, J., and Benner, S. A. (1995) Reconstructing the evolutionary history of the artiodactyl ribonuclease superfamily. *Nature* **374**: 57–59.

Kellis, M., Birren, B.W., and Lander, E.S. (2004) Proof and evolutionary analysis of ancient genome duplication in the yeast Saccharomyces cerevisiae. *Nature* **428**: 617–624.

Kreitman, M. and Akashi, H. (1995) Molecular evidence for natural selection. *Ann. Rev. Ecol. Syst.* **26**: 403–422.

Lehman, W.P. (1973) *Historical Linguistics*. Holt, Reinhard, Winston, New York.

Levins, R. and Lewontin, R. (1985) *The Dialectical Biologist*. Harvard University Press, Cambridge, MA.

Lewontin, R.C. (1974) *The Genetic Basis of Evolutionary Change*. Columbia University Press, New York.

Lynch, M. and Conery, J.S. (2000) The evolutionary fate and consequences of duplicate genes. *Science* **290**: 1151–1155.

McGeehan, G.M. and Benner, S.A. (1989) An improved system for expressing pancreatic ribonuclease in *Escherichia coli*. *FEBS Lett.* **247**: 55–56.

McGovern, P.E. (2004) Fermented beverages of pre- and proto-historic China. *Proc. Natl. Acad. Sci. USA* **101**: 17593–17598.

Nambiar, K.P., Stackhouse, J., Stauffer, D.M., Kennedy, W.G., Eldredge, J.K., and Benner, S.A. (1984) Total synthesis and cloning of a gene coding for the ribonuclease S protein. *Science* **223**: 1299–1301.

Nambiar, K.P., Stackhouse, J., Presnell, S.R., and Benner, S.A. (1987) Expression of bovine pancreatic ribonuclease A in *E. coli*. *Eur. J. Biochem.* **163**: 67–71.

Nielsen, R. (2002) Mapping mutations on phylogenies. *Syst. Biol.* **51**: 729–739.

Pagel, M. (1999) Inferring the historical patterns of biological evolution. *Nature* **401**: 877–884.

Pauling, L. and Zuckerkandl, E. (1963) Chemical paleogenetics molecular restoration studies of extinct forms of life. *Acta. Chem. Scand.* **17**: S9–S16.

Powers, D.A. and Schulte, P.M. (1998) Evolutionary adaptations of gene structure and expression in natural populations in relation to a changing environment: a multidisciplinary approach to address the million-year saga of a small fish. *J. Exp. Zool.* **282**: 71–94.

Pretorius, I.S. (2000) Tailoring wine yeasts for the new millennium: Novel approaches to the ancient art of winemaking. *Yeast* **16**: 675–729.

Robson, B. and Garnier, J. (1993) Protein structure prediction. *Nature* **361**: 506.

Schaaff, I., Heinisch, J., and Zimmerman, F.K. (1989) Overproduction of glycolytic enzymes in yeast. *Yeast* **5**: 285–290.

Segel, I.H. (1975) *Enzyme Kinetics*. John Wiley and Sons, New York.

Somero, G.N. (1995) Proteins and temperature. *Annu. Rev. Physiol.* **57**: 43–68.

Stackhouse, J., Presnell, S.R., McGeehan, G.M., Nambiar, K.P., and Benner, S.A. (1990) The ribonuclease from an extinct bovid. *FEBS Lett.* **262**: 104–106.

Stryer, L. (1995) *Biochemistry*, 4th edn. W.H. Freeman and Company, New York.

Sun, G. (2002) Archaefructaceae, a new basal angiosperm family. *Science* **296**: 899–904.

Suppe, F. (1977) *The Structure of Scientific Theories*, 2nd ed. University of Illinois Press, Urbana, IL.

Swofford, D.L. (1998) *PAUP* Phylogenetic Analysis Using Parsimony Version 4*. Sinauer Associates, Sunderland, MA.

Thomson, J.M. (2002) *Interpretive Proteomics: Experimental Paleogenetics as a Tool to Analyze Function and Discover Pathways in Yeast*. Dissertation, University of Florida.

Thomson, J.M., Gaucher, E.A., Burgan, M.F., De Kee, D.W., Li, T., Aris, J.P., and Benner, S.A. (2005) Resurrecting ancestral alcohol dehydrogenases from yeast. *Nat. Genet.* **37**: 630–635.

Wills, C. (1976) Production of yeast alcohol dehydrogenase isoenzymes by selection. *Nature* **261**: 26–29.

Wolfe, K.H. and Shields, D.C. (2001) Molecular evidence for an ancient duplication of the entire yeast genome. *Nature* **387**: 708–713.

Yang, Z. (1997) PAML: a program package for phylogenetic analysis by maximum likelihood. *Comput. Appl. Biosci.* **13**: 555–556.

Yang, Z., Kumar, S., and Nei, M. (1995) A new method of inference of ancestral nucleotide and amino acid sequences. *Genetics* **141**: 1641–1650.

Zhang, J.Z. and Nei, M. (1997) Accuracies of ancestral amino acid sequences inferred by the parsimony, likelihood, and distance methods. *J. Mol. Evol.* **44**: S139–S146.

CHAPTER 2

Ancestral sequence reconstruction as a tool to understand natural history and guide synthetic biology: realizing and extending the vision of Zuckerkandl and Pauling

Eric A. Gaucher

2.1 Historical context

The recent accumulation of DNA sequence data, combined with advances in evolutionary theory and computational power, have paved the way for innovative approaches to understanding the origins, evolution, and distribution of life and its constituent biomolecules (Pauling and Zuckerkandl, 1963; Benner et al., 2002; Gaucher et al., 2004). One approach to understanding ancestral states follows a present-day-backwards strategy, whereby genomic sequences from extant (modern) organisms are incorporated into evolutionary models that estimate the extinct (ancient) character states of genes no longer present on Earth (Fitch, 1971; Shih et al., 1993; Benner, 1995; Koshi and Goldstein, 1996; Schultz et al., 1996; Cunningham, 1999; Omland, 1999; Pagel, 1999; Schultz and Churchill, 1999; Chang and Donoghue, 2000; Thornton, 2004; Hall, 2006). These inferred ancestral gene sequences act as hypotheses that can be tested in the laboratory through the resurrection of the ancestral proteins themselves (paleogenetics). Results from functional assays of the protein products from these ancient genes permit us to accept or reject null hypotheses about the sequences themselves, or about their interactions, binding specificities, environments, etc. And beyond narratives relating ancient phenotypes and environments, paleogenetics provides the unique opportunity to "replay" the molecular tape of life (Gould, 1989).

It is probably appropriate that any discussion of ancestral sequence reconstruction be placed within a historical framework. By doing so in this chapter, I hope to convey how the field has made recent advances, and emphasize the need to continue this progress so that our concepts enjoy wider recognition and create greater impact than the achievements to date. The chapter will begin with a brief account of the field during the 1990s. I will continue with a discussion on one of the resurrected gene families that has been studied, and then conclude with a discussion on future directions in the field, with a particular emphasis on evolutionary synthetic biology.

In 1963, Emile Zuckerkandl and Linus Pauling published an intriguing article entitled "Chemical paleogenetics molecular restoration studies of extinct forms of life" (Pauling and Zuckerkandl, 1963). In it, they put forward the notion of reconstructing amino acid sequences of ancestral proteins by virtue of a comparison between sequences of related proteins found in contemporary organisms and subsequent synthesis (and thereby resurrection) of these sequences in the laboratory. While limits in technology prohibited the actual resurrection, Zuckerkandl and Pauling presented a sequence reconstruction of ancient mammalian hemoglobins. The duo then suggested that a future resurrection of ancient hemoglobins assayed for dioxygen affinity and pH dependence would generate higher-order inferences of biological interpretation. Or, more broadly, the joining of chemical, biological, and structural models to natural history would provide a more accurate description of macromolecular behavior beyond that supplied by studying individual molecules disconnected from the selective forces governing their evolution (this approach is also termed planetary biology).

It is important to note the significance of Zuckerkandl and Pauling's proposal. Molecular reconstructions and experimental resurrections are fundamentally exercises in theory/computation/algorithmic development and technological advances, respectively. Proposing these approaches would have been sufficiently significant. But Zuckerkandl and Pauling realized the potential of creating a paradigm that attempted to connect chemical structure to natural selection at the molecular, cellular, organismal, population, and planetary levels.

The first examples of molecular resurrections were performed on artiodactyl pancreatic ribonucleases and lysozymes (Malcolm et al., 1990; Stackhouse et al., 1990; Jermann et al., 1995; see Chapters 1 and 18 in this volume for further discussion). The former study was particularly notable in that it not only put Zuckerkandl and Pauling's theory into practice, it did so in a manner that connected chemical reactivity (RNA hydrolysis), molecular biology (single- versus double-stranded RNA-binding affinities), organismal evolution (bacterial fermentation and foregut digestion in ruminants), and planetary biology (diversification of grasses during the Oligocene cooling, and the ability of artiodactyls to digest and extract nutrients from these grasses via bacterial fermentation). These connections were the first to exemplify Zuckerkandl and Pauling's proposed paradigm, and by the late 1990s we were intent on extending the paradigm as far back in history as possible—ideally to the earliest life forms on Earth.

Successful studies on ancestral reconstructions require a sufficient base of knowledge in evolutionary theory and models. Computational simulations and experimental phylogenetics were used to differentiate accuracy and consistency between distance-based (e.g. neighbor-joining) and character-based (e.g. parsimony, likelihood, and Bayesian) approaches to phylogenetics. Further, several discussions about reconstructing ancestral character states were also appearing in the literature during this time. For instance, a symposium lead by Cunningham, Oakley, Omland, and others on behalf of the Society of Systematic Biology focused on this topic (Omland, 1999). Meanwhile, Pagel, Goldstein, Yang, and others were independently developing maximum-likelihood approaches based on Bayesian statistics (called ML or empirical Bayesian reconstruction; Yang et al., 1995; Koshi and Goldstein, 1996; Pupko et al., 2000). These authors advocated the use of maximum likelihood over parsimony for inferring ancestral states (see Pagel et al., 2004, and Schluter, 1995, for a criticism of previous work on resurrected proteins). Similarly, Ronquist, Huelsenbeck, Schultz, and others were developing hierarchical Bayesian algorithms for character-state inferences (Schultz and Churchill, 1999; Huelsenbeck and Bollback, 2001; Ronquist, 2004). A fuller discussion of methods for ancestral sequence reconstruction is presented in Chapters 4–9 in this volume.

Regardless of the growing literature, it remained unclear at the time which approach would best guide an experimental design of ancestral sequences. Maximum likelihood's superior performance under certain models of evolution was apparent from numerous simulation studies and laboratory evolution experiments with viruses (Bull et al., 1993; Zhang and Nei, 1997). Hierarchical Bayesian methods could theoretically outperform likelihood approaches, although only Yang's empirical approach was available to us at the time. Ideally, a hierarchical Bayesian approach accounts for uncertainty from phylogenetic hypotheses (these include uncertainty in the topology, branch lengths, and any other parameter estimates). A heuristic work-around, however, could take advantage of an empirical Bayesian approach in which alternative models, parameters, and topologies could be analyzed separately. The output from these separate analyses could then be combined with the expectation of resolving uncertainty or bias arising from any of the individual analyses alone.

The heuristic hierarchical Bayesian approach was initially applied to the alcohol dehydrogenase (ADH) gene family (the same approach was later applied to elongation factors (EFs) and seminal ribonucleases). Ancestral character states were inferred from the ADH phylogeny using multiple models each for DNA-, codon-, and amino acid-based evolution, all for two competing topologies (Thomson et al., 2005). The results from these separate analyses required that 12 putative ancestral genes be synthesized to account for the differences (or uncertainty) at individual sites among the various analyses. Although this approach did not eliminate all uncertainty and bias associated with the ancestral reconstruction methods, it was an improvement in the ability to capture the true ancestral state.

During our gene synthesis experiments of ancestral ADHs, Chang and colleagues had formulated a similar heuristic hierarchical Bayesian approach towards the resurrection of ancestral rhodopsins (Chang et al., 2002), and later with fluorescent proteins (Ugalde et al., 2004). Although their analysis did not account for alternative topologies, they did consider various models of sequence evolution and their effects on the ancestral state inferences.

We should note that the choice of individual amino acid residues at sites in an ancestral sequence remains under debate to this day. For instance, Pollock, Goldstein, and colleagues have suggested that selecting the ancestral residue with the highest probability at each site (most-probabilistic ancestral sequence, MPAS) can lead to erroneous inferences of the ancestral states (Williams et al., 2006; see Chapter 8). These authors advocate the synthesis of random sequences weighted from the posterior distribution. For instance, consider a pentapeptide in which each site in the ancestral sequence has a posterior probability for alanine of 80% and serine of 20%. The MPAS would be a penta-alanine (i.e. AAAAA), whereas the average weighted sample from the distribution would generate four alanines and one serine (e.g. ASAAA).

Recent discussions over which of these two sequences introduces less bias during the reconstruction procedure have resulted in lively exchanges, and will undoubtedly continue to do so for the next few years. The center issue remains unresolved: under what phylogenetic and evolutionary conditions will the two sequences be biased and thus no longer able to capture the true ancestral behavior/function? To what extent do sequence divergence, episodes of adaptive evolution (e.g. high nonsynonymous/synonymous ratios or heterotachy), heterogeneous processes, etc., lead to bias during the reconstruction process?

For instance, consider an ancient gene-duplication event resulting in neofunctionalization whereby the novel behavior is determined by amino acid replacements at two sites only. Each of the two sites experiences an alanine-to-serine replacement along the branch leading to the new function. Reconstructing the ancestral sequence at the node representing the common ancestor of the paralogs infers alanine at 80% and serine at 20%. The MPAS would correctly have alanines at these two positions. Sequences sampled from the posterior distribution, however, would only have alanines at these two positions 64% of the time (0.8×0.8). Therefore, one-third of the sampled sequences would have a behavior not analogous to the ancestral behavior, making it nearly impossible to generate accurate interpretations from the sampled sequences.

Alternatively, lack of phylogenetic signal and parallel/convergent evolution may cause the MPAS to be biased away from the true ancestral behavior. Here, ancestral states can be driven by the evolutionary models themselves when the data fail to provide sufficient signal to extract the true ancestral states. Any bias associated with the evolutionary models (e.g. preponderance of hydrophobic residues) may be propagated throughout the inference process, ultimately influencing the inferred protein behavior away from the true ancestral state. We tend to expect this for sites that are more rapidly evolving (e.g. coils on the protein surface) although simulations are required to confirm this notion. We anticipate that many of these issues will be addressed in the near future through extensive computational simulations that exploit biologically realistic models of protein evolution as well as experimental phylogenetics studies.

By no means do the above paragraphs serve as a definitive account of the field. Many other people were involved in the development of reconstruction theory, evolutionary models, algorithms, and software development. Rather, it serves as one historical account intended to convey the dynamic growth occurring within the field.

In fact, approximately 20 narratives have emerged to date where specific molecular systems from extinct organisms have been resurrected for study in the laboratory (Sassi et al., 2007). These include digestive proteins (ribonucleases, proteases, and lysozymes) in ruminants and primates to illustrate how digestive function arose from non-digestive function in response to a changing global ecosystem, fermentive enzymes from fungi to illustrate how molecular adaptation supported mammals as they displaced dinosaurs as the dominant large land animals, pigments in the visual system adapting to different environments, steroid hormone receptors adapting to changing function in steroid-based regulation of metazoans, fluorescent proteins from ocean-dwelling invertebrates, enzyme cofactor evolution, and proteins from very ancient bacteria helping to define environments where the earliest forms of bacterial life lived.

2.2 Temperature conditions of early life

By the end of the last decade, the most ancient paleomolecular resurrections had traveled back in time only c.200–300 million years. This had left untouched many of the most widely discussed questions about the nature of early life on Earth. We identified and concluded that the EF-Tu family had the greatest potential to address these questions. Further, this gene family could generate a robust statistical reconstruction in conjunction with biological interpretations (correlation between protein thermostability and optimal growth temperature of the host organism) and planetary integrations (correlating the temperature histories of early life from molecular and geologic records; Gaucher et al., 2003).

2.3 EFs

EFs are G-proteins that present charged aminoacyl-tRNAs to the ribosome during translation. Because of their relatively slow rates of sequence divergence, most character states of ancient EF sequences can be robustly reconstructed for proteins from deep nodes in the Bacterial phylogeny. Further, the optimal thermal stabilities of EFs correlate with the optimal growth temperature of the host organism. Thus, EFs from mesophiles, thermophiles, and hyperthermophiles, defined as organisms that grow at 20–40, 40–80, and >80°C, respectively, and represented by species of *Escherichia*, *Thermus*, and *Thermotoga*, have temperature optima in their respective ranges. This is consistent with a previous study based on a large set of proteins in which a correlation coefficient of 0.91 was calculated between environmental temperatures of the host organisms and protein-melting temperatures.

Figures 2.1a and 2.1b show the two topologies used to reconstruct ancestral sequences at the node representing the hypothetical organism lying at the stem of the Bacterial tree. The number of sequences in the outgroup, 3–20, did not affect the amino acid reconstructions at these nodes: ML-stem (maximum-likelihood tree for the stem bacteria) and Alt-stem (alternative tree for the stem bacteria). The ancestral sequence at the node representing the most recent common ancestor of only mesophilic bacterial lineages was also reconstructed, and named ML-meso (maximum-likelihood tree for mesophiles only). This node captures one feature of models that have concluded that the last common ancestor of the Bacteria was mesophilic. In all, these reconstructed ancestral sequences did not appear to be influenced by long-branch attraction or non-homogeneous modes of molecular evolution, such as changes in the mutability of individual sites in different branches of the bacterial subtree (see Chapter 9 for further discussion of this).

A BLAST search was then performed to identify the most similar extant sequences to the inferred ancestors. ML-stem and Alt-stem are most similar to the sequences of EFs from *Thermoanaerobacter tengcongensis* (a thermophile) and *Thermotoga*

24 ANCESTRAL SEQUENCE RECONSTRUCTION

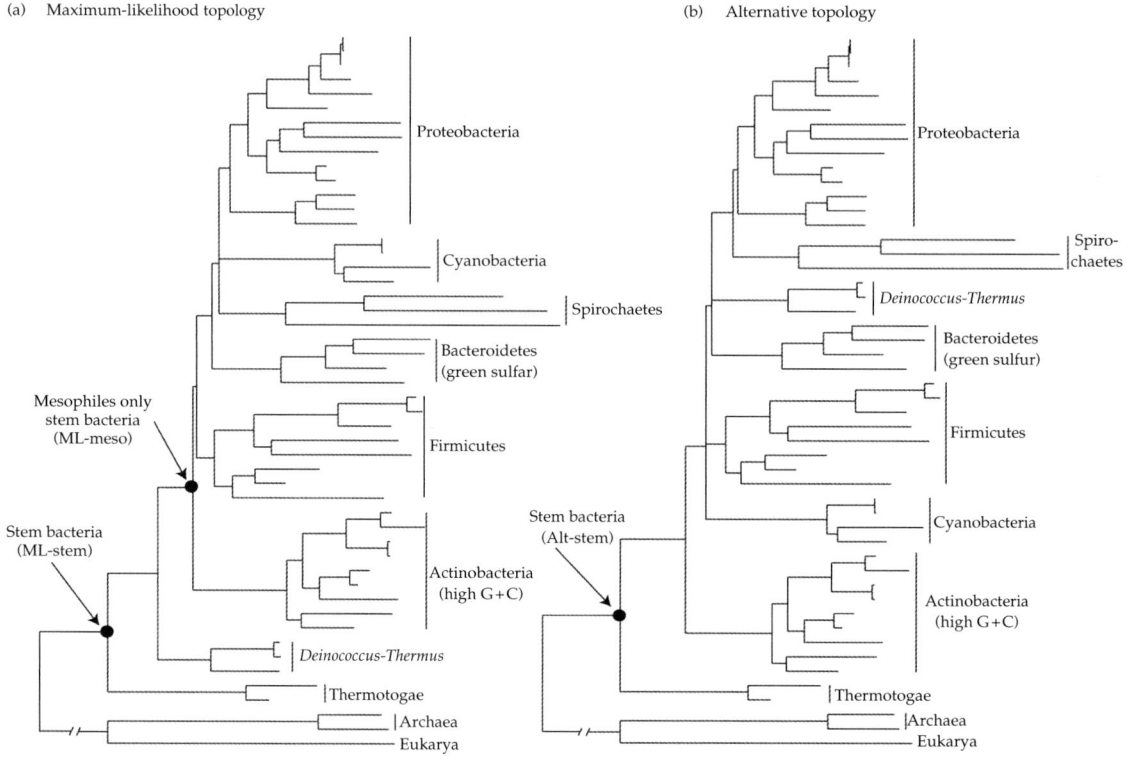

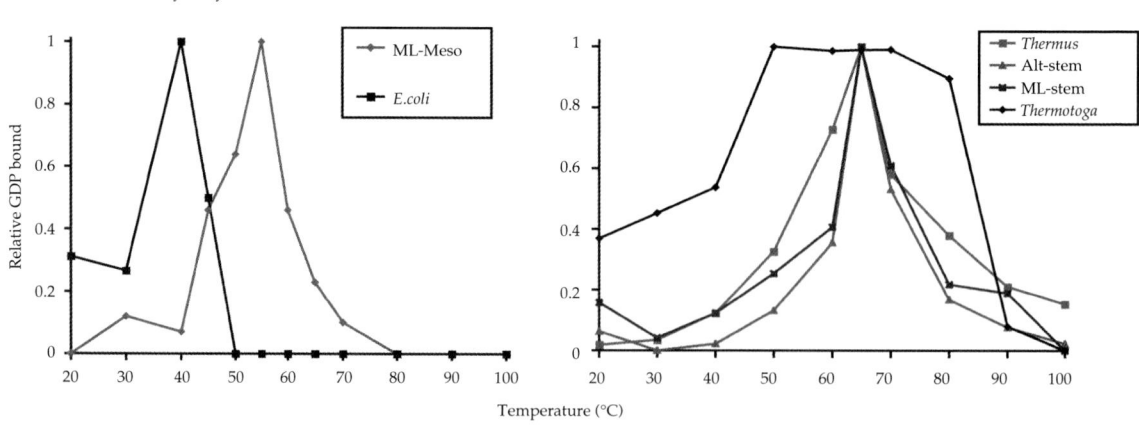

Figure 2.1 The two un-rooted universal trees used to reconstruct ancestral bacterial sequences. Archaea and Eukarya serve to provide a node within the bacterial subtree from which ancient sequences can be inferred. Lineages containing thermophiles (as known at the time of the original study) are highlighted in Italic. (a) Maximum-likelihood topology used to reconstruct the stem elongation factors (EFs) from bacteria (ML-stem), or most recent common ancestor of bacteria, and the ancestral sequence for mesophilic lineages only (ML-meso). (b) Alternative topology used to reconstruct the stem elongation factors from bacteria (Alt-stem). (c) GDP-binding assay to test thermostability of ancestral and modern EF proteins. The amount of tritium-labeled GDP bound at 0°C was subtracted from all other temperature values for a given protein. Shown is the relative amount of GDP bound compared to the amount bound at the optimal temperature for each protein. EF thermostability profiles for *Escherichia coli*, *Thermus aquaticus*, and *Thermotoga maritima* are shown with the three ancestral EF profiles. Reprinted from Gaucher *et al.* (2003) Inferring the paleoenvironment of ancient bacteria on the basis of resurrected proteins. *Nature* **425**: 285–288.

maritima (a hyperthermophile), respectively, and differ from each other by 28 amino acids. ML-meso is most similar to the sequence of EF from *Neisseria meningitides* (a mesophile). We expect that the identities of these best-hits will change as the databases themselves expand. If we had assumed, however, that similarity in sequence implies similarity in thermostability, it might have been predicted that the stem bacterium was thermophilic or hyperthermophilic, whereas the ancestral node constructed without considering thermophiles was a mesophile. To test these predictions based on this under-substantiated assumption, genes encoding the ancestral sequences were synthesized, expressed in an *Escherichia coli* host, and purified. The thermostabilities of these ancestral EFs, and three representative EFs from contemporary organisms, were then assessed by measuring the ability of each to bind GDP across a range of temperatures (Figure 2.1c).

Each resurrected protein behaved similarly. Both ML-stem and Alt-stem EFs bound GDP with a temperature profile similar to that of the thermophilic EF from modern *Thermus aquaticus*, with optimal binding at approximately 65°C. Although the sequence similarity was higher between Alt-stem and the modern hyperthermophilic *T. maritima*, the temperature profile of Alt-stem was not similar to that from *T. maritima*, which is maximally active up to approximately 85°C and has a broad optimal temperature range typical of hyperthermophiles. The observation that the amino acid sequences of ML-stem and Alt-stem shared only 93% identity but displayed the same thermostability profiles suggests that inferences of this ancestral property are robust with respect to both varying topologies and ancestral character-state predictions. Based on these given evolutionary models, this suggests that the paleoenvironment of the ancient bacterium was approximately 65°C.

Inferences were then drawn from a resurrected EF whose sequence was reconstructed from the last common ancestral sequences of contemporary organisms which, for the most part, grow optimally at mesophilic temperatures. The temperature profile of the ancestral protein, which displayed a maximum at 55°C, suggests that the ancestor of modern mesophiles lived at a higher temperature than any of its descendants (Figure 2.1c). This result shows that the behavior of an ancestor need not be an average of the behaviors of its descendants.

The observation that tree-based ancestral sequence reconstructions can give results different from consensus-sequence reconstructions may be general. It underscores a fact, well known in Structure Theory, that physical behavior in a protein is not a linear sum, or even a simple function, of the behavior of its parts. This, in turn, implies that an experiment in paleobiochemistry can yield information beyond that yielded by analysis of descendent proteins alone.

A short side note on the names of the ancestral EF proteins used in this study: an examination of the GenBank entries for ML-stem, Alt-stem and ML-meso will reveal their alternative names: BLANKET1–3. This acronym for bacterial-lineage ancestral reconstructions served as a small tribute to Pauling. It remains to be determined, however, if this was indeed a service to his name.

2.4 Conclusions from EF studies

This study pushed the experimental paleogenetics research strategy back to at least 3 billion years, to the most primitive ancestors from which descendants can be traced. Accordingly, the ambiguity encountered is substantial, and available sequence data may not be sufficient to manage it convincingly. Here, the ambiguities do not depend primarily on the details of the model used to infer ancestral states. They seem to arise rather from the uncertainty of the phylogenetic tree joining the protein family members.

It is noteworthy, therefore, that reconstructions can be made at all. Further, if the large-scale sequencing of random bacterial genomes, as undertaken by Venter and his group (Venter *et al.*, 2004), continues, there is good reason to hope that the reconstructions will improve. Indeed, the temperature history of Bacteria is already beginning to be defined by follow-up studies.

Adding interpretations from the geologic and molecular records to the results of ancestral EF proteins provides the necessary planetary integrations to fulfill the goals of Zuckerkandl and Pauling's molecular-restoration paradigm. Here, geologic evidence based on low $\delta^{18}O$ isotopic ratios in 3.5–3.2 billion-year-old cherts from the Barberton greenstone belt in South Africa suggests that the ocean temperature of early Earth was 55–85°C (Knauth, 2005). This result is remarkably consistent with the inferred optimal growth temperature of the microorganisms hosting the ancestral EF proteins. Inferences from the evolution of amino acid frequencies, as well as other studies, suggest that the last universal ancestor lived in a hot environment, further supporting the thermophily notion of early life (see Chapter 17).

It is important to note that these studies *do not* address issues related to the origins of life. The origins of bacteria and the last universal ancestor were undoubtedly complex microorganisms far removed from life's origins. These studies do, however, provide clues to the environment that hosted life's most recent common ancestors.

2.5 Ancestral resurrections and evolutionary synthetic biology

Zuckerkandl and Pauling's proposal to connect chemical structure and reactivity to cellular, organismal, and planetary interpretations using resurrected ancestral sequences was remarkably prophetic. Nearly all implementations of this proposal have followed the natural history paradigm laid out in Zuckerkandl and Pauling's seminal discussion (1963), and will undoubtedly soon be extended to developmental biology (Colosimo *et al.*, 2005). But exceptions to this paradigm exist. Most notable is the application of resurrected proteins to connect chemical and molecular reactivity to human health by Kodra, Liberles, and coworkers (*truly* applied molecular evolution; Skovgaard *et al.*, 2006; see Chapter 3). The driving goal in their study is not to elucidate the evolutionary history of the insulin-response pathway per se, but rather to search ancestral functional sequence-space in hopes of developing innovative therapeutics. Regardless of the outcome from this study, the prospects of similar research aimed at integrating molecular evolution and biomedicine using ancestral resurrections are highly energizing.

An additional application of ancestral reconstruction tools for biomedicine consists of generated predictions regarding the tolerability of amino acid replacements from human single-nucleotide-polymorphism data for disease-causing genes (Gaucher *et al.*, 2006).

The next logical extension of molecular resurrections beyond natural history and biomedicine is to biotechnology and synthetic biology. Not surprisingly, synthetic biology means different things to different scientific disciplines (Benner and Sismour, 2005; Endy, 2005). Surprisingly, however, biologists seem to have taken a back seat to chemists and engineers in the development of this field. It seems apparent that synthetic biology would stand to benefit if so-called molecular reconstructionists contributed to its progress. In this way, an evolutionary synthetic biology is formed. A few examples come to mind: synthetic DNA/protein libraries, tools for gene integration, cellular machines, and recombinant genomes.

2.5.1 Synthetic DNA/protein libraries

As the cost of DNA-synthesis technology diminishes, the constraint of resurrecting a few ancestral nodes on a phylogeny or a few variants at any node is released. Suppose, for instance, that the novel protein identified by Skovgaard *et al.* (2006) only existed in sequence space that was slightly removed from any of the given nodes connecting humans and gila monsters. Ancestral resurrections of these particular nodes could have failed to reveal a protein with interesting properties. An exercise in directed evolution intended to capture phylogenetic information, however, may be more successful in revealing the desired properties.

Directed evolution is a powerful technique for improving the activity, specificity and/or stability of proteins. It has been applied to a wide range of proteins with uses in therapeutics, agriculture, and chemistry.

My research group have developed a new method for designing libraries to include the functional diversity of large, highly divergent protein families while still maintaining a high proportion of active members. The strategy is to include only those changes that are associated with

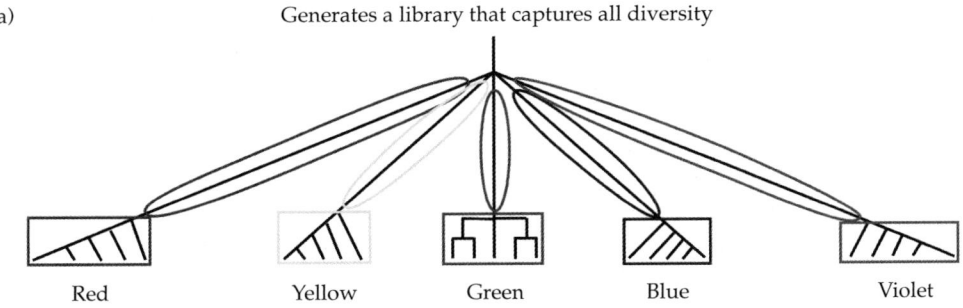
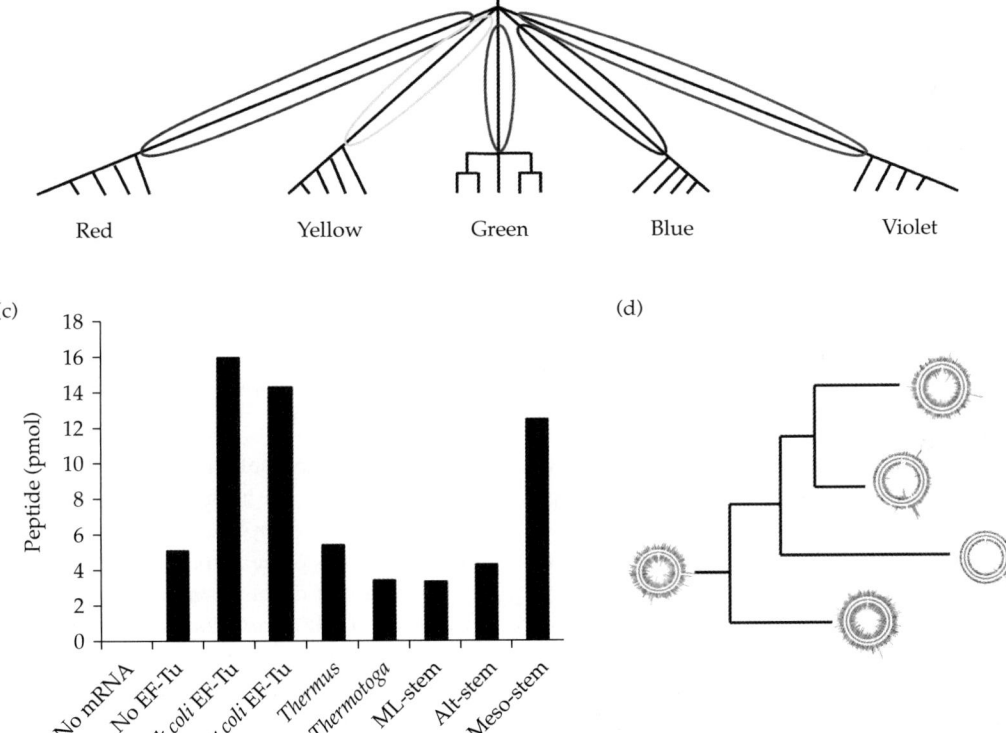

Figure 2.2 Ancestral sequence reconstruction as a tool for synthetic biology. (a) Accepted-diversity approach for the synthesis of DNA libraries intended to expand protein functionality. A library built using the approach integrates all the diversity found within extant sequences. (b) REAP approach for library synthesis. Note that a library built using this approach attempts to capture only sequence diversity responsible for differences between the five fluorescent subfamilies. (c) The amount of peptide synthesized from a reconstituted *in vitro* translation system that makes use of modern and ancestral EF-Tu proteins. *E. coli* EF-Tu was purified from both the Jack Szostak (Harvard Medical School) and Gaucher laboratories to compare purification protocols. (d) Resurrection of the ancestral genome.

new activities during the evolution of a protein family and to exclude the much larger set of changes that do not lead to new functions (phenotypically neutral). In other words, the approach identifies the amino acid changes that were associated with functional divergence during the evolutionary history of a family and builds libraries containing only those specific changes. The new method is referred to as reconstructing evolutionary adaptive paths (REAP).

To illustrate this strategy, consider a hypothetical case of fluorescent proteins. There are five families with individual fluorescent spectra. Each family contains five sequences and all five families share a common ancestor (polytomy). We would like to generate a fluorescent protein with novel properties by incorporating information contained within these 25 sequences.

The standard approach (accepted diversity) takes advantage of processes governed by natural selection. Here, amino acids present in modern (extant) sequences are combinatorially used to generate a library. Highlighted in Figure 2.2a, patterns of amino acid residues that evolved either within a family (boxed regions) or along the branches that gave rise to the individual families (circled regions) are recorded. Most of the amino acid patterns observed in modern proteins presumably arose within the families (boxed regions) and thus have little to offer in terms of generating novel properties. These amino acid patterns arose mostly from neutral evolution, assuming a lack of selective pressure to diversify within a given family.

The REAP approach is based on a model of molecular evolution that attempts to eliminate amino acid patterns predicted to have minimal contributions to novel protein behaviors. This is achieved by reconstructing the ancestral patterns that arose during the evolution of 'unique' properties compared to the last common ancestor of the fluorescent proteins (circled branches only), and ignoring the amino acid patterns that arose within a family (see Figure 2.2b). In doing so, we increase the unique behaviors captured using the accepted-diversity approach, while decreasing the noise associated with its approach.

The REAP methodology is not predicted to be ideal for all library designs. The approach requires numerous homologous sequences to generate an articulated phylogeny. Further, the phylogeny needs to represent a family of sequences with diverse behaviors, otherwise the extracted amino acid patterns may not generate novel behaviors. When the appropriate information is available, however, REAP is predicted to have a number of significant advantages including the incorporation of information from highly diverse family members into a highly *functional* library and no requirements for information regarding protein structure or the mutability of sites (mutagenesis experiments) to guide the library design.

We anticipate that the REAP approach will make substantial contributions to synthetic biology. For instance, DNA libraries based on the REAP method may generate polymerases with high fidelity towards non-standard nucleotides and/or protein variants capable of supporting unnatural amino acid incorporation during protein synthesis. The resulting biopolymers will then serve as the information (novel coding systems) and catalysis (novel side-chain chemistry) components of an expanded biology.

2.5.2 Ancient transposons as tools for gene integration

Transposable elements are pieces of DNA that have the ability to jump around an organism's genome. In their simplest form, these mobile elements code for a protein that excises their encoded DNA from one chromosomal position and reintegrates it at different location in the genome. Terminal repeats flank the coding sequence and provide the necessary regulatory information for transposition. The mobile properties of these elements have considerable potential as tools for molecular biology and therapeutic gene delivery.

For instance, a member of the Tc1/mariner family has been resurrected from salmonid fish pseudogenes using a consensus approach, as opposed to a true phylogenetic reconstruction of ancestral states (Figure 2.3a; Ivics *et al.*, 1997). The aptly named Sleeping Beauty (SB) element was shown to be active in transposition, and, equally important, it was active in non-fish species. Follow-up mutagenesis studies have produced a

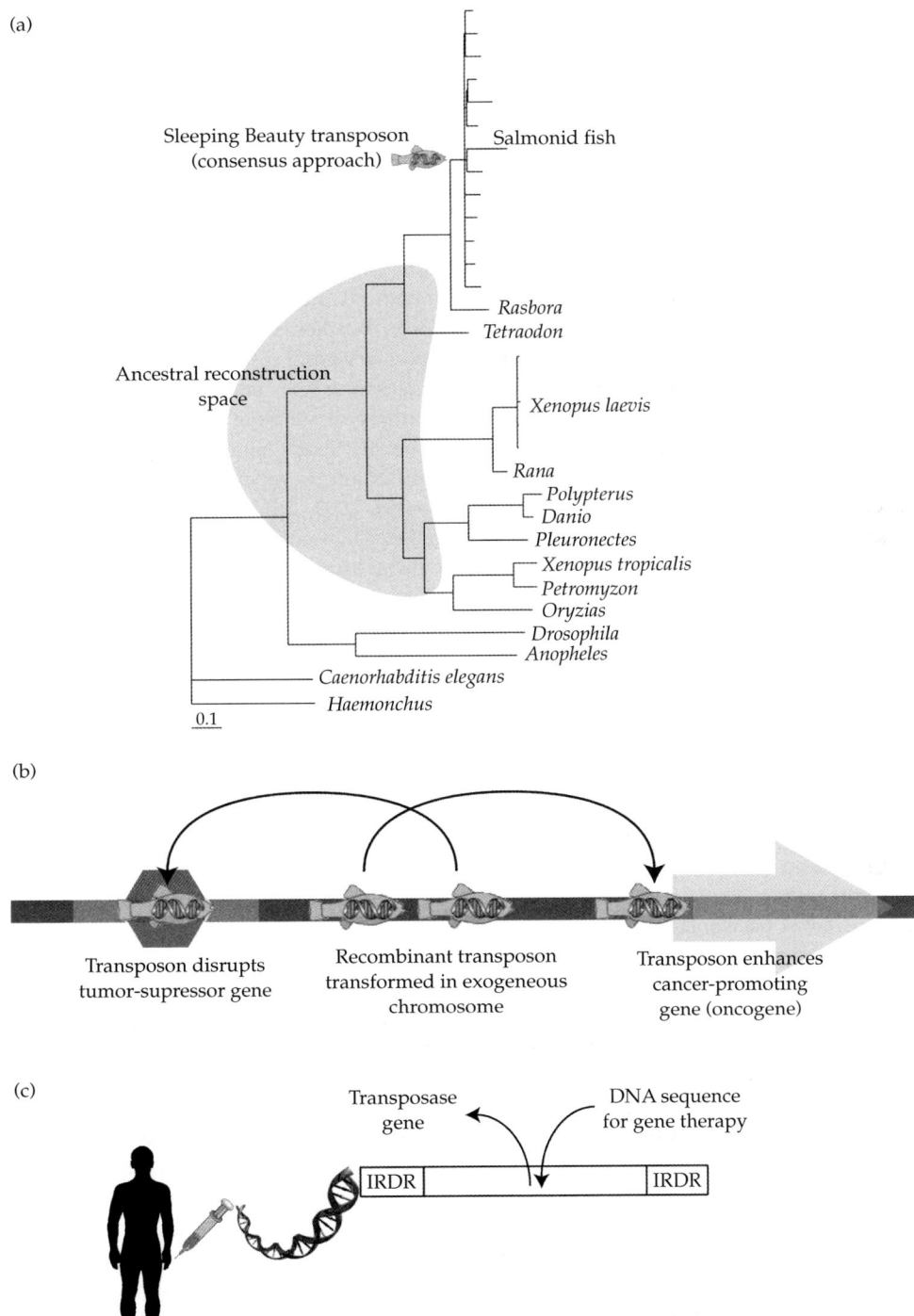

Figure 2.3 Ancestral transposons as tools. (a) Phylogeny of Tc1/mariner family of transposable elements. Sleeping Beauty (Ivics *et al.*, 1997) transposon generated by the consensus approach is shown along with ancestral sequence space of interest to our laboratory. (b) Schematic of an engineered Sleeping Beauty transposon that serves as a tool to identify genes involved in tumor suppression/growth. Adapted from Weiser and Justice (2005). (c) Transposons can be engineered as a vector for gene therapy. Inverted-repeat/direct-repeat elements (IRDRs) are regulatory/binding regions for transposition.

variant of the SB element that displays enhanced transposition activity.

One application of the SB variant has identified genes involved in mouse tumor suppression/promotion (Collier et al., 2005). Here, the transposase gene was replaced with a murine stem-cell-virus (MSCV) promoter, a string of poly(A)s, and slice-acceptor/-donor sites flanking the promoter and terminator signals. This transposon construct was then integrated in the mouse genome. Transposition was carried out by a transposase gene supplied in trans. The arrangement of splice sites in the transposon, and the location of the transposon either upstream or within an endogenous mouse gene, dictated whether the mouse gene was truncated or displayed increased expression. In this way, the transposon was used to disrupt tumor suppressors or enhance expression of cancer-promoting genes (Figure 2.3b).

The utility of the resurrected SB transposon is apparent. We are currently probing whether other ancestral sequences of the Tc1/mariner family have expanded utility (Figure 2.3a). Our motivation stems from the notion that a consensus approach may only capture evolutionary information in a heuristic manner. Reconstructions that utilize phylogenetic methods with explicit models of sequence evolution are more likely to recapitulate the ancestral form.

Overall, resurrected transposable elements have demonstrated their value by identifying genes involved in tumor growth. The next step is to develop a transposon capable of serving as a vector for gene therapy (Figure 2.3c). This will require that researchers overcome hurdles associated with the current limitations of these transposons (size of gene insert, site-integration specificity, etc.).

2.5.3 Cellular machines and recombinant (ancestral) genomes

In the interest of space limitations, let us briefly introduce the two remaining topics.

Cellular machines have a broad range of potentials; from simple expression of heterologous genes for laboratory analysis or environmental detection of explosive compounds such as TNT, to the synthesis of minimal artificial cells for small-compound drug or protein synthesis (Martin et al., 2003; Noireaux and Libchaber, 2004). We anticipate that ancestral reconstructed sequences will provide much of the foundation of genetic information for these machines in the future. As a first step, we have demonstrated that ancestral EF proteins can participate in a reconstituted translation system in vitro designed to incorporate unnatural amino acids (Figure 2.2c; Josephson et al., 2005). Further, experimental evolution studies of these ancestral genes introduced into laboratory organisms will enhance our biological understanding of adaptive and sequence landscapes and allow us to ask whether replaying the molecular tape of life is repetitive (Gould, 1989; Elena and Lenski, 2003; Lunzer et al., 2005; Weinreich et al., 2006). This work will have obvious extensions to natural history and the origins of (early) life.

Ancestral reconstructions will undoubtedly include larger and larger segments of a genome in the near future. As a first step in developing the appropriate technology to support this vision, the Venter Institute is in the process of constructing a minimal synthetic *Mycoplasma* genome (H. Smith, personal communication). Once the technology is available, why not construct a complete ancestral biochemical pathway (e.g. operon) or an ancestral genome (Figure 2.2d)? The ancestral reconstruction field would no longer be confined to single-gene reconstructions (Blanchette et al., 2004). Developing narratives based on resurrected pathways and genomes will have profound effects on our understanding of evolution (natural selection/biodiversity) and biomedicine. It is also quite possible that our understanding of what constitutes a sustaining minimal genome required to support life will be altered through ancestral reconstructions. In this way, homologous genes performing two different, but related, functions may share a single common ancestor that performed both of these functions, albeit with less efficiency or specificity.

We anticipate that the ancestral reconstruction field will drive synthetic biology once the technology permits us. We will then find ourselves in the center of the debate on artificial life, which will raise even more debate both within and outside of our field. Until then...

2.6 Conclusions

The reconstruction field has made tremendous strides since 1963 (Thornton, 2004). We anticipate that the past 40 years will go down in history as our lag-phase period. It is necessary then, as we enter the exponential phase and as technology harnesses our power, that we do not forget the importance of biological integration and natural history that Zuckerkandl and Pauling promoted. Some scientists have noted that the field of molecular biology had this connection but lost it in recent decades. For example, Woese has noted that *empirical* reductionism is suitable in our quest to understand the broader context of biological systems (Woese, 2004). *Fundamentalist* reductionist approaches toward molecular biology, however, trivialize the natural world and are therefore paralyzing. It is important that we do not reach such a stationary phase in our quest to understand biological systems and diversity through ancestral reconstructions. In a field full of narratives, consider this as merely one more.

2.7 Acknowledgments

This work was funded in part by a National Research Council/NASA Astrobiology fellowship and grants from the NASA Exobiology program and NIH.

I would like to thank David Liberles, David Ardell, and Giorgio Matassi for organizing the inaugural ancestral reconstruction conference held in Sweden in 2005, and for overseeing the publication of this book. I would further like to thank all of the contributors to this book for their enthusiasm to make our field as rigorous and fascinating as possible. Thanks to Mike Thomson, Slim Sassi, Michelle Burgan, Jack Szostak, and Kris Josephson for their assistance with our research, and to my undergraduate advisor (George P. Smith) for introducing me to Bayesian statistics in the early 1990s.

A special thanks to Steven Benner for adopting me into his laboratory as a graduate student. The parallels between Pauling and Benner in their adroit abilities to integrate physical sciences and natural history should not go unnoticed. Any discussion with Steve about his work on organic reactivity, biochemical pathways, stereoselectivity, non-standard nucleotides, or protein-structure prediction always lead to biological interpretations within a Darwinian framework.

References

Benner, S.A. (1995) Reconstructing ancient forms of life. *J. Cell. Biochem.* 200–200, suppl. 19A.

Benner, S.A. and Sismour, A.M. (2005) Synthetic biology. *Nat. Rev. Genet.* **6**: 533–543.

Benner, S.A., Caraco, M.D., Thomson, J.M., and Gaucher, E.A. (2002) Planetary biology–paleontological, geological, and molecular histories of life. *Science* **296**: 864–868.

Blanchette, M., Green, E.D., Miller, W., and Haussler, D. (2004) Reconstructing large regions of an ancestral mammalian genome in silico. *Genome Res.* **14**: 2412–2423.

Bull, J.J., Cunningham, C.W., Molineux, I.J., Badgett, M.R., and Hillis, D.M. (1993) Experimental molecular evolution of bacteriophage-T7. *Evolution* **47**: 993–1007.

Chang, B.S.W. and Donoghue, M.J. (2000) Recreating ancestral proteins. *Trends Ecol. Evol.* **15**: 109–114.

Chang, B.S.W., Jonsson, K., Kazmi, M.A., Donoghue, M.J., and Sakmar, T.P. (2002) Recreating a functional ancestral archosaur visual pigment. *Mol. Biol. Evol.* **19**: 1483–1489.

Collier, L.S., Carlson, C.M., Ravimohan, S., Dupuy, A.J., and Largaespada, D.A. (2005) Cancer gene discovery in solid tumours using transposon-based somatic mutagenesis in the mouse. *Nature* **436**: 272–276.

Colosimo, P.F., Hosemann, K.E., Balabhadra, S., Villarreal, G., Dickson, M., Grimwood, J. *et al.* (2005) Widespread parallel evolution in sticklebacks by repeated fixation of Ectodysplasin alleles. *Science* **307**: 1928–1933.

Cunningham, C.W. (1999) Some limitations of ancestral character-state reconstruction when testing evolutionary hypotheses. *Syst. Biol.* **48**: 665–674.

Elena, S.F. and Lenski, R.E. (2003) Evolution experiments with microorganisms: the dynamics and genetic bases of adaptation. *Nat. Rev. Genet.* **4**: 457–469.

Endy, D. (2005) Foundations for engineering biology. *Nature* **438**: 449–453.

Fitch, W. (1971) Towards defining the course of evolution. Minimum change for a specific tree topology. *Syst. Zool.* **20**: 406–416.

Gaucher, E.A., Thomson, J.M., Burgan, M.F., and Benner, S.A. (2003) Inferring the palaeoenvironment of ancient bacteria on the basis of resurrected proteins. *Nature* **425**: 285–288.

Gaucher, E., Graddy, L., Li, T., Simmen, R.C., Simmen, F.A., Schreiber, D.R. et al. (2004) The planetary biology of cytochrome P450 aromatases. *BMC Biol.* **2**: 19.

Gaucher, E.A., De Kee, D.W., and Benner, S.A. (2006) Application of DETECTER, an evolutionary genomic tool to analyze genetic variation, to the cystic fibrosis gene family. *BMC Genomics* **7**: 44.

Gould, S.J. (1989) *Wonderful Life: The Burgess Shale and the Nature of History.* W.W. Norton & Company: New York.

Hall, B.G. (2006) Simple and accurate estimation of ancestral protein sequences. *Proc. Natl. Acad. Sci. USA* **103**: 5431–5436.

Huelsenbeck, J.P. and Bollback, J.P. (2001) Empirical and hierarchical Bayesian estimation of ancestral states. *Syst. Biol.* **50**: 351–366.

Ivics, Z., Hackett, P.B., Plasterk, R.H., and Izsvak, Z. (1997) Molecular reconstruction of Sleeping Beauty, a Tc1-like transposon from fish, and its transposition in human cells. *Cell* **91**: 501–510.

Jermann, T.M., Opitz, J.G., Stackhouse, J., and Benner, S.A. (1995) Reconstructing the evolutionary history of the artiodactyl ribonuclease superfamily. *Nature* **374**: 57–59.

Josephson, K., Hartman, M.C.T., and Szostak, J.W. (2005) Ribosomal synthesis of unnatural peptides. *J. Am. Chem. Soc.* **127**: 11727–11735.

Knauth, L.P. (2005) Temperature and salinity history of the Precambrian ocean: implications for the course of microbial evolution. *Palaeogeogr. Palaeocl.* **219**: 53–69.

Koshi, J.M. and Goldstein, R.A. (1996) Probabilistic reconstruction of ancestral protein sequences. *J. Mol. Evol.* **42**: 313–320.

Lunzer, M., Milter, S.P., Felsheim, R., and Dean, A.M. (2005) The biochemical architecture of an ancient adaptive landscape. *Science* **310**: 499–501.

Malcolm, B.A., Wilson, K.P., Matthews, B.W., Kirsch, J.F., and Wilson, A.C. (1990) Ancestral lysozymes reconstructed, neutrality tested, and thermostability linked to hydrocarbon packing. *Nature* **345**: 86–89.

Martin, V.J.J., Pitera, D.J., Withers, S.T., Newman, J.D., and Keasling, J.D. (2003) Engineering a mevalonate pathway in Escherichia coli for production of terpenoids. *Nat. Biotechnol.* **21**: 796–802.

Noireaux, V. and Libchaber, A. (2004) A vesicle bioreactor as a step toward an artificial cell assembly. *Proc. Natl. Acad. Sci. USA* **101**: 17669–17674.

Omland, K.E. (1999) The assumptions and challenges of ancestral state reconstructions. *Syst. Biol.* **48**: 604–611.

Pagel, M. (1999) Inferring the historical patterns of biological evolution. *Nature* **401**: 877–884.

Pagel, M., Meade, A., and Barker, D. (2004) Bayesian estimation of ancestral character states on phylogenies. *Syst. Biol.* **53**: 673–684.

Pauling, L. and Zuckerkandl, E. (1963) Chemical paleogenetics molecular restoration studies of extinct forms of life. *Acta Chem. Scand.* **17**: S9–S16.

Pupko, T., Pe'er, I., Shamir, R., and Graur, D. (2000) A fast algorithm for joint reconstruction of ancestral amino acid sequences. *Mol. Biol. Evol.* **17**: 890–896.

Ronquist, F. (2004) Bayesian inference of character evolution. *Trends Ecol. Evol.* **19**: 475–481.

Sassi, S.O., Benner, S.A., and Gaucher, E.A. (2007) Molecular paleosciences. Systems biology from the past. In *Advances in Enzymology and Related Areas of Molecular Biology: Protein Evolution*, vol. 75, pp. 1–132 (Toone, E., ed.). Wiley, Chichester.

Schluter, D. (1995) Uncertainty in ancient phylogenies. *Nature* **377**: 108–109.

Schultz, T.R. and Churchill, G.A. (1999) The role of subjectivity in reconstructing ancestral character states: a Bayesian approach to unknown rates, states, and transformation asymmetries. *Syst. Biol.* **48**: 651–664.

Schultz, T.R., Cocroft, R.B., and Churchill, G.A. (1996) The reconstruction of ancestral character states. *Evolution* **50**: 504–511.

Shih, P., Malcolm, B.A., Rosenberg, S., Kirch, J.F., and Wilson, A.C. (1993) Reconstruction and testing of ancestral proteins. molecular evolution: producing the biochemical data. *Molecular Evolution: Producing the Biochemical Data. Methods in Enzymology*, vol. 224, pp. 3–725 (Zimmer, E.A., White, T.J., Cann, R.L., and Wilson, A.C., eds). Academic Press, San Diego.

Skovgaard, M., Kodra, J.T., Gram, D.X., Knudsen, S.M., Madsen, D., and Liberles, D.A. (2006) Using evolutionary information and ancestral sequences to understand the sequence-function relationship in GLP-1 agonists. *J. Mol. Biol.* **363**: 977–988.

Stackhouse, J., Presnell, S.R., Mcgeehan, G.M., Nambiar, K.P., and Benner, S.A. (1990) The ribonuclease from an extinct bovid ruminant. *FEBS Lett.* **262**: 104–106.

Thomson, J.M., Gaucher, E.A., Burgan, M.F., De Kee, D. W., Li, T., Aris, J.P., and Benner, S.A. (2005) Resurrecting ancestral alcohol dehydrogenases from yeast. *Nat. Genet.* **37**: 630–635.

Thornton, J.W. (2004) Resurrecting ancient genes: experimental analysis of extinct molecules. *Nat. Rev. Genet.* **5**: 366–375.

Ugalde, J.A., Chang, B.S., and Matz, M.V. (2004) Evolution of coral pigments recreated. *Science* **305**: 1433.

Venter, J.C., Remington, K., Heidelberg, J.F., Halpern, A.L., Rusch, D., Eisen, J.A. et al. (2004) Environmental genome shotgun sequencing of the Sargasso Sea. *Science* **304**: 66–74.

Weinreich, D.M., Delaney, N.F., Depristo, M.A., and Hartl, D.L. (2006) Darwinian evolution can follow only very few mutational paths to fitter proteins. *Science* **312**: 111–114.

Weiser, K.C. and Justice, M.J. (2005) Cancer biology: Sleeping Beauty awakens. *Nature* **436**: 184–186.

Williams, P.D., Pollock, D.D., Blackburne, B.P., and Goldstein, R.A. (2006) Assessing the accuracy of ancestral protein reconstruction methods. *PLoS Comput. Biol.* **2**, e69.

Woese, C.R. (2004) A new biology for a new century. *Microbiol. Mol. Biol. Rev.* **68**: 173–186.

Yang, Z.H., Kumar, S., and Nei, M. (1995) A new method of inference of ancestral nucleotide and amino-acid-sequences. *Genetics* **141**: 1641–1650.

Zhang, J.Z. and Nei, M. (1997) Accuracies of ancestral amino acid sequences inferred by the parsimony, likelihood, and distance methods. *J. Mol. Evol.* **44**: S139–S146.

CHAPTER 3

Linking sequence to function in drug design with ancestral sequence reconstruction

Janos T. Kodra, Marie Skovgaard, Dennis Madsen, and David A. Liberles

3.1 Introduction

As we have seen in the previous two chapters, ancestral sequence reconstruction is a powerful technique with utility in several fields of basic science. Ancestral sequence reconstruction also holds increasing promise in applied science. Examples of its potential utility to the pharmaceutical industry will be presented here.

When developing protein and peptide natural products as pharmaceutical candidates, an understanding of the link between sequence and bioactivity (function) is usually needed. To this end, techniques such as combinatorial chemistry and alanine-scanning mutagenesis (Morrison and Weiss, 2001; Liou and Khosla, 2003) have traditionally been applied. For large peptides and proteins, sampling of sequence space is limited. Combinatorial approaches lead to random and sparse sampling of global sequence space, whereas alanine-scanning mutagenesis samples only a selected and frequently very local region of single-mutation points in sequence space. However, the mapping of functional fitness landscapes on sequence space is not random, and is subject to complex covarion processes that result in peaks that are not accessible through examination of single mutation points, where the optimal amino acid at a given position is context-dependent (Lopez *et al.*, 2002).

Designing experiments based upon evolution is one way to explore complex sequence-fitness landscapes in order to discover drug candidates with new and desirable properties. Phage display and evolution *in vitro* may be viewed as a forward-looking experimental approach to assessing sequence space (Barbas *et al.*, 1991). Ancestral sequence reconstruction involves the backward-looking complement to this. Given a strong selective pressure to retain a functional molecule, one expects the ancestral nodes in phylogenetic trees also to have been functional. Examining such sequences and the substitutional mapping to the phylogenetic tree enables a fuller appreciation of both sequence space and the fitness landscape.

The purpose of the analysis is often to understand sequence-function mapping rather than the evolutionary history. This involves identification of homologous molecules with different desirable properties which have not previously been found in related molecular species. Here mapping of the correlative substitutions to branches where the properties emerged can increase the power to detect candidate causative substitutions. Combining such substitutions in different ancestral sequence backgrounds and testing activities experimentally will enable researchers to understand better what it is possible to generate and what is subject to pleiotropic constraint. Branches in gene families of protein-encoding genes that have been subjected to positive selection are good candidates for relevant functional changes (see for example Liberles *et al.*, 2001).

Designing and/or optimizing proteins for selected pharmaceutical or biological purposes such as stability, selectivity, bioavailability, and potency have proven to be challenging tasks. The mere size of proteins often renders the *ab initio* approach of describing protein–protein interactions impossible. Once a protein drug is delivered *in vivo*, its efficacy is a complex function of the interactions with its targets (such as a receptor or enzyme) as well as the interactions with a number of known and unknown proteins involved in the clearance, possible transportation mechanism, and other properties that determine the efficiency of a protein as a drug.

Animal venoms have proved to be a valuable source of ligands for a wide variety of pharmacological targets. Crude venoms usually contain a diverse array of different peptides/proteins, many of which are bioactive. The pharmaceutical benefits of venoms have been known since early human history. Several peptides/proteins isolated from venoms are now being tested in the clinic for treatment of disease, including metabolic disorders and pain relief. Recently, the US Food and Drug Administration (FDA) has approved Ziconotide, a peptide isolated from the venom of marine snails for the treatment of pain relief. Exenatide, a peptide isolated from the venom of the gila monster *Heloderma suspectum*, has been marketed for use in treatment of type 2 diabetes, and several peptides isolated from snakes are currently being tested in the clinic (see Table 3.1).

Venoms are often associated with an envenomation apparatus for the delivery of bioactive molecules into the soft tissue of the prey via subcutaneous or intravenous routes. For the venom peptides to be efficient, they need to reach their pharmaceutical targets, which results in the high bioavailability seen in venom peptides/proteins. Furthermore, venom peptides/proteins need to be sufficiently stable to survive chemical degradation in solution at ambient temperature via enzymatic degradation by proteases present within the venom itself, as well as in the tissues of the prey species. Biopharmaceuticals often have similar delivery routes and are therefore faced with similar challenges. Biopharmaceuticals most often require injection and need high bioavailability as well as sufficient stability in the face of protease degradation. In that context, it might not be so surprising that venoms are a valuable source of new leads for various pharmaceutical targets. Venom peptides/proteins are already optimized by nature with regard to several of the parameters often required for biopharmaceuticals.

Peptides/proteins from animal venoms typically have three major physiological targets—nerves, muscles, and blood circulation—in which they alter the host pathways for the benefit of the venomous animal. Nature often reuses scaffolds within protein frameworks to develop new protein

Table 3.1 Some of the venom peptides or truncated versions tested in the clinic and on the market.

Name	Target/related protein	Disease	Organism	Clinical stage	Company
Prialt/Ziconotide	Voltage-gated Ca^{2+} channel antagonist	Pain relief	Cone snail	Granted FDA approval (December 2004)	Elan Corporation
Exendin-4	Glucagon-like peptide-1 receptor agonist	Type 2 diabetes	Gila monster	Granted FDA approval (April 2005)	Amylin and Eli Lilly
Alfimeprase	Metallo proteinase	Stroke (peripheral arterial occlusion)	Southern copperhead viper	Phase III	Nuvelo
TM-601	Chloride-channel agonist	Brain tumor	Death stalker scorpion	Phase II	TransMolecular
ACV1	Nicotinic acetylcholine receptor antagonist	Pain relief	Cone snail	Phase I	Metabolic Pharmaceuticals
Q8010	Prothrombinase	Hemostasis	Australian common brown snake	Preclinical	QRxPharma

properties. The binding core of the peptides/proteins from venomous animals is often conserved through species (built on a small number of common permissive scaffolds). The same scaffolds are often found in nature in proteins performing other non-toxic functions and it is likely that such conserved motifs are the results of divergent evolution in most cases.

3.2 Glucagon-like peptide-1 (GLP-1) and exendin-4

One example which is likely the result of divergent evolution is provided by the venom peptide exendin-4, related to the physiologically important hormone GLP-1. GLP-1 is involved in stimulation of glucose-dependent insulin secretion and insulin biosynthesis, inhibition of glucagon secretion, gastic emptying, and inhibition of food intake (Toft-Nielsen *et al.*, 2001; Zander *et al.*, 2002). This physiological role has suggested that GLP-1 is a promising candidate for treatment of type 2 diabetes. However, GLP-1 has a very short half-life in humans, thereby limiting its potential as a drug (Kieffer *et al.*, 1995). From the venom of the gila monster, a 39-amino acid peptide called exendin-4 has been isolated (Eng *et al.*, 1992). The gila monster and the Mexican bearded lizard are the only known venomous lizards. Compared to native GLP-1, exendin-4 is a long-acting agonist at the GLP-1 receptor but otherwise it displays similar properties to GLP-1 in regulating gastric emptying, insulin secretion, food intake, and glucagon secretion. Exendin-4 was recently approved by the FDA for the treatment of type 2 diabetes.

GLP-1 shares 53% amino acid identity with exendin-4 (see Figure 3.1;). This suggests that the peptides originated from a common ancestral peptide coded by a common ancestral gene that has been duplicated one or more times during the course of evolution. GLP-1 sequences are highly conserved in tetrapods and completely conserved at the amino acid level in all mammalian species investigated so far. A larger degree of variation can be observed in amphibian sequences where multiple copies of GLP-1 are found. Surprisingly, exendin-4 groups with these amphibian peptides (Figure 3.2). Given the phylogenetic informational content in a short peptide this may indicate an ancient duplication event leading to exendin-4 or convergent evolution under functional selective pressure, or may simply be a phylogenetic artefact of a short sequence.

From a pharmaceutical viewpoint, exendin-4 is a GLP-1 analog with superior drug-like properties, due primarily to improved stability towards enzymatic degradation. However, when compared to human GLP-1 the 14 amino acid substitutions found in exendin-4 clearly make them more diverged analogs than are typically obtained by the traditional design of biopharmaceuticals. Given the large number of combinations by which one could substitute 14 positions with 20 amino acids, it is unlikely that exendin-4 would result from human GLP-1 by using one of the traditional approaches. Hence, assuming an origin by divergent evolution, 350 million years of evolution over an unknown fitness landscape is the difference between having an efficient drug and not. Examining the evolutionary history of the two peptides would present a systematic way of exploring the amino acids combinations that introduced the various properties in exendin-4. Along these lines, use of substitutional mapping of residues to create a human GLP-1 with residues that mapped uniquely to exendin-4 (mapped to the most recent branch leading to exendin-4) generated a molecule that showed enhanced protease stability *in vitro*, but not in a mouse model, when compared to endogenous human GLP-1 (Skovgaard *et al.*, 2006).

Sequence of exendin-4   HGEGTFTSDLSKQMEEEAVRLFIEWLKNGGPSSGAPPS
Sequence of GLP-1   HAEGTFTSDVSSYLEGQAAKEFIAWLVKGR

Figure 3.1 The sequences of exendin-4 and human GLP-1 are shown.

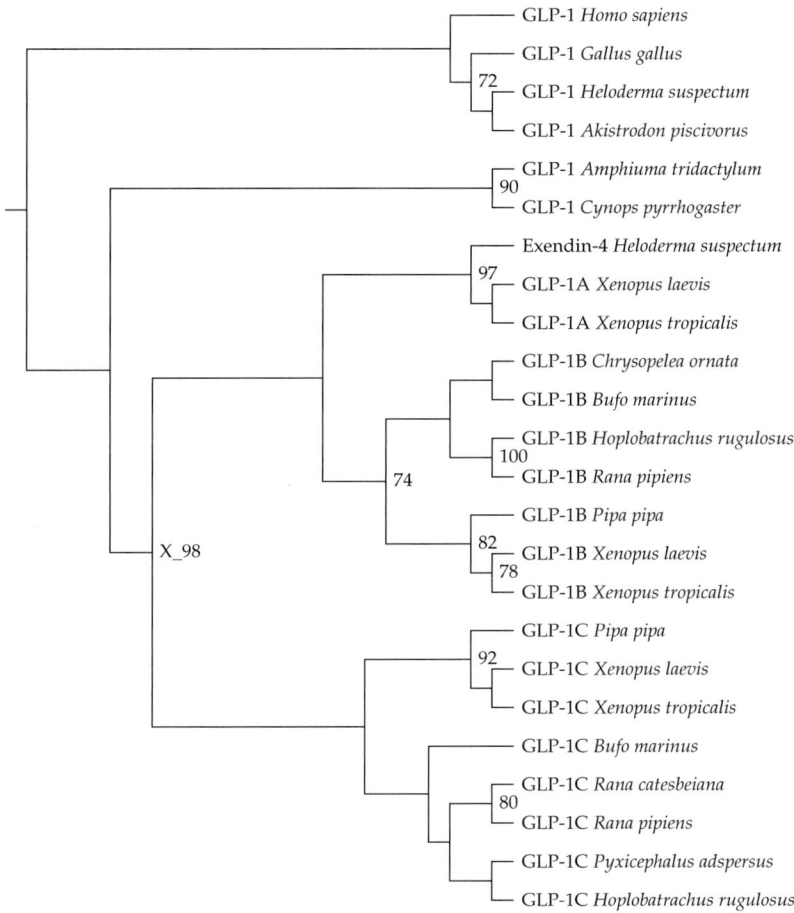

Figure 3.2 The phylogenetic relationship derived from a multiple sequence alignment including exendin-4 and GLP-1 homologs is shown, as derived from Skovgaard *et al.* (2006) Using Evolutionary Information and Ancestral Sequences to Understand the Sequence-Function Relationship in GLP-1 Agonists. *Journal of Molecular Biology*, **393**: 977–988. With permission from Elsevier. As all currently known mammalian sequences are identical, only the sequence from *Homo sapiens* is included. The exon encoding GLP-1 has been duplicated in frogs and toads: these peptides can be recognized in the alignment as they are named GLP-1A, B, or C. The tree is annotated with bootstrap values and X indicates a node where the activity has been sampled by ancestral sequence reconstruction. As described in the text, synthetic peptides were also designed by placing substitutions that occurred along the branches leading to exendin-4 and to the GLP-1A clade in a human GLP-1 context.

3.3 Snake venom and hemostasis

A bite by a venomous snake results in severe damage mediated through various physiological pathways. This would typically include blocking of nerve–muscle junctions by neurotoxins, changes in hemostasis, and frequent hemorrhaging. Snake venom utilizes the blood-coagulation process as part of a strategy to immobilize prey and/or increase the target-tissue permeability for other toxins. This evolutionary history has resulted in snake venom that contains many proteins and peptides that act selectively on different blood coagulation factors, blood cells, and tissues. Hence there appears to be a venom protein that can activate or inactivate almost every factor involved in coagulation or fibrinolysis.

Venom proteins acting as coagulation factors include Factor V activators, Factor X activators, prothrombin activators, and thrombin-like enzymes. Other venom components are involved in anti-coagulation, including Factor IX-/X-binding proteins, protein C activators, thrombin inhibitors, and phospholipase A_2. Finally, some act on fibrinolysis, including fibrinolytic enzyme and plasminogen activator.

Australian snakes are among the most venomous in the world and have evolved a unique approach to dealing with the blood-clotting pathway in their prey (Broad et al., 1979; Fry, 1999). The prothrombin activators from Australian snakes have evolved from Factor X during at least two different gene-duplication events (see Figure 3.3). The duplication events also represent two functional groups, namely those depending on the functionality of prey Factor V and those distributing their own Factor V homolog with the venom (Kini, 2005). Prothrombin activator from the common brown snake (*Pseudonaja textilis*) carries a non-enzymatic Factor V domain. It is currently under preclinical studies by QRxPharma as an agent that rapidly stops bleeding (thus minimizing blood loss) and reducing the length of surgical procedures (see Table 3.1). Differences in the activities of venom prothrombin activators and Factor X can be pinpointed using mutational mapping along the relevant branches during the evolution of venom proteins and along other branches where functions have changed. This mutational mapping will enable a linkage between sequence and function. Ancestral sequence reconstruction can also pinpoint changes in sequence scaffold that may be necessary for the function of individual substitutions. In the case of prothrombin activators, post-translational modifications must be considered in addition to amino acid-level changes, adding to the complexity of this example. Another group of snake-venom proteins that has gained attention as having possible therapeutic properties are the snake-venom thrombin-like enzymes, which share sequence similarity with human thrombin (Castro et al., 2004). This is yet another example of a protein family where ancestral sequence reconstruction could be of interest.

3.4 Conclusion

Design of biopharmaceuticals with desired biological profiles remains a challenging task. First, the size of protein drugs complicates calculations *ab initio* when addressing protein–protein and ligand–receptor interactions. Second, the number of possible chemical variations within a protein of a given size is extremely large and in most cases is out of reach for a complete screen. Despite intensive study and much progress, the parameters involved in determining structure and stability remain incompletely understood. This is also true regarding the effects of scaffold and allosteric changes on functional activity changes (and their relationship in the mutational landscape).

Most biopharmaceuticals are either natural proteins or slightly modified versions of natural proteins. Nature recycles good ideas and presents sets of homologous proteins sometimes performing divergent functions in different organisms. In these cases, ancestral sequence reconstruction and mutational mapping can be used as valuable tools in protein drug design. Basically, nature has provided a library of functional molecules, and ancestral sequence reconstruction may be used

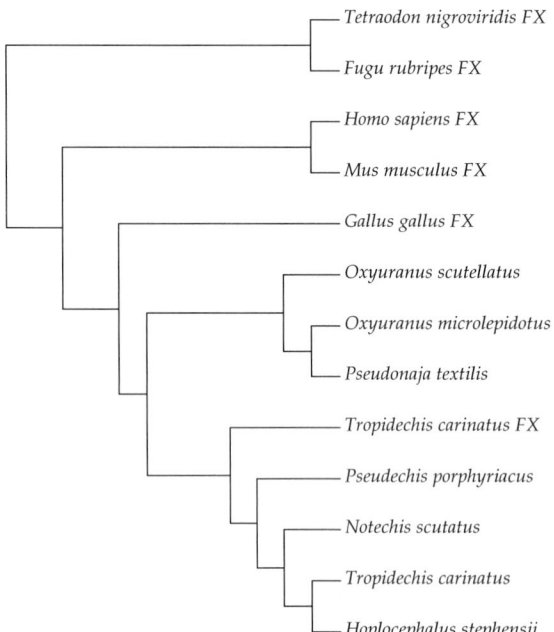

Figure 3.3 The phylogenetic relationship of Factor X (FX) homologs from Australian snakes calculated using Mr. Bayes is shown (Ronquist and Huelsenbeck 2003). Factor X has been sequenced from a wide variety of vertebrate species. Only one snake Factor X from the homeostatic system has been sequenced (*Tropidechis carinatus* Factor X). This sequence clearly shows that the Factor X-like venomous peptides have evolved recently from vertebrate Factor X.

to expand this library. This allows the protein designer to discover combinations of functional mutations that are otherwise difficult to screen for. When additional functionalities have arisen in homologous proteins, such as stability or bioavailability, the reconstruction of the common ancestors enables determination of when the different properties arose, and shows which substitutions drove the changes.

Some challenges remain. Most critically, the bioactive proteins and peptides isolated from venoms often come from reptiles, which are poorly represented in genomic databases. Another problem is the possible immune response. Biopharmaceuticals originating from venom can, like other proteins unknown to the human body, cause the generation of antibodies and an immune response. This problem, however, is not expected to be greater for a reconstructed protein/peptide compared to biopharmaceuticals engineered by any other method. In fact, substitutional mapping can be used to incorporate residues from the human protein that may lessen the immune response. Nevertheless, the general idea of looking at a protein as member of a family of homologous proteins which together contain more information than any individual sequence is as true for a protein drug candidate as it is for any other protein.

References

Barbas, C.F., Kang, A.S., Lerner, R.A., and Benkovic, S.J. (1991) Assembly of combinatorial antibody libraries on phage surfaces: the gene III site. *Proc. Natl. Acad. Sci. USA* **88**: 7978–7982.

Broad, A.J., Sutherland, S.K., and Coulter, A.R. (1979) The lethality in mice of dangerous Australian and other snake venom. *Toxicon* **17**: 661–664.

Castro, H.C., Zingali, R.B., Albuquerque, M.G., Pujol-Luz, M., and Rodrigues, C.R. (2004) Snake venom thrombin-like enzymes : from reptilase to now.*Cell. Mol. Life Sci.* **61**: 843–856.

Eng, J., Kleinman, W.A., Singh, L., Singh, G., and Raufman, J.-P. (1992) Isolation and characterization of exendin-4, an exendin-3 analogouge, from *Heloderma suspectum*. *J. Biol. Chem.* **267**: 7402–7405.

Fry, B.G. (1999) Structure-function properties of venom components from Australian elapids. *Toxicon* **37**: 11–32.

Kieffer, T.J., McIntosh, C.H., and Pederson, R.A. (1995) Degradation of glucose-dependent insulinotropic polypeptide and truncated glucagon-like peptide 1 in vitro and in vivo by dipeptidyl peptidase IV. *Endocrinology* **136**: 3585–3596.

Kini, R.M. (2005) The intriguing world of prothrombin activators from snake venom. *Toxicon* **45**: 1133–1145.

Liberles, D.A., Schreiber, D.R., Govindarajan, S., Chamberlin, S.G., and Benner, S.A. (2001) The Adaptive Evolution Database (TAED). *Genome Biology* **2**: research0028.1–0028.6.

Liou, G.F. and Khosla C. (2003) Building-block selectivity of polyketide synthases. *Curr. Opin. Chem. Biol.* **7**: 279–284.

Lopez, P., Casane, D., and Philippe, H. (2002) Heterotachy, an important process of protein evolution. *Mol. Biol. Evol.* **19**: 1–7.

Morrison, K.L. and Weiss, G.A. (2001) Combinatorial alanine-scanning. *Curr. Opin. Chem. Biol.* **5**: 302–307.

Ronquist, F. and Huelsenbeck, J.P. (2003) Mr Bayes 3: Bayesian phylogenetic inference under mixed models. *Bioinformatics* **19**: 1572–1574.

Skovgaard, M., Kodra, J.T., Gram, D.X., Knudsen, S.M., Madsen, D., and Liberles, D.A. (2006) Using evolutionary information and ancestral sequences to understand the sequence-function relationship in GLP-1 agonists. *J. Mol. Biol.* **363**: 977–988.

Toft-Nielsen, M.B., Madsbad, S., and Holst, J.J. (2001) Determinants of the effectiveness of glucagon-like peptide-1 in type 2 diabetes. *J. Clin. Endocrinol. Metab.* **86**: 3853–3860.

Zander, M., Madsbad, S., Madsen, J.L., and Holst, J.J. (2002) Effect of 6-week course of glucagon-like peptide 1 on glycaemic control, insulin sensitivity, and beta-cell function in type 2 diabetes: a parallel-group study. *Lancet* **359**: 824–830.

II
Computational methodology and concerns

CHAPTER 4

Probabilistic models and their impact on the accuracy of reconstructed ancestral protein sequences

Tal Pupko, Adi Doron-Faigenboim, David A. Liberles, and Gina M. Cannarozzi

4.1 Probabilistic evolutionary models

Recent large-scale sequencing efforts are changing the focus of biological research. To understand and utilize the ever-increasing sequence databases, one must use sequence evolutionary models (reviewed in Whelan et al., 2001). These models are fundamental in various bioinformatics endeavors, such as prediction of protein structure and function, sequence-motif finding, active-site prediction, evolutionary studies, gene prediction, comparative genomics, RNA-structure prediction, tree reconstruction, and ancestral sequence reconstruction (ASR). When using evolutionary models, we have to be careful in choosing the model assumptions. There are many examples in which the use of unrealistic models of sequence evolution leads to erroneous conclusions (Sullivan and Swofford, 1997; Pupko et al., 2002a). Novel models of sequence evolution are continuously being developed in terms of both modeling choices and computational tools. Advanced statistical techniques are being used to learn the parameters of these models and to predict with them. Special effort is being directed to take into account better the realistic biological phenomena, removing possible sources of error from existing oversimplified models. In this chapter, the effect of the model assumptions on ASR will be discussed. Existing methods for sequence reconstruction are based on the maximum-parsimony criterion or on probabilistic models. ASR using probabilistic models is based on either the maximum-likelihood or the Bayesian paradigms. This chapter will focus on probabilistic-based methods for ASR of protein-coding sequences. However, for completeness, in the following section we briefly describe ASR based on the maximum-parsimony criterion, which was widely used before probabilistic ASR methodology was developed.

4.2 ASR based on the maximum-parsimony criterion

The idea of maximum parsimony is to identify the ancestral states at each node of a tree that minimize the number of character changes needed to explain the observed differences among the sequences at the leaves. Algorithms for ASR based on this criterion were developed by Fitch (1971), Sankoff (1975), and Sankoff and Rousseau (1975). These algorithms use dynamic programming, ensuring efficient reconstruction. The Fitch algorithm, introduced with nucleotide sequence data, penalizes equally any change among the four character states (*A*, *C*, *G*, and *T*). For the reconstruction of a specific position, the algorithm proceeds by assigning to each node of the tree a set of character states that are compatible with the minimum number of changes. The algorithm processes the tree in post-order; that is, each tree node is visited only after its descendants are visited. Thus, the algorithm starts by assigning character sets to the leaves of the tree. If, for example, a leaf is labeled by the character *A*, a set {*A*} is assigned to that leaf. Next, an internal node for which both descendents have already been visited is evaluated. The set assigned to this internal node is the intersection of the sets at its two descendant nodes if this intersection is not empty, or the union of the two sets if the intersection is empty. If the new set is a union, one change is counted so that the number of changes is the number of the union operations. The next step is to traverse the tree from root to leaves, in pre-order, to determine the ancestral states for internal nodes. Initially the ancestral state at the root is equal to the character state in its set. If this set includes more than one character then different, equally parsimonious reconstructions exist. Then, each descendent of the root is evaluated as follows: if the ancestral state at the root is a member of the set of the descendent node the same ancestral state is assigned to the descendent node, otherwise another state from the set in the descendent node is chosen. This procedure is applied for each node in the tree. This procedure will find some of the most parsimonious reconstructions but not all. To guarantee that all most parsimonious reconstructions are found, comparisons involving the outgroups of a node must be performed (Maddison *et al.*, 1984; Harvey and Pagel, 1991).

To exemplify the algorithm of ASR using Fitch's algorithm consider the simple five-taxon tree in Figure 4.1. For the character illustrated, the data observed are $X_1 = A$, $X_2 = C$, $X_3 = G$, $X_4 = C$, and $X_5 = T$. At the leaves the character sets are simply: $X_1 = \{A\}$, $X_2 = \{C\}$, $X_3 = \{G\}$, $X_4 = \{C\}$, and $X_5 = \{T\}$. At node X_6, the intersection of the sets of its two descendants X_1 and X_2 is $\{A\} \cap \{C\} = \phi$. Hence, the union set is assigned: $\{A\} \cup \{C\} = \{A, C\}$. Likewise, the union set of X_3 and X_4 is assigned at node X_7; that is, $\{G\} \cup \{C\} = \{G, C\}$. Now the set at node X_8 can be determined, since the intersection of sets X_5 and X_7 is again empty, the union of these sets $\{G, C, T\}$ is assigned. Finally the set in the root (X_9) is the intersection of the sets X_8 and X_6: $\{A, C\} \cap \{G, C, T\} = \{C\}$. Three union operations were needed; thus, a minimum of three changes is needed for this reconstruction. In the next step, the ancestral states are determined (marked in bold type in Figure 4.1) by traversing the tree in pre-order (from the root to the leaves). First the state C is determined at the root; the state at X_8 is also set to C since this state is the ancestral state in the parent (X_9) and is a member of the set at that node (X_8). Similarly, the state at X_6 is C since this state is the ancestral state in the parent, as well as a member of the set at node X_6. Finally, the state at node X_7 is assigned and is also equal to C.

The Sankoff (1975) algorithm is a generalization of Fitch's algorithm. Instead of assuming that all state changes are equally likely, it allows different costs for different character changes. Similarly to the Fitch algorithm, the tree is visited in post-order followed by pre-order steps.

Both algorithms may reconstruct more than one ancestral state for each node. When the results are ambiguous two different methods of assignment—acceleration transformation (ACCTRAN) or delayed transformation (DELTRAN)—can be applied (Swofford and Maddison, 1987). ACCTRAN assumes that the character changes happen at the earliest possible point and thus prefer reversals over convergences. DELTRAN tries to delay the changes and thus maximizes parallelism.

4.3 ASR using probabilistic models

In phylogeny, probabilistic models (both maximum-likelihood and Bayesian approaches) are considered the state-of-the-art methods for tree reconstruction (Holder and Lewis, 2003). Felsenstein's (1981) seminal work showing how to efficiently compute the likelihood of a tree, together with the efficient computer program PHYLIP

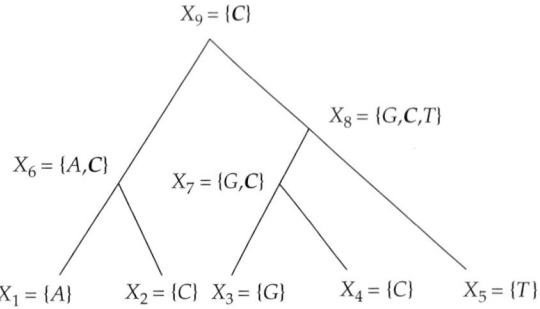

Figure 4.1 Illustration of the maximum-parsimony algorithm of Fitch for ancestral sequence reconstruction. The post-order (leaves to root) step of the algorithm is shown.

(Felsenstein, 2005; distributed from 1980) boosted interest in probabilistic models for phylogeny. During the next two decades, such probabilistic approaches replaced the previously more common maximum-parsimony approach as numerous studies demonstrated the many shortcomings of the latter (e.g. Holder and Lewis, 2003). For instance, maximum parsimony is inherently biased toward overestimating the number of "common to rare" changes (Eyre-Walker, 1998). Furthermore, this method does not supply statistically robust means for discriminating among equally parsimonious reconstructions (Yang et al., 1995). Unlike maximum parsimony, the probabilistic approach also allows the statistical testing of various hypotheses, such as testing whether two tree topologies are significantly different or testing for a monophyletic origin of a clade (reviewed in Goldman et al., 2000).

The history of ASR followed that of tree reconstruction, albeit with a delay. Until the concept of probabilistic sequence reconstruction was introduced in the early 1990s (Gonnet and Benner, 1991; Schluter, 1995; Yang et al., 1995; Koshi and Goldstein, 1996), maximum parsimony was the method of choice (e.g. Jermann et al., 1995; Stewart, 1995). However, it is clear that the same shortcomings of maximum parsimony that have been mentioned in the context of tree reconstruction are also valid for ASR that is parsimony based. Later Koshi and Goldstein (1996) and Pupko et al. (2000) developed efficient algorithms for both joint and marginal reconstruction using the probabilistic approach (see below for a detailed explanation of these concepts).

A vital advance in the development of evolutionary models was the consideration of heterogeneity of evolutionary rates among sequence sites (Yang, 1993). Yang has shown that a model that takes into account such among-site rate variation significantly increases the tree likelihood. In the case of ASR, Pupko et al. (2002b) showed via simulation that failure to account for among-site rate variation also reduces the accuracy of ASR and results in lower likelihood for the reconstructed sequences. As can be expected, a general pattern emerges: models that better fit data for tree reconstruction are also better for ASR. Areas in which model improvements have been attempted include accounting for rate variation between different amino acids (the substitution matrix) and the variation of this substitution matrix among different sites of a protein, and among different branches of the phylogenetic tree. The tree and its associated branches are also considered as part of the probabilistic model. Thus the impact of taking into account uncertainties in tree topology and model parameters within a Bayesian approach on the accuracy of ASR was also explored. In the subsequent sections we describe each aspect in detail.

4.4 How ancestral sequences are computed using probabilistic models

To exemplify the concept of probabilistic ASR consider the simple four-taxon tree in Figure 4.2. For simplicity we consider a two-state (0 or 1) alphabet (for example, polar and non-polar amino acids). The ancestral character assignments x_5 and x_6 at the internal nodes X_5 and X_6 are unknown. Numbers above branches indicate branch lengths; that is, the average number of substitutions per sequence site. In all probability-based models, the probabilities are expressed in terms of summations and multiplications of $P_{ij}(t)$ factors, the probability that character i will be replaced by character j along a branch of length t. The $P_{ij}(t)$ factors are usually expressed in a matrix form $P(t)$, so that $[P(t)]_{ij} = P_{ij}(t)$. The matrix $P(t)$ can be computed by $P(t) = e^{Qt}$, where Q is the instantaneous rate matrix (Felsenstein, 2004). The probability model also contains initial probabilities: for each character x, $P(x)$ denotes the probability of observing x at the root of the tree. The likelihood of the data

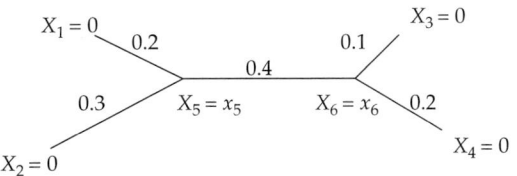

Figure 4.2 An example of probabilistic-based ancestral sequence reconstruction. The example is for the simple case of a two-states alphabet (0 and 1). Numbers above branches reflect evolutionary distances. In joint reconstruction, the pair of characters that is assigned to the two internal nodes X_5 and X_6 is that which maximizes the likelihood of the data.

describes the probability of observing the characters at the leaves given the tree topology, the branch lengths, and the $P_{ij}(t)$ factors. Thus, the following expression represents the likelihood of the tree in Figure 4.2:

$$\sum_{x_5}\sum_{x_6} P(x_5) P_{x_5,x_1}(0.2) P_{x_5,x_2}(0.3) P_{x_5,x_6}$$
$$(0.4) P_{x_6,x_3}(0.1) P_{x_6,x_4}(0.2) \qquad (4.1)$$

This likelihood is a sum of four different terms, each corresponding to a specific ancestral sequence assignment ($x_5 = x_6 = 0$, $x_5 = x_6 = 1$, $x_5 = 0$ and $x_6 = 1$, and $x_5 = 1$ and $x_6 = 0$). In this case, internal node X_5 was arbitrarily chosen as the root of the tree. Most evolutionary models used are time-reversible. In mathematical terms, a model is time-reversible if $P(i)P_{ij}(t) = P(j)P_{ji}(t)$ for all pairs of characters, i and j. Felsenstein (1981) showed that for time-reversible models, the position of the root of the tree does not affect the likelihood score. The joint character assignment (x_5, x_6) which contributes the most to the above likelihood is called the joint ASR and is explicitly given by the expression:

$$\mathrm{argmax}_{x_5,x_6} [P(x_5) P_{x_5,x_1}(0.2) P_{x_5,x_2}(0.3) P_{x_5,x_6}$$
$$(0.4) P_{x_6,x_3}(0.1) P_{x_6,x_4}(0.2)] \qquad (4.2)$$

The maximum of the above expression is the likelihood of the joint reconstruction. In essence, x_5 and x_6 are two non-independent random variables over the {0, 1} set. The term of eqn 4.1 above, in which $x_5 = 0$ and $x_6 = 0$, corresponds to $P(x_5 = 0, x_6 = 0, data)$, whereas $data$ refers to $x_1 = 0$, $x_2 = 0$, $x_3 = 1$, and $x_4 = 0$. The probability of $x_5 = 0$ and $x_6 = 0$ given the data is simply

$$P(x_5 = 0, x_6 = 0 \mid data) = \frac{P(x_5 = 0, x_6 = 0, data)}{P(data)} \qquad (4.3)$$

$P(data)$ is exactly the expression given in eqn 4.1. The four possible values $P(x_5, x_6 \mid data)$ can be expressed in tabular form: see table 4.1. From these joint probabilities, one can easily compute the marginal probabilities. For example, the probability that character 0 was the ancestral state at node X_6, given the available current sequence, is $P(x_6 = 0 \mid data) = P(x_5 = 0, x_6 = 0 \mid data) + P(x_5 = 1, x_6 = 0 \mid data)$. Thus, if we are interested in the best

Table 4.1 Joint and marginal probabilities for the ancestral sequence reconstruction of the tree in Figure 4.2

	$x_6 = 0$	$x_6 = 1$	Sum
$x_5 = 0$	$P(x_5 = 0, x_6 = 0 \mid data)$	$P(x_5 = 0, x_6 = 1 \mid data)$	$P(x_5 = 0 \mid data)$
$x_5 = 1$	$P(x_5 = 1, x_6 = 0 \mid data)$	$P(x_5 = 1, x_6 = 1 \mid data)$	$P(x_5 = 1 \mid data)$
Sum	$P(x_6 = 0 \mid data)$	$P(x_6 = 1 \mid data)$	1

character assignment to node X_6, we should compare $P(x_6 = 0 \mid data)$ and $P(x_6 = 1 \mid data)$—if the former is higher, 0 is the most likely reconstruction at this node, otherwise it is 1. As was shown by Pupko et al. (2000), joint and marginal reconstructions are not always the same. If for example $P(x_5 = 0, x_6 = 0 \mid data) = 0.4$, $P(x_5 = 0, x_6 = 1 \mid data) = 0.3$, $P(x_5 = 1, x_6 = 0 \mid data) = 0.05$, and $P(x_5 = 1, x_6 = 1 \mid data) = 0.25$, the highest would be the first, indicating character 0 at both internal nodes. However, the marginal probabilities of assigning 1 to node 6 would be 0.55, indicating that this is the most likely marginal reconstruction. A similar explanation for such ancestral reconstruction probabilities can be found in Koshi and Goldstein (1996) and Yang et al. (1995).

4.5 ASR taking into account among-site rate variation

Whereas heterogeneity in the number of amino acid replacements at different sites can result from the stochastic nature of the substitution process, the observed pattern of variability in the number of replacements significantly deviates from the expected pattern under a model which assumes a homogenous rate at all sites (Uzzell and Corbin, 1971). This rate heterogeneity stems from the fact that not all sites in a protein are subject to the same evolutionary constraints. Sites that are important to maintaining the structure or function of a protein, such as the active-site residues, are usually highly conserved, whereas other sites, such as those on the protein surface, undergo substitution at much higher rates. Among-site rate variation (ASRV) models aim to express mathematically this heterogeneity of evolutionary rates across sites.

In ASRV models it is assumed that each site has a fixed rate r, which indicates how fast this position evolves relative to the average rate over all positions (Yang, 1993). Thus, a site with a rate of 2 evolves twice as quickly as the average; that is, the expected number of substitutions at this site is twice that of the average. More formally, when a position evolves at a rate r, we assume that the rate matrix Q underlying the site's Markovian process is multiplied by r. We note that since $P(t) = e^{Qt}$, the same replacement probabilities will be obtained by either multiplying the rate matrix by r or by multiplying the branch length t by the same factor r. This equivalency shows that the likelihood of a position that evolves at a rate r can be computed by first multiplying all the branch lengths of the tree by r, and then computing the likelihood of that position with Q. This allows the computation of the likelihood of all sites with the same Q, rather than with a different Q matrix for each rate, thus significantly reducing computation time.

The rate at each site is in general unknown. One can try to estimate the most likely rate at each site (e.g. Nielsen, 1997; Pupko et al., 2002c) but this requires a separate parameter at each site. Another biologically relevant model is to assume that a percentage of sites is invariant; that it has zero rate of change (Hasegawa et al., 1987). More often, a distribution of rates is assumed. Uzzell and Corbin (1971), the first to model rate variation, used a gamma distribution because it was analytically tractable, varied from 0 to infinity, and had a parameter that allows controlling the extent of the rate variation (see also Felsenstein, 2001). Olsen (1987) modeled ASRV using a lognormal distribution. Although continuous-probability density functions can be numerically integrated for any rate density function, for computational ease discrete models are more often used. A discrete approximation of the gamma distribution (Yang et al., 1994) is by far the most widely used rate distribution.

When the goal is to reconstruct the tree topology or the ancestral characters, it is preferable to include the various possible rates in the computation in a probabilistic manner (Felsenstein, 2001). Thus it is usually assumed that several rates are allowed at each site, each rate with a specific probability. Given such a rate distribution, the likelihood of the data is computed by summing the likelihood over all possible rates, taking into account their probabilities.

In Yang's paper on probabilistic ASR (Yang, 1995), ASRV was not taken into account. Yang suggested that "the relative contributions to the likelihood by different reconstructions at a site [are] unlikely to change significantly when the branch lengths are multiplied by a constant..." Although this statement may be true regarding one most likely character at a specific node at a specific site, it is not true for the probability vector: the probabilities of each character at each node and site (see Chapter 5 in this volume). The most likely character is just the one that maximizes the probability vector. This probability vector is often taken to represent the confidence interval of the reconstruction, and moreover it is used in various applications such as detecting co-evolving substitutions and substitution mapping (Dimmic et al., 2005; Dutheil et al., 2005; Bollback, 2006) and detecting radical replacements (Pupko et al., 2003). It is intuitively expected that the most likely reconstruction will be less sensitive to model assumptions. Yet this is clearly not the case for probability vectors and for applications which use such vectors as part of their computations. Furthermore, it was later shown that taking into account ASRV significantly increases the accuracy of the most likely reconstruction, especially for sequences that are highly divergent (Pupko et al., 2002b). Thus, ASRV should be especially critical when highly divergent sequences are reconstructed, which is often the case (e.g. Thornton, 2001; Chang et al., 2002).

In some cases, computing ancestral sequences using ASRV model is not trivial. When marginal reconstruction is needed, computational time is linear with the number of sequences whether or not ASRV is assumed. However, computing joint reconstruction assuming ASRV is exponential with the number of sequences. To this end, Pupko et al. (2002b) developed an efficient branch-and-bound algorithm that, although exponential in the worst case, can handle dozens of sequences in most practical cases. It should be noted that this algorithm *guarantees* finding the joint maximum-likelihood reconstruction; it is not an heuristic approach.

One limitation of all ASRV models discussed above is that they assume a constant rate throughout evolution, which is not always the case. This issue is discussed in section 4.8.

4.6 Model selection and ASR

Which rate distribution is best for ASR? As in phylogenetic reconstruction, the best distribution cannot be determined *a priori* and should be determined based on the available data. The most widely used distribution for modeling ASRV is the discrete gamma distribution. Is the discrete gamma distribution the best? In terms of likelihood it seems that allowing a proportion of the sites to be invariable and having the rest of the rates sampled from a gamma distribution results in an improved likelihood (Gu *et al.*, 1995). However, this gamma + invariant model is also not ideal, as the proportion of invariant sites seems to be highly sensitive to the number of sequences used in the analysis. Recently, Mayrose *et al.* (2005) suggested using a mixture of gamma distributions to better account for the complicated pattern of ASRV. This model significantly increases the likelihood and it is expected that it will also increase the accuracy of ASRs and methods that rely on them. Complex evolutionary models are usually more realistic from the biological point of view. Yet they often require the estimation of additional parameters from the same amount of data, and hence the standard error of each estimated parameter is increased. This tradeoff between rich models with many parameters and simple models with a few parameters is the topic of extensive research in statistics, which is known as model selection (Burnham and Anderson, 2003). In general, criteria such as the Akaike Information Criterion (AIC) and the likelihood ratio test (LRT; Posada and Buckley, 2004) are often used to compare different models. Hence, it is advisable to search for the best-fitting model for the given data using programs such as ProTest (Abascal *et al.*, 2005), and only then to perform ASR using this model. Theoretically, it is best to estimate both the ancestral sequences and the model parameters simultaneously. However, a two-step approach, consisting of first finding the parameters that maximize the likelihood of the data (average over all reconstructions), and then fixing these parameters for the ASR, is more efficient and should provide essentially identical results as the simultaneous approach.

4.7 The instantaneous rate matrix and its impact on ASR

As stated above, current likelihood models are based on two components: the rate matrix Q which determines the substitution probabilities and the ASRV parameters. The determination of Q is critical to any evolutionary model. As it turns out, different approaches for the determination of Q were developed for DNA (coding or non-coding), RNA, amino acids, and codons. As the focus of this book is on ancestral protein sequences, we will describe in detail only amino acid- and codon-based matrices.

Most amino acid matrices are empirical. Empirical matrices are derived from large data-sets and hence accurate estimates of amino acid-replacement probabilities can be obtained. A few empirical amino acid-replacement matrices have been proposed previously (Dayhoff *et al.*, 1978; Gonnet *et al.*, 1992; Jones *et al.*, 1992; Adachi and Hasegawa, 1996; Whelan and Goldman, 2001). These matrices are used extensively in amino acid-based applications such as programs for multiple sequence alignment, detecting distant homologs, phylogenetic tree reconstruction, and ASR (e.g. Thompson *et al.*, 1994; Altschul *et al.*, 1997; Penny and Hasegawa, 1997). The strength of empirical matrices stems from the fact that they are derived from averaging over many genes from various organisms, and thus the rate-matrix entries usually correspond to accurate estimation for the *average* replacement probabilities. However, in these matrices, no information about replacement probabilities can be learned from the specific data analyzed; that is, Q has no free parameters. This concern can be a major difficulty when the protein analyzed evolves in a manner distinct from the hypothetical average protein.

Should the same amino acid matrix be used to model the evolution of all positions of a given protein? It is well known that not all regions within a protein evolve under the same evolutionary constraints. For example, transmembrane regions of proteins are known to evolve under different evolutionary constraints than non-transmembrane regions. Jones *et al.* (1994) have computed specific amino acid-replacement matrices for transmembrane and non-transmembrane domains. Other context-dependent matrices were developed for secondary structures (α-helices, β-sheets and loops) or for buried and exposed structural elements (Koshi and Goldstein, 1995). Unfortunately, these matrices are not commonly used for ASR, although it is expected that using an α-helix-based matrix when reconstructing the ancestral characters of an α-helix region will result in more accurate reconstruction compared to a general matrix such as the widely used JTT matrix of Jones *et al.* (1992). In addition, specific amino acid matrices were developed for the mitochondrial (Adachi and Hasegawa, 1996) and chloroplast (Adachi *et al.*, 2000) genomes. These matrices reflect both the different mutation pattern in these genomes (correlated with the different genetic codes used in these genomes, the different replication machinery, etc.) and the different selection pressures on these organelles. With the advent of more sophisticated algorithms for constructing amino acid-replacement models (e.g. Muller and Vingron, 2000; Muller *et al.*, 2002), it is expected that more such context-dependent matrices will be developed.

An effort was made to create mechanistic models for amino acid-replacement probabilities—models which include parameters that are fit to each data-set analyzed. For example, in some models Q_{ij} depends on the difference in fitness between the pair of amino acids, which is based on some physicochemical property, such as α-helical propensity or hydrophobicity (Koshi *et al.*, 1997; Koshi and Goldstein, 1998). Another approach is to construct amino acid-based matrices from codon models (Yang *et al.*, 1998). These models provide an intriguing alternative to the commonly used empirical matrices. More research is needed to test their applicability for inference of phylogenetic trees and for ASR.

All the models suggested above assume that the same Q matrix models all sites. These models thus do not allow heterogeneity of substitution pattern among sites. Models that take into account among-site substitution variation are similar in concept to discrete ASRV models, in the sense that in the former each site can evolve according to some predefined rate category, and in the latter each site can evolve according to some predefined Q matrix. Such an approach was used by Dimmic *et al.* (2000), in which several Q matrices were used to model mitochondrial proteins; each such matrix constructed so that it can model different selection forces underlying the evolution of the various sites of these proteins.

Goldman and Yang (1994) and Muse and Gaut (1994) were among the first to suggest mechanistic codon-based evolutionary models. The more sophisticated model by Goldman and Yang takes into account the transition/transversion bias, the codon frequencies, and the different replacement probabilities between amino acids based on the Grantham (1974) physicochemical distance matrix. However, these models did not account for the heterogeneity of the evolutionary selection pressure among protein sites. Nielsen and Yang (1998) and Yang *et al.* (2000) further developed mechanistic Bayesian models that account for such selection heterogeneity. In their model, a prior distribution of the ratio of the non-synonymous substitution rate (K_a) to the synonymous substitution rate (K_s) is assumed. Sites showing K_a/K_s values significantly lower than 1 are regarded as undergoing purifying selection and therefore may have a functionally or structurally important role. Sites showing K_a/K_s values significantly higher than 1 are indicative of positive Darwinian selection, suggesting adaptive evolution. However, unlike the model of Goldman and Yang (1994), these models ignore the fact that distinct amino acids differ in their replacement rates. Recently, an empirical codon-substitution matrix was developed by Schneider *et al.* (2005). This model was used to estimate synonymous distance between coding sequences (Schneider *et al.*, 2006). Another direction is the derivation of codon matrices from empirical amino acid matrices (Doron-Faigenboim and Pupko, 2006).

Current efforts in codon models focus on their applicability for detecting positive selection. However, it is clear that for coding sequences these models can be very helpful for phylogenetic reconstruction and ASR. In this respect, the different substitution rates between amino acids must be taken into account, as in the original paper of Goldman and Yang (1994).

4.8 Covarion models and their impact on ASR

Several new approaches show promise in generating better models of protein evolution. The realization that different sites in a protein evolve at different rates has led to the use of the above mentioned ASRV models. However, these models assume that over the course of evolution, the rate of substitution remains unchanged at a given protein site. Recent studies have shown this may not always be the case (e.g. Lopez et al., 2002). Sites that are highly conserved in one part of the tree may be variable in the rest of the tree and vice versa. Although this notion, termed covarion, was described decades ago by Fitch and Markowitz (1970), an evolutionary probabilistic model for the covarion process was only developed relatively recently (Galtier, 2001; see also Chapter 5). Galtier's covarion model assumes that the rate itself is not fixed for each site, but rather follows a continuous-time Markov process along the tree, with a specific rate-change parameter. The higher this rate-switching parameter, the more common the rate jumps. In his pioneering work Galtier applied such a covarion model to infer the ancestral GC content of the most recent common ancestor of extant life forms. The moderate GC content found in this study brings into question the previously accepted hypothesis of a hyperthermophilic (or even thermophilic) origin of the most recent common ancestor. Gaucher et al. (2003) also used ASR to study whether the ancient bacteria were thermophiles. In this work, sequences of ancestral elongation factors were inferred, synthesized in the laboratory and, finally, their activity measured empirically as a function of temperature. Interestingly, the optimal temperature of the ancestral protein found strengthens the view that ancient bacteria were thermophiles (although not hyperthermophiles).

It is difficult to draw many conclusions about the effect of covarion on ASR, since in Galtier's study the covarion model was used to study ancestral GC content rather than to infer ancestral sequences. Thus, the effect of the covarion assumption on the accuracy of the reconstructed sequences was not tested. Furthermore, the model used was a non-homogenous one (see Chapter 9 in this volume, where this is discussed in more detail), and thus no analysis was performed to determine the effect of accounting for the covarion in standard homogenous models. However, especially when highly divergent sequences are analyzed, ASR is expected to be much more accurate when covarion models are introduced. This is especially true as it was shown that refraining from taking into account the covarion process underestimates the number of multiple substitutions (Galtier, 2001).

4.9 Deviations from homogenous stationary reversible models

Standard evolutionary models usually make the following three assumptions: (1) the process is homogenous, so that the same Q matrix models the evolution in all branches; (2) the process is stationary, so that, on average, the same character frequencies hold for all branches; (3) the process is reversible, so $P(i)Q_{ij} = P(j)Q_{ji}$. These concepts are explained elsewhere, for example by Galtier and Gouy (1998). All these assumptions are known to be violated in many biological examples (e.g. Lockhart et al., 1994; Galtier and Gouy, 1995; Chang and Campbell, 2000). The main reason for these assumptions is to avoid the introduction of too many parameters, and thus risking over-fitting the data (see Chapter 5). In addition, models that do not make these assumptions are computationally intensive for some tasks, such as finding the maximum-likelihood tree. Thus these models may not be applicable even for moderately sized datasets. When the tree topology is known, however, as is often the case in many ASR studies, this computational limitation is not a real hurdle. This enables the use of such complex models, and the subsequent gains of insights from these models regarding the function of the ancestral sequences.

One such example is the non-homogenous model developed by Galtier and Gouy (1998).

In their model, the G+C content is allowed to change over time, so that each branch along the tree can have a distinct G+C content. They have shown via simulation studies that maximum-likelihood inference with this model can accurately infer the ancestral G+C content. They then used a variant of this model (allowing also ASRV) to reconstruct the G+C content in the most recent common ancestor (see section 4.8).

Yap and Speed (2005) compared three models of nucleotide substitutions: the standard reversible one (REV), a stationary non-reversible one (STAT), and a non-stationary, non-reversible one (NONSTAT). They found that the NONSTAT model significantly improved the likelihood and could be used to root simple trees. In their NONSTAT model, the nucleotide frequencies at the root can be different from the stationary frequencies. Although the authors did not explore the possible effect of the NONSTAT model on ASR, their model suggests a way to test whether the character frequencies at extant sequences reliably reflect those of the ancestral sequence. The effects and impact of the different assumptions—non-reversibility and non-stationarity—on ASR await further investigation.

4.10 Using additional information

Accurate ASR depends on how well the model fits the data. Each model contains parameters such as tree topology, branch lengths, and the transition/transversion ratio, which are estimated from the data. This estimation depends on the amount of information the data contain: in general, the more sequences and positions available for analysis, the higher the accuracy of each estimated parameter. However, one can distinguish between two types of data. The first is the protein alignment whose ancestral sequences are to be inferred. We term this the ASR data. The second type of data is any information that is not directly related to the protein sequence analyzed, yet can contribute to the accuracy of the ASR. We term this second type additional information.

Assume our goal is to reconstruct the ancestral cytochrome b of human, chimpanzee, gorilla, and baboon. A simple approach would consider adding an outgroup (e.g. mouse), and using these ASR data to search for the maximum-likelihood tree topology and branch lengths and then to compute the ancestral sequence at the ancestor of these primates. An alternative to this approach is instead of searching the tree topology based on cytochrome b sequences alone, to estimate the species tree based on a large set of orthologous sequences available for all these organisms. In this case, the species tree is assumed to reflect the topology of the gene trees, and this tree should be used for the ASR, rather than the maximum-likelihood tree based on the ASR data alone.

Such an approach was used for example in Krishnan *et al.* (2004) when reconstructing ancestral primate mitochondrial DNA. Even when a Bayesian approach, which takes into account many alternative trees, is considered, one should compute the trees' posterior probabilities not from the ASR data alone, but rather from all other available additional information. When using information external to the sequences to build a gene tree, one should be careful to evaluate the fit of the gene tree to the imposed species tree. Techniques for such evaluation have been developed previously using the parsimony (Berglund-Sonnhammer *et al.*, 2006) or the Bayesian (Arvestad *et al.*, 2003) frameworks.

Assuming the tree topology is known, should we infer branch lengths from the ASR data or can we use side information for more accurate branch-length estimation? In other words, for the above example, assume that in addition to the cytochrome b sequences we also have the sequences of cytochrome c oxidase subunit 1 from these four primates and mouse: how can this information be used to more accurately estimate the branch lengths of the cytochrome b tree? One approach is to concatenate these two data-sets and infer the branch lengths from the concatenated data. However, as has been shown previously, concatenation is by far the least-correct method for combining different data-sets (Yang, 1996; Pupko *et al.*, 2002a). In essence, concatenation assumes that all genes evolve at the same rate, an assumption that is known to be wrong: some genes are highly conserved (slow evolving) whereas others are highly variable (fast evolving). An alternative approach (the separate model) is to assume that branch lengths of each data-set are independent of each other. This assumes that no information is shared regarding evolutionary rates of different species.

It is known that mice evolve faster than humans for all genes, so that the branch leading to human should be, on average, shorter than the branch leading to mouse. When assuming the separate model, this information is ignored, and a possible increase in branch-length accuracy is overlooked. Furthermore, the separate model assumes a free parameter for each branch for each data-set: a very large number of parameters resulting in decreased accuracy for each branch length.

An intermediate approach, first suggested by Yang (1996), is to assume a proportional model, in which a base tree topology and branch lengths are assumed to be shared among all members of the data-set (see also Pupko et al., 2002a; Bevan et al., 2005). The tree for each gene in the data-set has the same topology as this base tree, but its branches are multiplied by a fixed rate factor, the evolutionary rate of this gene. To exemplify this point consider a hypothetical case in which the analysis of many genes from human and chimpanzee resulted in an estimate that the chimpanzee evolves 1.05 times faster than human. It can then be assumed that this same ratio exists between the branch lengths leading to human and chimpanzee in the analyzed ASR data. Knowing such relative evolutionary rates between organisms can significantly reduce the number of free parameters. We note that for some genes the proportionality assumption can be rejected if, for example, rapid evolution of the gene in a specific lineage has occurred. Therefore, we suggest first testing which model best fits the entire data analyzed (concatenation, proportional, or separate analysis) and then use the best model for branch-length estimation.

Finally, should we use other vertebrate cytochrome b sequences if our goal is only to reconstruct the human/chimpanzee/gorilla/baboon ancestral sequence? In general, the answer is yes. All model parameters (tree topology and branch lengths, the α-parameter of the gamma distribution, the transition/transversion ratio in the case of a DNA-based model, amino acid frequencies, etc.) are more accurately inferred in cases when more data are available. One should always be cautious to specific cases in which this general assumption of more data equating to more accuracy fails. The identification of such cases is strongly linked with methods aimed at identifying functional shifts in proteins, and is an active area of current research (e.g. Gaucher et al., 2002; Pupko and Galtier, 2002; Gu, 2003).

4.11 Taking into account uncertainties in the tree topology, branch lengths, and model parameters

In the empirical Bayesian approach of ASR, one computes the most likely tree topology, branch lengths, and model parameters in the first computational step. Then the ancestral sequences are reconstructed using these maximum-likelihood estimates as fixed values. This approach fails to consider the uncertainty in the maximum-likelihood estimate and may result in overestimated confidence.

Bayesian methods for ASR were suggested to overcome this uncertainty (Schultz and Churchill, 1999; Huelsenbeck and Bollback, 2001). Schultz and Churchill (1999) studied the effect of different prior distributions on the posterior probabilities of ancestral states in the simple case of a two-character-state model. Their approach was fully Bayesian: the parameters of the prior distribution were fixed and were not influenced by the data.

Huelsenbeck and Bollback (2001) reconstructed ancestral DNA sequences using the HKY85 substitution model (Hasegawa et al., 1985) with ASRV. Uncertainties in the tree topology, branch lengths, ASRV, and substitution-model parameters were considered. Prior distributions on model parameters were assumed. In the case of ASRV, the Bayesian method is hierarchical: the rate at each site is assumed to follow a gamma distribution with parameter α, and this α parameter is assumed to be derived from a uniform distribution between 0 and 10. Computing exact posterior probabilities is computationally infeasible, so the Markov chain Monte Carlo (MCMC) technique was used to efficiently approximate these probabilities (see Mau et al. (1999) for an example of applying this method for tree reconstruction).

Consider the goal of reconstructing the ancestral sequence at a specific node of a given tree. One problem arises: when integrating over the space of all possible trees, how trees in which this node does not exist should be considered in the MCMC computation? Huelsenbeck and Bollback (2001) considered only trees in which this node exists in their computation. However, Pagel *et al.* (2004) have shown that such a method of disregarding irrelevant trees introduces a bias in the estimation of the posterior probability. In essence, they suggest that the uncertainty in the existence of the node must also be taken into account in the ASR.

4.12 Gaps and unknown characters

All the models described thus far do not explicitly consider gapped positions and unknown characters. A common technique in phylogenetic tree reconstruction is to exclude from the analysis all positions in which at least one sequence contains a gap. However, for ASR, the goal is to infer the most likely ancestral sequence and thus it is essential to determine whether a character or gap is the ancestral state.

One approximation to escape this problem is to consider a gap as a missing character (e.g. Yang, 1997; Pupko *et al.*, 2000). In this approach, all possible character states are considered at the gapped position in the ASR computation. One problem with this approach is that the ancestral sequence is always longer than or equal in length to the longest sequence—clearly an unrealistic result.

An alternative approach is to represent a gap by adding an additional character to the model (thus creating an alphabet of size 21 for amino acids, or five for DNA/RNA). Since gaps are considered as all other characters, the probability of a gap being replaced by any other character and vice versa must be determined. There are two main difficulties with this approach. First, the probabilities of such hypothetical transitions from each amino acid to a gap and vice versa are unknown. More importantly, this approach assumes independence among sites. Thus, an insertion of two residues will be considered as two independent so-called character-to-gap transitions, rather than the more parsimonious explanation of a single insertion of two amino acids.

Clearly it is more realistic to consider insertions and deletions of more than one residue as components of the evolutionary model. To this end, the tree-based hidden Markov model (T-HMM) scheme was developed (Mitchison and Durbin, 1995; Mitchison, 1999). In this approach three hidden states are allowed in each position: match, insertion, and deletion. Each match state emits a character. A Markovian substitution scheme between these hidden states is assumed in two dimensions: the spatial dimension across the sequence and the temporal dimension along evolutionary times. Qian and Goldstein (2003) have used this approach for finding remote homologs. To this end, they applied ASR methodology to build sequence profiles that were then used in the homology search. However, the evaluation of the impact of these models on ASR awaits further studies.

A different approach to deal with gaps was suggested by Edwards and Shields (2004). This algorithm first approximates the probabilities of gaps at each position and internal node, using a two-states character model (0 for a gapped position, 1 for any other character). Once the ancestral state (0/1) for each node was determined, the non-gapped sites are estimated in an informal likelihood approach using probabilities derived from empirical substitution matrices. Although they show that their method is not as accurate as maximum likelihood, their novelty is in dividing the ASR algorithm into two separate tasks: first reconstruct gapped and non-gapped positions and then use this reconstruction for ASR of the un-gapped position.

Chapter 5 presents efficient algorithms for doing dynamic programming using probabilistic sequences. The dynamic-programming algorithm is guaranteed to find the highest-scoring alignment of two probabilistic sequences and can be used with any model of gap evolution, each of which implies a penalty for incurring a gap. The Affine gap penalty using one penalty for opening a gap and a smaller penalty to extend a gap is the most commonly used gap penalty as it simplifies the computation of the dynamic-programming scores. Benner *et al.* (1993) have argued for a gap penalty based on the logarithm of the length of the gap as this model better fits the distribution of gaps in empirical data. Given a multiple sequence alignment and a tree, dynamic programming can be used to build ancestral sequences at every node from the leaves to the root, as shown in Chapter 5.

4.13 Using structural and physicochemical information when reconstructing ancestral proteins

The different purifying selection forces resulting from different constraints on the structure, stability, foldability, and function of a protein should be considered as part of the evolutionary model. There are a few interesting directions towards this goal. For example, it is possible to include information on secondary structures or surface accessibility in the model. This is done by considering amino acid-substitution matrices for specific secondary structures or for buried and exposed residues.

Ideally, reconstructed proteins should be stable, foldable, and functional. With current computational techniques, it is difficult to predict whether this is the case for a reconstructed sequence. Towards this goal, biophysical models of protein evolution were developed that can be used to study the relationship between reconstructed ancestral proteins and their stability (Taverna and Goldstein, 2000, 2002; DePristo et al., 2005; Rastogi and Liberles, 2005). Recently, such a model was used to study the accuracy of various ASR methods. In a simulation study, the thermodynamic properties of the reconstructed sequence were compared with these properties in the "true" ancestral sequences (Williams et al., 2006). Such methods are likely to have impact on the detection of co-evolving substitutions, and as such on ASR, since non-independent evolution of residues in a protein is a direct result from purifying selection forces acting to maintain interactions between amino acid sites, and thus maintain protein stability and proper folding.

The ongoing effort to improve existing models for protein evolution, the endeavor to develop more realistic models, and the integration of these models with efficient ASR methods should increase our ability to accurately infer ancestral sequences and genomes, a vital element in evolutionary research.

In this chapter we have reviewed the evolutionary models used in ancestral reconstruction and the many modifications to the basic models and methodologies that can improve the accuracy of the reconstruction. In the next chapter, one of these models, a basic Markovian model, is discussed in mathematical detail. Computation of the probabilistic vectors describing the ancestral sequences under this model, as well as efficient algorithms for scoring alignments of these probabilistic sequences, are discussed and applied to various problems of sequence alignment.

References

Abascal, F., Zardoya, R., and Posada, D. (2005) ProtTest: selection of best-fit models of protein evolution. *Bioinformatics* **21**: 2104–2105.

Adachi, J. and Hasegawa, M. (1996) Model of amino acid substitution in proteins encoded by mitochondrial DNA. *J. Mol. Evol.* **42**: 459–468.

Adachi, J., Waddell, P.J., Martin, W., and Hasegawa, M. (2000) Plastid genome phylogeny and a model of amino acid substitution for proteins encoded by chloroplast DNA. *J. Mol. Evol.* **50**: 348–358.

Altschul, S.F., Madden, T.L., Schaffer, A.A., Zhang, J., Zhang, Z., Miller, W., and Lipman, D.J. (1997) Gapped BLAST and PSI-BLAST: a new generation of protein database search programs. *Nucleic Acids Res.* **25**: 3389–3402.

Arvestad, L., Berglund, A.C., Lagergren, J., and Sennblad, B. (2003) Bayesian gene/species tree reconciliation and orthology analysis using MCMC. *Bioinformatics* **19** (suppl. 1): i7–i15.

Benner, S.A., Cohen, M.A., and Gonnet, G.H. (1993) Empirical and structural models for insertions and deletions in the divergent evolution of proteins. *J. Mol. Biol.* **229**: 1065–1082.

Berglund-Sonnhammer, A.C., Steffansson, P., Betts, M.J., and Liberles, D.A. (2006) Optimal gene trees from sequences and species trees using a soft interpretation of parsimony. *J. Mol. Evol.* **63**: 240–250.

Bevan, R.B., Lang, B.F., and Bryant, D. (2005) Calculating the evolutionary rates of different genes: a fast, accurate estimator with applications to maximum likelihood phylogenetic analysis. *Syst. Biol.* **54**: 900–915.

Bollback, J.P. (2006) SIMMAP: stochastic character mapping of discrete traits on phylogenies. *BMC Bioinformatics* **7**: 88.

Burnham, K.P. and Anderson, D.R. (2003) *Model Selection and Multi-Model Inference*, 2nd edn. Springer, Berlin.

Chang, B.S. and Campbell, D.L. (2000) Bias in phylogenetic reconstruction of vertebrate rhodopsin sequences. *Mol. Biol. Evol.* **17**: 1220–1231.

Chang, B.S., Jonsson, K., Kazmi, M.A., Donoghue, M.J., and Sakmar, T.P. (2002) Recreating a functional ancestral archosaur visual pigment. *Mol. Biol. Evol.* **19**: 1483–1489.

Dayhoff, M.O., Schwartz, R.M., and Orcutt, B.C. (1978) A model of evolutionary change in proteins. In *Atlas of Protein Sequence and Structure* (Dayhoff, M.O., ed.), pp. 345–352. National Biomedical Research Foundation, Washington DC.

DePristo, M.A., Weinreich, D.M., and Hartl, D.L. (2005) Missense meanderings in sequence space: a biophysical view of protein evolution. *Nat. Rev. Genet.* **6**: 678–687.

Dimmic, M.W., Mindell, D.P., and Goldstein, R.A. (2000) Modeling evolution at the protein level using an adjustable amino acid fitness model. *Pac. Symp. Biocomput.* 18–29.

Dimmic, M.W., Hubisz, M.J., Bustamante, C.D., and Nielsen, R. (2005) Detecting coevolving amino acid sites using Bayesian mutational mapping. *Bioinformatics* **21** (suppl. 1): i126–i135.

Doron-Faigenboim, A. and Pupko, T. (2007) A combined empirical and mechanistic codon model. *Mol. Biol. Evol.*, **24**: 388–397.

Dutheil, J., Pupko, T., Jean-Marie, A., and Galtier, N. (2005) A model-based approach for detecting coevolving positions in a molecule. *Mol. Biol. Evol.* **22**: 1919–1928.

Edwards, R.J. and Shields, D.C. (2004) GASP: Gapped Ancestral Sequence Prediction for proteins. *BMC Bioinformatics* **5**: 123.

Eyre-Walker, A. (1998) Problems with parsimony in sequences of biased base composition. *J. Mol. Evol.* **47**: 686–690.

Felsenstein, J. (1981) Evolutionary trees from DNA sequences: a maximum likelihood approach. *J. Mol. Evol.* **17**: 368–376.

Felsenstein, J. (2001) Taking variation of evolutionary rates between sites into account in inferring phylogenies. *J. Mol. Evol.* **53**: 447–455.

Felsenstein, J. (2004) *Inferring Phylogenies.* Sinauer Associates, Sunderland, MA.

Felsenstein, J. (2005) *PHYLIP (Phylogeny Inference Package) version 3.6.* Department of Genome Sciences, University of Washington, Seattle.

Fitch, W.M. (1971) Toward defining course of evolution—minimum change for a specific tree topology. *Syst. Zool.* **20**: 406–416.

Fitch, W.M. and Markowitz, E. (1970) An improved method for determining codon variability in a gene and its application to the rate of fixation of mutations in evolution. *Biochem. Genet.* **4**: 579–593.

Galtier, N. (2001) Maximum-likelihood phylogenetic analysis under a covarion-like model. *Mol. Biol. Evol.* **18**: 866–873.

Galtier, N. and Gouy, M. (1995) Inferring phylogenies from DNA sequences of unequal base compositions. *Proc. Natl. Acad. Sci. USA* **92**: 11317–11321.

Galtier, N. and Gouy, M. (1998) Inferring pattern and process: maximum-likelihood implementation of a nonhomogeneous model of DNA sequence evolution for phylogenetic analysis. *Mol. Biol. Evol.* **15**: 871–879.

Gaucher, E.A., Gu, X., Miyamoto, M.M., and Benner, S.A. (2002) Predicting functional divergence in protein evolution by site-specific rate shifts. *Trends Biochem. Sci.* **27**: 315–321.

Gaucher, E.A., Thomson, J.M., Burgan, M.F., and Benner, S.A. (2003) Inferring the palaeoenvironment of ancient bacteria on the basis of resurrected proteins. *Nature* **425**: 285–288.

Goldman, N. and Yang, Z. (1994) A codon-based model of nucleotide substitution for protein-coding DNA sequences. *Mol. Biol. Evol.* **11**: 725–736.

Goldman, N., Anderson, J.P., and Rodrigo, A.G. (2000) Likelihood-based tests of topologies in phylogenetics. *Syst. Biol.* **49**: 652–670.

Gonnet, G.H. and Benner, S.A. (1991) *Computational Biochemistry Research at ETH*, pp. 1–18. ETH, Zurich.

Gonnet, G.H., Cohen, M.A., and Benner, S.A. (1992) Exhaustive matching of the entire protein sequence database. *Science* **256**: 1443–1445.

Grantham, R. (1974) Amino acid difference formula to help explain protein evolution. *Science* **185**: 862–864.

Gu, X. (2003) Functional divergence in protein (family) sequence evolution. *Genetica* **118**: 133–141.

Gu, X., Fu, Y.X., and Li, W.H. (1995) Maximum likelihood estimation of the heterogeneity of substitution rate among nucleotide sites. *Mol. Biol. Evol.* **12**: 546–557.

Harvey, P.H. and Pagel, M.D. (1991) *The Comparative Method in Evolutionary Biology*. Oxford University Press, Oxford.

Hasegawa, M., Kishino, H., and Yano, T. (1985) Dating of the human-ape splitting by a molecular clock of mitochondrial DNA. *J. Mol. Evol.* **22**: 160–174.

Hasegawa, M., Kishino, H., and Yano, T. (1987) Man's place in Hominoidea as inferred from molecular clocks of DNA. *J. Mol. Evol.* **26**: 132–147.

Holder, M. and Lewis, P.O. (2003) Phylogeny estimation: traditional and Bayesian approaches. *Nat. Rev. Genet.* **4**: 275–284.

Huelsenbeck, J.P. and Bollback, J.P. (2001) Empirical and hierarchical Bayesian estimation of ancestral states. *Syst. Biol.* **50**: 351–366.

Jermann, T.M., Opitz, J.G., Stackhouse, J., and Benner, S.A. (1995) Reconstructing the evolutionary history of the artiodactyl ribonuclease superfamily. *Nature* **374**: 57–59.

Jones, D.T., Taylor, W.R., and Thornton, J.M. (1992) The rapid generation of mutation data matrices from protein sequences. *Comput. Appl. Biosci.* **8**: 275–282.

Jones, D.T., Taylor, W.R., and Thornton, J.M. (1994) A mutation data matrix for transmembrane proteins. *FEBS Lett.* **339**: 269–275.

Koshi, J.M. and Goldstein, R.A. (1995) Context-dependent optimal substitution matrices. *Protein Eng.* **8**: 641–645.

Koshi, J.M. and Goldstein, R.A. (1996) Probabilistic reconstruction of ancestral protein sequences. *J. Mol. Evol.* **42**: 313–320.

Koshi, J.M. and Goldstein, R.A. (1998) Models of natural mutations including site heterogeneity. *Proteins* **32**: 289–295.

Koshi, J.M., Mindell, D.P., and Goldstein, R.A. (1997) Beyond mutation matrices: physical-chemistry based evolutionary models. In *Genome Informatics* (Miyano, S. and Takagi, T., eds), pp. 80–89. Universal Academy Press, Tokyo.

Krishnan, N.M., Seligmann, H., Stewart, C.B., De Koning, A.P., and Pollock, D.D. (2004) Ancestral sequence reconstruction in primate mitochondrial DNA: compositional bias and effect on functional inference. *Mol. Biol. Evol.* **21**: 1871–1883.

Lockhart, P.J., Steel, M.A., Hendy, M.D., and Penny, D. (1994) Recovering evolutionary trees under a more realistic model of sequence. *Mol. Biol. Evol.* **11**: 605–612.

Lopez, P., Casane, D., and Philippe, H. (2002) Heterotachy, an important process of protein evolution. *Mol. Biol. Evol.* **19**: 1–7.

Maddison, W.P., Donoghue, M.J., and Maddison, D.R. (1984) Outgroup Analysis And Parsimony. *Syst. Zool.* **33**: 83–103.

Mau, B., Newton, M.A., and Larget, B. (1999) Bayesian phylogenetic inference via Markov chain Monte Carlo methods. *Biometrics* **55**: 1–12.

Mayrose, I., Friedman, N., and Pupko, T. (2005) A Gamma mixture model better accounts for among site rate heterogeneity. *Bioinformatics* **21** (suppl. 2): ii151–ii158.

Mitchison, G. and Durbin, R. (1995) Tree-based maximal likelihood substitution matrices and hidden Markov models. *J. Mol. Evol.* **41**: 1139–1151.

Mitchison, G.J. (1999) A probabilistic treatment of phylogeny and sequence alignment. *J. Mol. Evol.* **49**: 11–22.

Muller, T. and Vingron, M. (2000) Modeling amino acid replacement. *J. Comput. Biol.* **7**: 761–776.

Muller, T., Spang, R., and Vingron, M. (2002) Estimating amino acid substitution models: a comparison of Dayhoff's estimator, the resolvent approach and a maximum likelihood method. *Mol. Biol. Evol.* **19**: 8–13.

Muse, S.V. and Gaut, B.S. (1994) A likelihood approach for comparing synonymous and nonsynonymous nucleotide substitution rates, with application to the chloroplast genome. *Mol. Biol. Evol.* **11**: 715–724.

Nielsen, R. (1997) Site-by-site estimation of the rate of substitution and the correlation of rates in mitochondrial DNA. *Syst. Biol.* **46**: 346–353.

Nielsen, R. and Yang, Z. (1998) Likelihood models for detecting positively selected amino acid sites and applications to the HIV-1 envelope gene. *Genetics* **148**: 929–936.

Olsen, G.J. (1987) Earliest phylogenetic branchings: comparing rRNA-based evolutionary trees inferred with various techniques. *Cold Spring Harb. Symp. Quant. Biol.* **52**: 825–837.

Pagel, M., Meade, A., and Barker, D. (2004) Bayesian estimation of ancestral character states on phylogenies. *Syst. Biol.* **53**: 673–684.

Penny, D. and Hasegawa, M. (1997) Molecular systematics. The platypus put in its place. *Nature* **387**: 549–550.

Posada, D. and Buckley, T.R. (2004) Model selection and model averaging in phylogenetics: advantages of akaike information criterion and bayesian approaches over likelihood ratio tests. *Syst. Biol.* **53**: 793–808.

Pupko, T., Pe'er, I., Shamir, R., and Graur, D. (2000) A fast algorithm for joint reconstruction of ancestral amino acid sequences. *Mol. Biol. Evol.* **17**: 890–896.

Pupko, T. and Galtier, N. (2002) A covarion-based method for detecting molecular adaptation: application to the evolution of primate mitochondrial genomes. *Proc. R. Soc. Lond. B.* **269**: 1313–1316.

Pupko, T., Huchon, D., Cao, Y., Okada, N., and Hasegawa, M. (2002a) Combining multiple data sets in a likelihood analysis: which models are the best? *Mol. Biol. Evol.* **19**: 2294–2307.

Pupko, T., Pe'er, I., Hasegawa, M., Graur, D., and Friedman, N. (2002b) A branch-and-bound algorithm for the inference of ancestral amino-acid sequences when the replacement rate varies among sites: application to the evolution of five gene families. *Bioinformatics* **18**: 1116–1123.

Pupko, T., Bell, R.E., Mayrose, I., Glaser, F., and Ben-Tal, N. (2002c) Rate4Site: an algorithmic tool for the identification of functional regions in proteins by surface mapping of evolutionary determinants within their homologues. *Bioinformatics* **18** (suppl. 1): S71–S77.

Pupko, T., Sharan, R., Hasegawa, M., Shamir, R., and Graur, D. (2003) Detecting excess radical replacements in phylogenetic trees. *Gene* **319**: 127–135.

Qian, B. and Goldstein, R.A. (2003) Detecting distant homologs using phylogenetic tree-based HMMs. *Proteins* **52**: 446–453.

Rastogi, S. and Liberles, D.A. (2005) Subfunctionalization of duplicated genes as a transition state to neofunctionalization. *BMC Evol. Biol.* **5**: 28.

Sankoff, D. (1975) Minimal mutation trees of sequences. *Siam J. Appl. Math.* **28**: 35–42.

Sankoff, D. and Rousseau, P. (1975) Locating vertices of a Steiner tree in an arbitrary metric space. *Math. Program.* **9**: 240–246.

Schluter, D. (1995) Uncertainty in ancient phylogenies. *Nature* **377**: 108–110.

Schneider, A., Cannarozzi, G.M., and Gonnet, G.H. (2005) Empirical codon substitution matrix. *BMC Bioinformatics* **6**: 134.

Schneider, A., Gonnet, G.H., and Cannarozzi, G.M. (2006) Synonymous codon substitution matrices. ICCS 2006. *Lecture Notes Comput. Sci.* **3992**: 630–637.

Schultz, T.R. and Churchill, G.A. (1999) The role of subjectivity in reconstructing ancestral character states: a Bayesian approach to unknown rates, states, and transformation asymmetries. *Syst. Biol.* **48**: 651–664.

Stewart, C.B. (1995) Molecular evolution. Active ancestral molecules. *Nature* **374**: 12–13.

Sullivan, J. and Swofford, D.L. (1997) Are guinea pigs rodents? The importance of adequate models in molecular phylogenetics. *J. Mammal. Evol.* **4**: 77–86.

Swofford, D.L. and Maddison, W.P. (1987) Reconstructing ancestral character states under Wagner parsimony. *Math. Biosci.* **87**: 199–229.

Taverna, D.M. and Goldstein, R.A. (2000) The distribution of structures in evolving protein populations. *Biopolymers* **53**: 1–8.

Taverna, D.M. and Goldstein, R.A. (2002) Why are proteins so robust to site mutations? *J. Mol. Biol.* **315**: 479–484.

Thompson, J.D., Higgins, D.G., and Gibson, T.J. (1994) CLUSTAL W: improving the sensitivity of progressive multiple sequence alignment through sequence weighting, position-specific gap penalties and weight matrix choice. *Nucleic Acids Res.* **22**: 4673–4680.

Thornton, J.W. (2001) Evolution of vertebrate steroid receptors from an ancestral estrogen receptor by ligand exploitation and serial genome expansions. *Proc. Natl. Acad. Sci. USA* **98**: 5671–5676.

Uzzell, T. and Corbin, K.W. (1971) Fitting discrete probability distributions to evolutionary events. *Science* **172**: 1089–1096.

Whelan, S. and Goldman, N. (2001) A general empirical model of protein evolution derived from multiple protein families using a maximum-likelihood approach. *Mol. Biol. Evol.* **18**: 691–699.

Whelan, S., Lio, P., and Goldman, N. (2001) Molecular phylogenetics: state-of-the-art methods for looking into the past. *Trends Genet.* **17**: 262–272.

Williams, P.D., Pollock, D.D., Blackburne, B.P., and Goldstein, R.A. (2006) Assessing the accuracy of ancestral protein reconstruction methods. *Plos Computat. Biol.* **2**: 598–605.

Yang, Z. (1993) Maximum-likelihood estimation of phylogeny from DNA sequences when substitution rates differ over sites. *Mol. Biol. Evol.* **10**: 1396–1401.

Yang, Z. (1996) Maximum-likelihood models for combined analyses of multiple sequence data. *J. Mol. Evol.* **42**: 587–596.

Yang, Z. (1997) PAML: a program package for phylogenetic analysis by maximum likelihood. *Comput. Appl. Biosci.* **13**: 555–556.

Yang, Z., Goldman, N., and Friday, A. (1994) Comparison of models for nucleotide substitution used in maximum-likelihood phylogenetic estimation. *Mol. Biol. Evol.* **11**: 316–324.

Yang, Z., Kumar, S., and Nei, M. (1995) A new method of inference of ancestral nucleotide and amino acid sequences. *Genetics* **141**: 1641–1650.

Yang, Z., Nielsen, R., and Hasegawa, M. (1998) Models of amino acid substitution and applications to mitochondrial protein evolution. *Mol. Biol. Evol.* **15**: 1600–1611.

Yang, Z., Nielsen, R., Goldman, N., and Pedersen, A.M. (2000) Codon-substitution models for heterogeneous selection pressure at amino acid sites. *Genetics* **155**: 431–449.

Yap, V.B. and Speed, T. (2005) Rooting a phylogenetic tree with nonreversible substitution models. *BMC Evol. Biol.* **5**: 2.

CHAPTER 5

Probabilistic ancestral sequences based on the Markovian model of evolution: algorithms and applications

Gina M. Cannarozzi, Adrian Schneider, and Gaston H. Gonnet

5.1 Introduction

Biological sequence analysis has brought the understanding of life to unprecedented heights. New genome projects are continually being announced as new technologies promise to increase sequencing ability 100-fold (Margulies et al., 2005). Never again will there be a dearth of sequence data for present-day organisms. For the understanding of the evolution of life, however, analysis of the biological sequences of ancient organisms is desirable. Unfortunately, with rare exceptions, there is no remaining organic matter from which to extract DNA, so the only recourse is to infer the ancestral sequences from the present-day sequences. Here we describe a formalism based on maximum likelihood and Markovian evolution to infer ancestral DNA or protein sequences at the internal nodes of a phylogenetic tree.

In the context of biological sequences, the sequences of the leaves of a phylogenetic tree are known sequences and those at the internal nodes of the tree are reconstructed to give probabilistic ancestral sequences (PASs). Often the PAS is desired at a node in the tree representing an important evolutionary event such as the last common ancestor of mammals, vertebrates, or metazoans. A PAS is a sequence which is defined by probability vectors in each position and is associated with an internal node in a phylogenetic tree. This is illustrated by the example in Table 5.1.

We see that in a probabilistic sequence each position is described by a vector of probabilities (all ≥ 0 and adding to 1). A position is fully determined when one probability is 1 and the others are 0, but could also be partially informative (as in positions 3, 5, and 6) or could indicate a complete lack of information (as in position 7).

A probabilistic sequence defines a probability for each instance of the sequence. For this example, the sequence ACCGAACT has probability 1/48. This probability can be computed as the product of the corresponding entries for each position:

$$1 \times 1 \times \frac{1}{2} \times 1 \times \frac{1}{3} \times \frac{1}{2} \times \frac{1}{4} \times 1 = \frac{1}{48}$$

Reconstruction of ancestral sequences has found uses in homology searching as well as prediction of function. Motif or profile search methods are often more sensitive at detecting long-distance homologs than single-sequence searches alone (Altschul et al., 1997). Ancestral reconstruction also allows the computation of branch-specific evolutionary measures used to detect selection as well as the prediction of residues conferring functional specificity within families of related proteins.

This chapter is organized as follows. The Markovian model of evolution upon which our reconstruction methods are based is summarized and applied to the reconstruction of probabilistic ancestral characters. Then the concept of Dayhoff

Table 5.1 An example of a probabilistic sequence of DNA

Base	Probability								
Position...	1	2	3	4	5	6	7	8	...
A	1	0	0	0	$\frac{1}{3}$	$\frac{1}{2}$	$\frac{1}{4}$	0	...
C	0	1	$\frac{1}{2}$	0	$\frac{1}{3}$	$\frac{1}{4}$	$\frac{1}{4}$	0	...
G	0	0	$\frac{1}{2}$	1	$\frac{1}{3}$	$\frac{1}{8}$	$\frac{1}{4}$	0	...
T	0	0	0	0	0	$\frac{1}{8}$	$\frac{1}{4}$	1	...

matrices for scoring sequence alignments is presented and extended to the application to probabilistic sequences. Efficient algorithms for optimally aligning probabilistic sequences are then described. Finally some applications of ancestral sequence reconstruction, pairwise sequence alignment and multiple sequence alignment (MSA) will be discussed and it will be shown how selected problems can be solved using the ancestral reconstruction package in the *Darwin* software system (Gonnet et al., 2000).

5.2 Model

The approach to reconstruct PASs is based on a Markovian model of evolution. Dayhoff *et al.* (1978) presented a method for estimating mutation matrices which describe the substitution probability between pairs of amino acids from empirical data. The Markovian process employed by the Dayhoff formalism is described in the form of a mutation matrix, denoted by M, describing the probabilities of amino acid mutations for a given period of evolution.

$$M_{ji} = Pr\{amino\ acid\ i \to amino\ acid\ acid\ j\} \quad (5.1)$$

Mutations at a given position are assumed to occur at random and independent from other positions as well as from their own history.

A 1-PAM mutation matrix (where PAM stands for point accepted mutations) describes an amount of evolution which will change, on average, 1% of the amino acids. In mathematical terms this is expressed as a matrix M such that

$$\sum_i f_i(1-M_{ii}) = 0.01 \quad (5.2)$$

where the sum goes over all elements i in the set Σ of all characters. The frequency of the ith character is denoted as f_i. The diagonal elements of M are the probabilities that a given character does not change. The set of characters for biological sequences can be the four nucleotides, the 20 amino acids or the set of 64 (or 61) codons. In this chapter, for simplicity, we usually refer to proteins for which Σ is a character set composed of the 20 amino acids. Clearly, however, the formalism is the same for other character sets.

If we have a probability or frequency vector p, the product Mp gives the probability vector after an amount of evolution equivalent to 1 PAM unit. Or, if we start with a given amino acid (a probability vector which contains a 1 in position i and 0s in all others) $M_{\cdot i}$ (the ith column of M) is the corresponding probability vector after one unit of random evolution. Because of the Markovian nature of the model, after k units of evolution (what is called k PAM) a frequency vector p will be changed into the frequency vector $M^k p$.

Because the natural frequencies f correspond to the steady-state frequencies of the Markovian process, the following equation holds for any distance k and characters A and B:

$$f_A(M^k)_{BA} = f_B(M^k)_{AB} \quad (5.3)$$

This symmetry also reflects the time-reversibility of this model.

5.3 Computation of PASs

5.3.1 Probability of an evolutionary configuration

An evolutionary configuration is a tree with defined branch lengths and probabilistic internal nodes over a given set of external nodes (the leaves). Following the theory of maximum likelihood, we choose the configuration of ancestors, branch lengths, and tree topology which maximizes the likelihood of the sequences. It is important to note that when computing the probability of an evolutionary configuration, each character (column of an MSA) is treated as independent. In other words, the probability of the entire MSA will be computed as the product of the probabilities of each aligned position. We assume that the phylogenetic tree and the MSA are given

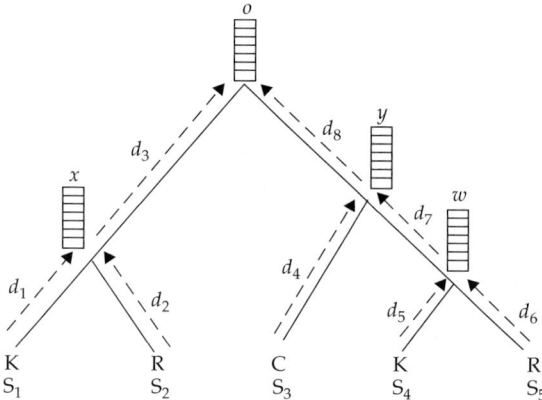

Figure 5.1 An example of an evolutionary configuration as a rooted tree over five amino acids in an MSA, with four unknown internal nodes.

and then, for each aligned position in the MSA, we will have a tree with one amino acid at each leaf. Sequences having a gap at this position are ignored as they do not contribute any information about possible ancestral characters at this position. The tree in Figure 5.1 represents a slice of an MSA at a position for which sequence 1 (S_1) has a K, sequence 2 (S_2) has an R, etc. In this case, o, w, x, and y are the unknown amino acids at the divergence points and are therefore represented as probability vectors.

We can compute the probability of this evolutionary configuration; that is, the probability of the tree and the alignment given the sequences. Let o be the point which identifies the root of the phylogenetic tree. To compute the probability of a given position, we sum over all possible unknown amino acids, o, w, x, and y, and evaluate all of the mutation probabilities.

$$Pr\{Config.\} = \sum_o f_o \sum_x M_{xo}^{d_3} M_{Kx}^{d_1} M_{Rx}^{d_2}$$
$$\sum_y M_{yo}^{d_8} M_{Cy}^{d_4} \sum_w M_{wy}^{d_7} M_{Kw}^{d_5} M_{Rw}^{d_6} \quad (5.4)$$

For each branch $A \to B$ (of branch length d) there is a term of the form M_{BA}^d, for each unknown internal amino acid v there is a sum over all of its possibilities, and for the root o we have the sum of each possible amino acid times its natural frequency of occurrence. The computation cost of the probability of an evolutionary configuration is only linear in the number of leaves (and not exponential, as it may seem from eqn 5.4). The reason behind this is that the probability can be computed bottom-up from the leaves towards the root. In order to do so, for each node x of the tree we introduce a temporary vector T^x. If x is a leaf—a known amino acid—then

$$T_i^x = \begin{cases} 1 & i=x \\ 0 & i \neq x \end{cases} \quad (5.5)$$

If x is an internal node, then it is computed from the two descendants of x, denoted y and z:

$$T_i^x = \left(\sum_j M_{ji}^{d_{xy}} T_j^y\right)\left(\sum_k M_{ki}^{d_{xz}} T_k^z\right) \quad (5.6)$$

The final computation of the probability is performed at the root of the tree and is simplified to

$$Pr\{Config.\} = \sum_i f_i T_i^o = f T^o \quad (5.7)$$

The computation of each T for an internal node requires $|\Sigma|(2|\Sigma|+1)$ multiplications and the vector product at the root needs additional $|\Sigma|$ multiplications. This means that for k sequences, each of length m, the total computation of the probability of the evolutionary configuration requires $m|\Sigma|((k-1)(2|\Sigma|+1)+1)$ multiplications.

The PAS at the root of the tree can be computed by normalizing T^o:

$$PAS(o) = T^o / \sum_i T_i^o \quad (5.8)$$

This is implemented in the *Darwin* package **Ancestor** by the two functions **ComputePAS** and **PASfromMSA**.

5.3.2 Independence of the position of the root

The probability of any evolutionary configuration is independent of the position of the root because the Markovian model of evolution is time-reversible, as shown in the previous section. To prove this, we first show that the probabilities are not affected if we move the root along the edge linking its two descendants (x and y). The length of the branch from x to y will remain constant, we will just change d_x and d_y so that $d_x + d_y = d_{xy}$ = constant. The probability of the evolutionary configuration is

$$\sum_i f_i \left(\sum_j M_{ji}^{d_x} T_j^x \right) \left(\sum_k M_{ki}^{d_y} T_k^y \right) \quad (5.9)$$

Moving the summations around, and using the symmetry relation of f and M and noticing that the summation over i is a matrix product, we obtain

$$\sum_k \sum_j T_k^y T_j^x \sum_i f_i M_{ki}^{d_y} M_{ji}^{d_x} = \sum_k \sum_j T_k^y T_j^x f_j \left(M^{d_x+d_y} \right)_{kj}$$
$$(5.10)$$

and the formula is independent of d_x and d_y, but only depends on their sum. By continuity, this can be extended to any position in the phylogenetic tree. Consequently, for the purposes of evaluating the probability of a particular evolutionary configuration, we can place the root wherever we want. This is convenient as the correct placement of the root requires an outgroup which is not always present.

Extending this argument, the PAS for every internal node of a tree can be computed, reusing previously computed values of T. As an example we reconsider Figure 5.1, but we now want to compute the PAS o' at the (ternary) internal node y. Following eqns 5.6 and 5.7 and envisioning o' slightly above y (but with distance 0), $T^{o'}$ can be computed as:

$$T_i^{o'} = f_i \left(\sum_j M_{ji}^{d_{o'y}} T_j^{yo'} \right) \left(\sum_k M_{ki}^{d_{o'x}} T_k^z \right)$$
$$= f_i T_i^y \left(\sum_k M_{ki}^{d_3+d_8} T_k^x \right) \quad (5.11)$$

The PAS is then the normalized $T^{o'}$. When the PASs for all internal nodes need to be computed, the T vectors can be reused. However, it is important to note that each T can be computed from three different directions, depending on which node is taken to be its parent. For the reconstruction of the PASs at internal nodes, the correct T has to be used; that is, the one with the correct parent node once the tree is rearranged for a new root o'.

5.4 Scoring

Molecular sequence analysis very often depends on an alignment between two or more sequences. Alignments can efficiently be computed using dynamic programing, which requires scores that describe the similarities between characters. Aligning sequences by dynamic programming using the Dayhoff formalism is equivalent to finding the alignment which maximizes the probability that the two sequences evolved from an ancestral sequence as opposed to being random sequences. More precisely, we are comparing two events

1 that the two sequences have evolved from some common ancestor after some amount, t, of evolution;
2 that the two sequences are independent of each other. This corresponds to the alignment by random occurence of the two sequences.

In this section we will first summarize the often-used theory about scoring matrices for two known sequences (the usual case) and then explain how this can be extended to match a probabilistic sequence against a sequence or even how to match two probabilistic sequences.

5.4.1 Scoring known sequences

The easiest case is defining a score between two known amino acids. Following the theory developed by Dayhoff et al. (1978), the score between two characters A and B is defined as

$$S(A,B) = 10 \log_{10} \frac{Pr\{A \text{ and } B \text{ have a common ancestor}\}}{Pr\{\text{random alignment of } A \text{ and } B\}}$$
$$(5.12)$$

The probability that the two characters have evolved from some common ancestral sequence after some amount, t, of evolution is

$$Pr\{A \text{ and } B \text{ from a common ancestor}\}$$
$$= \sum_x f_x Pr\{x \to A\} Pr\{x \to B\}$$
$$= \sum_x f_x (M^t)_{Ax} (M^t)_{Bx}$$
$$= \sum_x f_B (M^t)_{Ax} (M^t)_{xB} = f_B (M^{2t})_{AB} \quad (5.13)$$
$$= f_A (M^{2t})_{BA}$$

The probability of a random alignment of an arbitrary position with amino acid A and another arbitrary position with amino acid B is equal to the product of the individual frequencies:

$$Pr\{random\ alignment\ of\ A\ and\ B\} = f_A f_B \quad (5.14)$$

Therefore the score for every pair of characters and a given evolutionary distance t can be pre-computed and stored in a matrix (called a Dayhoff matrix) as

$$S_t(A, B) = 10 \log_{10} \frac{M^t_{BA}}{f_B} \quad (5.15)$$

Since dynamic programming maximizes the sum of the similarity measure, it maximizes the sum of the logarithms, which is equivalent to maximizing the product of these quotients of probabilities. As a consequence, using Dayhoff matrices as the scoring matrices for the dynamic programming finds the alignment which maximizes the probability of the sequences having evolved from a common ancestor and is called a maximum-likelihood alignment.

This formalism has long been applied to protein sequences to obtain amino acid scoring matrices (Gonnet et al., 1992). In addition, matrices have been built on such properties as secondary structure, dipeptides, and hydrophobicity. More recently, codon-substitution matrices have been constructed from vertebrate DNA (Schneider et al., 2005). Single-nucleotide mutation matrices contain less information but can be built from an appropriate data-set for a particular application; for example, for aligning mitochondrial DNA.

5.4.2 Scoring a probabilistic against a known sequence

Scoring an amino acid against a probabilistic sequence is not trivial if it is done according to the standard Markovian model of evolution. The first approach of taking the expected value of the score for each entry is not correct. Again we will use the logarithm of the quotient between the probability of being homologous—coming from a common ancestor x—divided by the probability of being matched by chance.

Let v be the probability vector to be scored against amino acid A, d_1 the distance between x and v, and d_2 the distance between x and A as shown in the left-hand panel of Figure 5.2. The quotient of probabilities is:

$$\frac{Pr\{common\ ancestor\}}{Pr\{random\ alignment\ of\ v\ and\ A\}}$$

$$= \frac{\sum_j f_j \left(\sum_i Pr\{j \to i\} v_i \right) Pr\{j \to A\}}{\left(\sum_i v_i f_i \right) f_A} \quad (5.16)$$

The probability is summed over all possible values j of the ancestor x and all possible values i of the probabilistic character v. We will develop the simplification of this formula in detail, as there are several important steps in its computation.

Notice that the event of having a common ancestor is the product of three probabilities which are visualized in Figure 5.2. First we have the natural frequency of the root, f_j, then the probability that a j mutates to v_i, and finally that j, independently of the previous, mutates to an A. Using the theorems introduced in the previous section, the probabilities quotient becomes:

$$\frac{Pr\{common\ ancestor\}}{Pr\{random\ alignment\ of\ v\ and\ A\}}$$

$$= \frac{\sum_j \sum_i f_j (M^{d_1})_{ij} v_i (M^{d_2})_{Aj}}{fvf_A} \quad (5.17)$$

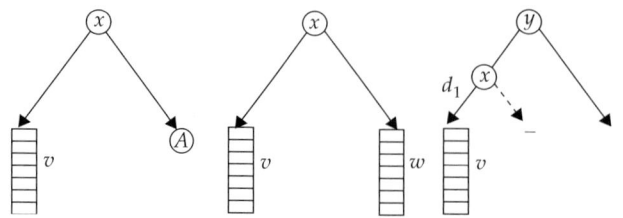

Figure 5.2 Evolutionary models used in the scoring of a probabilistic sequence against a sequence (left), two probabilistic sequences against each other (middle), and a position involving a gap (right).

Using the symmetry relation on the second M and inverting the summation we obtain:

$$\frac{\sum_i \sum_j v_i (M^{d_1})_{ij} (M^{d_2})_{jA}}{fv} \quad (5.18)$$

The inner summation is a term of the matrix product:

$$\frac{\sum_i v_i (M^{d_1+d_2})_{iA}}{fv} \quad (5.19)$$

which finally becomes

$$S(v, A) = 10 \log_{10} \frac{u_A}{fv} \quad (5.20)$$

where $u = v^T M^{d_1+d_2}$. Observe that the final score does not depend on d_1 and d_2 individually, but only on their sum. This confirms the findings of a previous section, that the placement of the root is inconsequential.

$S(v, A)$ can be precomputed for every amino acid A at the cost of a matrix-vector multiplication (computing u) and the same amount of space as v itself. With this precomputation, scoring each position is $O(1)$ and extremely efficient. This has a very practical implication for profile searching, when a given probability profile (e.g. describing a protein family) is compared against an entire database (to search for other members of that family). The consequence of the above result is that dynamic programming alignments of a probabilistic sequence against a known sequence can be done in the same run-time order as two plain sequences. Therefore a database search using a probabilistic sequence is almost as fast as the search with only one specific sequence.

5.4.3 Scoring two probabilistic sequences

Scoring one probabilistic sequence against another follows a similar procdure. This situation is shown in the middle panel of Figure 5.2. This time we have a triple sum, over k in the unknown ancestor and over i and j in the two probabilistic descendants, thus:

$$\frac{Pr\{v \text{ and } w \text{ have a common ancestor}\}}{Pr\{\text{random alignment of } v \text{ and } w\}}$$
$$= \frac{\sum_k \sum_i \sum_j f_k Pr\{k \to i\} v_i Pr\{k \to j\} w_j}{fv\, fw} \quad (5.21)$$

Using the same techniques as before we obtain:

$$S(v, w) = 10 \log_{10}(w^* v^*) \quad (5.22)$$

where $v^* = \frac{v}{fv}$ can be precomputed for the first sequence, $w^* = \left(\frac{(w^T M^{d_1+d_2} F)}{(fw)}\right)$ can be precomputed for the second sequence and F is a diagonal matrix containing the natural frequencies; that is, $F_{ii} = f_i$ (multiplying a vector by F can be done in linear time, so this multiplication does not affect the complexity). The precomputation takes as much space as the probabilistic sequences and a matrix-vector multiplication for each position w. Clearly we choose w to be in the shorter of the two probabilistic sequences. The computation of the score, after the precomputation is finished, requires an internal product of two vectors of size $|\Sigma|$.

5.5 Applications

5.5.1 Pairwise alignments of probabilistic sequences

Once scores between probabilistic sequences have been defined as shown in the previous section, the concept of dynamic programming can easily be extended to this kind of sequence. The dynamic programming itself can be executed as usual (Smith and Waterman, 1981; Gotoh, 1982), but instead of a single Dayhoff matrix with all possible scores, the vectors v^* and w^* have to be precomputed as described above. This precomputation is linear in the length of the sequences and is dwarfed by the actual matching phase (which is $O(mn)$ for sequences of lengths m and n). Instead of one lookup in the Dayhoff matrix, the vector product and then the logarithm of $v^* \cdot w^*$ have to be computed. This is a factor of $|\Sigma|$ more, but still very efficient considering that probabilistic sequences will be aligned without any shortcuts according to the Markovian model of evolution. Insertions and deletions are handled as in dynamic programming, for example, with empirical affine gap costs (Gonnet et al., 1992).

5.5.2 Handling of gaps in PAS

As a sequence alignment may contain indels (gaps), it may be necessary to reconstruct a PAS at a position involving gaps. The situation of gap against a gap results in a gap, and is not interesting. The situation which requires analysis is a probabilistic sequence against a gap (Figure 5.2, right-hand panel).

A gap on the right of x means that there is no information about amino acids and that any further computation, for example at y, should behave as if using v directly. This is achieved by "evolving" v up to x (using the symmetry of Markovian evolution):

$$T^x = M^{d_1} v. \tag{5.23}$$

This way the gap is effectively ignored (as it does not contribute any information) while at the same time the character v at the other leaf and the distance d_1 to the internal node x are properly treated. This also implies that although the only character under x is v, x is not exactly equal to v, but an additional uncertainty exists as x could have mutated over the distance d_1.

5.5.3 MSAs

MSAs are frequently used to understand the structure, function, and evolution of protein families. Optimal pairwise alignments can be computed exactly using dynamic programming between sequences and or probabilistic sequences. For more than two sequences, however, the problem is much harder. The complexity of dynamic programming for n sequences of length m is $O(m^n)$ in time, which is in practice impossible for n greater than 4. It can be proven to be NP complete and thus not solvable exactly for real problems. Thus approximate algorithms are a necessity.

An algorithm for the construction of MSAs falls naturally from the construction of probabilistic sequences from pairs of sequences. Although dynamic programming of n sequences quickly becomes intractable, we can use a phylogenetic tree and the scoring developed in the previous section to align probabilistic sequences at nodes of the tree from the leaves to the root. We assume that a phylogenetic tree is given and that for each external node (leaf) of the phylogenetic tree we have biological sequences of codons, amino acids, or nucleotides.

The algorithm for constructing MSAs using dynamic programming of probabilistic sequences is based on the observation that an MSA can be summarized as a probabilistic sequence. At conserved positions only one character has a high probability, while variable positions will result in several non-negligible probabilities, appropriately weighted according to the phylogeny of the underlying sequences. The MSA for all the sequences at the leaves of a tree is computed bottom-up, from the leaves to the root. At each internal node, the following steps have to be executed.

- Assume that the two children of this node have probabilistic sequences assigned. If a child is a leaf, the corresponding sequence can be converted to a probabilistic sequence by setting at each position the probability to 1 for the character at that position and to 0 for all other characters. If the child is an internal node, it has a probabilistic sequence from previous steps of the algorithm.
- Apply the dynamic-programming algorithm for probabilistic sequences from the previous section to the probabilistic sequence from the children nodes.
- Construct a probabilistic sequence for this node by computing the ancestral sequence of the two aligned probabilistic sequence as shown previously.
- Assume further that both children also have a partial MSA assigned. Partial means that it is an MSA over all sequences under that particular node. For leaves, the MSA is just the plain sequence.

Construct a new (partial) MSA for this node. This is based on the probabilistic sequence alignment. From the way the probabilistic sequence are constructed, the probabilistic sequence and MSA at each node have the same length and therefore the mapping between the positions is trivial. For each aligned position in the probabilistic sequence alignment, the two corresponding positions (columns) in the MSA are joined to a column in the new MSA. If one of the aligned probabilistic sequence has a gap at that position, the new MSA column consists of the MSA column from the non-gap sequence and all characters in the rows corresponding to the gap sequence are set to gaps.

In summary, at each internal node, two partial MSAs are joined to a MSA over all sequences under this node by mapping the MSAs to the aligned probabilistic sequences. This algorithm is implemented in *Darwin*'s **MAlign** package and can be accessed by choosing the option **prob**.

5.5.4 Rate variation among sites

Much evidence indicates that substitution rates vary across sites. For example, sites on the surface of a protein are more likely to undergo mutations than sites in the core of the protein. There are generally two approaches to treat this phenomenon. Either a distribution of the rates among sites is assumed (typically a discrete gamma distribution; Yang, 1994), or the most likely rate for each site is estimated. The latter approach is implemented in the *Darwin* system as **ScaleIndex** and works as follows: after a MSA and the phylogenetic tree, including the branch lengths are determined, a scale factor r is estimated for each site (i.e. a position in the MSA) using maximum likelihood. The factor r expands or contracts the distances d_i between the nodes of the tree, but maintains the topology of the tree and the relative branch lengths (see Figure 5.3). r is estimated such that the probability of the evolutionary configuration at this particular site (eqn 5.4) is maximized. The construction of the PAS can then be performed as in eqn 5.8.

This approach has the property that a separate factor r is estimated for each position, which could lead to overfitting. Therefore, typically a different approach is taken. Instead of estimating a fixed rate for each site individually, a rate distribution is assumed. A distribution has only very few parameters (e.g. only 1 in the case of the gamma distribution). Typically, the ancestral characters are not reconstructed by using the most likely rate at each site, but by computing the expected value given the rate distribution.

Reconstruction methods that search for the most likely configuration of all nodes of the tree result in exponential computation time, since the expected likelihood of every configuration has to be computed separately (Pupko *et al.*, 2002). However, in order to reconstruct the ancestor only at the root, bottom-up approaches are known that scale linearly with the number of sequences (Yang, 1997). The probabilistic reconstruction presented here also falls into that category. Therefore, incorporating rate variance among sites for the reconstruction of PASs could be done by reconstructing $PAS(o \mid r_i)$ according to eqn 5.8 for each individual rate separately. The computation given a rate r_i is performed following eqn 5.6, but all distances d_j are multiplied by r_i. In the end, the expected value of the ancestral vector given different rates can be computed. For k discrete rate categories r_i, each having probability $P\{r = r_i\}$ according to the rate distribution, the PAS is computed as:

$$PAS(o) = \sum_{i=1}^{k} PAS(o \mid r_i) \times Pr\{r = r_i\}.$$

It follows that the computation of a PAS under the assumption of rate variation can be performed in time proportional to the number of sequences. In fact, it is only a factor k more expensive than under the assumption of a constant rate.

5.6 Example

We present an example of probabilistic ancestral reconstruction on the alcohol dehydrogenase class III protein family. The purpose of this example is to illustrate the advantages of using ancestral sequences for homology search and also to introduce the **Ancestor** package of the *Darwin* software. The sequences are all from the SwissProt database. Figure 5.4 shows the *Darwin* distance tree of these sequences with plants in the left subtree and metazoan sequences in the right.

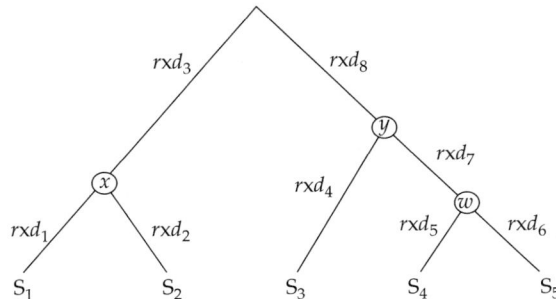

Figure 5.3 Use of a scale factor r.

The first task after starting *Darwin* is to prepare the Dayhoff matrices, loading the SwissProt database and getting the list of entries of the protein family. The latest release of the SwissProt database can be downloaded from the Internet and then converted to the *Darwin* database format using the **DbToDarwin** command. We use release 45.0 in this example.

```
CreateDayMatrices();
ReadDb('SwissProt');
fam := [SearchDb('ADHX_')];
```

The SwissProt IDs and sequences are extracted into lists and a phylogenetic tree is constructed using the **PhylogeneticTree** package with option **DISTANCE**.

```
seqs := [seq(Sequence(z),z=fam)]:
ids  := [seq(SearchTag('ID',z),z=fam)]:
tree := PhylogeneticTree(seqs,ids,
        DISTANCE);
```

The PASs at the internal nodes A and B, which are the roots of the plant and metazoan subtrees, respectively, can be reconstructed using the **PASfromTree** function:

```
A := PASfromTree(seqs, tree['Left']):
B := PASfromTree(seqs, tree['Right']):
```

The **print(A)** command displays the most probable amino acids at each position of the ancestor. Because the plant sequences show little variation, many positions are determined with almost absolute certainty. Only at 52 of the 381 positions has the most probable amino acid, a probability of less than 99.9%. In the ancestral sequence of the metazoans, this is the case at 149 of 384 positions.

```
pos Most probable chars
  1 M 0.94 L 0.03 I 0.01 V 0.00 A 0.00
  2 A 0.99 S 0.00 V 0.00 G 0.00 T 0.00
  3 S 0.96 A 0.01 T 0.01 G 0.00 N 0.00
  4 A 0.71 S 0.26 P 0.03 T 0.00 G 0.00
  5 T 1.00 S 0.00 A 0.00 V 0.00 K 0.00
  6 Q 1.00 E 0.00 K 0.00 R 0.00 A 0.00
...
```

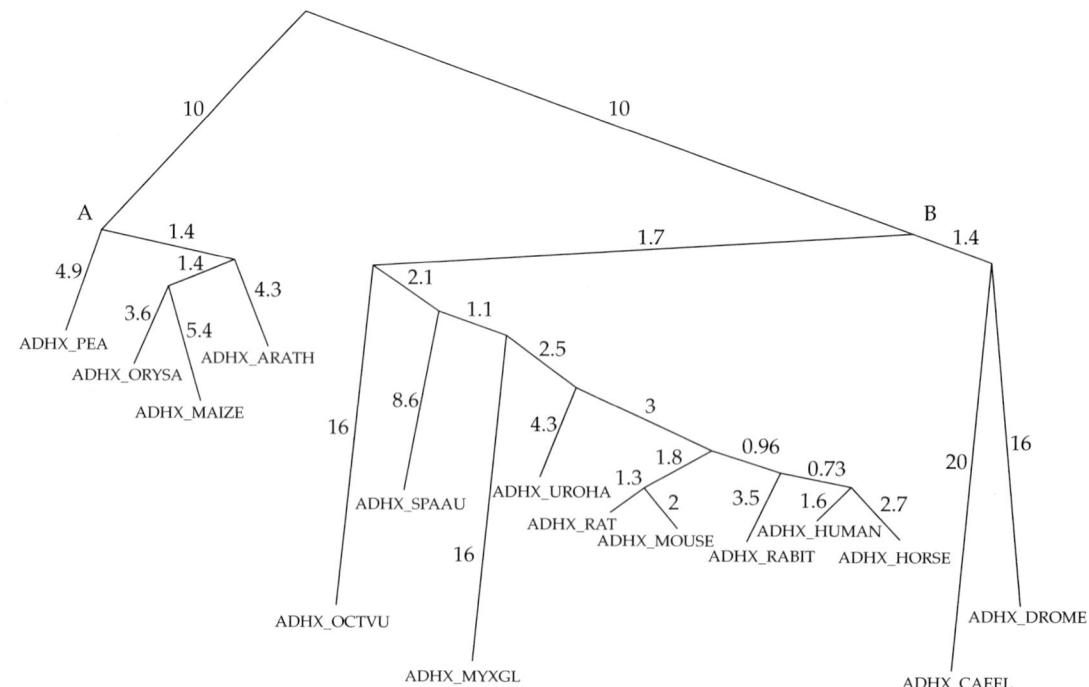

Figure 5.4 *Darwin* tree of the alcohol dehydrogenase class III proteins in SwissProt.

In this example we illustrate that aligning the common ancestor of a group of proteins to other members of the protein family achieves a higher score than aligning the individual sequences against those same family members. Specifically, we will compare the scores of alignments between the probabilistic ancestor at internal node A and all sequences in the right subtree against the scores of pairwise protein alignments between sequences in the left subtree and proteins in the right subtree.

In order to do so, we first compute all pairwise protein alignments and compute statistics over the scores. Useful functions are the **Leaves** iterator, which loops through all leaves of a (sub)tree, the **Align** function, which performs the dynamic programming between two proteins, and the **Stat** function, which can be used to compute and display univariate statistical information.

```
S := Stat('Pairwise Scores'):
for l1 in Leaves(tree['Left']) do
    s1 := seqs[l1[3]];
    for l2 in Leaves(tree['Right']) do
        s2 := seqs[l2[3]];
        dist := abs(l1['Height']) + abs
        (l2['Height']);
        al := Align(s1,s2,DayMatrix
        (dist));
        S+al[Score];
    od od:
```

With **print(S)** the collected statistics can be viewed. The average alignment score between the proteins in the plants and those in the metazoans amounts to 2334.

```
Pairwise Scores: number of sample
points = 44
mean = 2334 +- 34
variance = 13034 +- 4631
skewness = -0.374038, excess = -0.551367
minimum = 2074.3, maximum = 2524.34
```

This is result is now compared to the average score that is achieved when the ancestral sequence at B (the ancestor of all the metazoans) is aligned against all proteins in the left subtree. Alignments between probabilistic sequences are computed using the **PSDynProg** function.

```
S2 := Stat('B vs. leaves'):
for l1 in Leaves(tree['Left']) do
    s1 := seqs[l1[3]];
    dist := abs(tree['Right',
    'Height']) + abs(l1['Height']);
    psa1 := PSDynProg(B,
    ProbSeq(s1, IntToA),dist):
    S2+psa1[1];
od:
```

With **print(S2)** the score statistics can be viewed and compared with the pairwise score:

```
B vs. leaves: number of sample points = 4
mean = 2531 +- 51
variance = 2702 +- 1788
skewness = -0.374528, excess = -2.15185
minimum = 2467.68, maximum = 2585.08
```

The average score of 2531 is clearly higher than the result from the pairwise comparison. This confirms that the ancestor at node B is indeed closer to the family members in the other subtree and therefore better suited to detect additional members of the protein family. When done in the other direction (i.e. A against all metazoan sequences), the average score is 2407, which is less, but still higher than the 2334 from the pairwise alignments. That the score in this direction is lower comes most likely from the fact that the plant sequences are more homogeneous than those from the metazoan set. Therefore the difference between the single plant sequences and their common ancestor is relatively small.

References

Altschul, S.F., Madden, T.L., Schaffer, A.A., Zhang, J., Zhang, Z., Miller, W., and Lipman, D.J. (1997) Gapped BLAST and PSI-BLAST: a new generation of protein database search programs. *Nucleic Acids Res.* **25**: 3389–3402.

Dayhoff, M.O., Schwartz, R.M., and Orcutt, B.C. (1978) A model for evolutionary change in proteins. In *Atlas of Protein Sequence and Structure,* vol. 5 (Dayhoff, M.O., ed.), pp. 345–352. National Biomedical Research Foundation, Washington DC.

Gonnet, G.H., Cohen, M.A., and Benner, S.A. (1992) Exhaustive matching of the entire protein sequence database. *Science* **256**: 1443–1445.

Gonnet, G.H., Hallett, M.T., Korostensky, C., and Bernardin, L. (2000) Darwin v. 2.0: an interpreted computer language for the biosciences. *Bioinformatics* **16**: 101–103.

Gotoh, O. (1982) An improved algorithm for matching biological sequences. *J. Mol. Biol.* **162**: 705–708.

Margulies, M., Egholm, M., Altman, W.E., Attiya, S., Bader, J.S., Bemben, L.A. *et al.* (2005) Genome sequencing in microfabricated high-density picolitre reactors. *Nature* **437**: 376–380.

Pupko, T., Pe'er, I., Hasegawa, M., Graur, D., and Friedman, N. (2002) A branch- and-bound algorithm for the inference of ancestral amino-acid sequences when the replacement rate varies among sites: application to the evolution of five gene families. *Bioinformatics* **18**: 1116–1123.

Schneider, A., Cannarozzi, G.M., and Gonnet, G.H. (2005) Empirical codon substitution matrix. *BMC Bioinformatics* **6**: 134.

Smith, T.F. and Waterman, M.S. (1981) Identification of common molecular subsequences. *J. Mol. Biol.* **147**: 195–197.

Yang, Z. (1994) Statistical properties of the maximum likelihood method of phylogenetic estimation and comparison with distance matrix methods. *Syst. Biol.* **43**: 329–342.

Yang, Z. (1997) PAML: a program package for phylogenetic analysis by maximum likelihood. *Comput. Appl. Biosci.* **13**: 555–556.

CHAPTER 6

Estimating the history of mutations on a phylogeny

Jonathan P. Bollback, Paul P. Gardner, and Rasmus Nielsen

6.1 Introduction

Evolution is a historical process that has left its signature on the molecules and morphology of living organisms. Attempts to better understand the specific and general features of evolution involve making inferences about the past from these tell-tale signs (Felsenstein, 1985; Brooks and McLennan, 1991; Harvey and Pagel, 1991; Pagel, 1999). Ancestral state reconstruction is a powerful tool in this endeavor, as exemplified by its application to a wide array of questions (e.g. Langley and Fitch, 1974; Gillespie, 1991; Templeton, 1996; Messier and Stewart, 1997; Bishop et al., 2000). Traditionally, ancestral reconstruction has relied on well-understood approaches such as parsimony. However, the last decade has seen an excited flurry of research into statistical approaches as exemplified by the contents of this volume and the primary literature.

In a general sense, most methods for ancestral reconstruction have focused on reconstructing the ancestral states at the internal nodes of a phylogeny. Often we are not interested in particular nodes of the phylogeny but the whole history of a character. In this chapter we focus on a Bayesian method for estimating these histories on phylogenies (we refer to a complete description of a character's history as its mutational path). Mutational path methods differ most notably from other approaches in their ability to estimate not only the ancestral states at the internal nodes of a phylogeny but also the order and timing of mutational changes on the phylogeny. Our goal here is to provide a concise introduction to the statistical tools necessary for estimating mutational histories and making inferences from these histories, and to provide some examples of the power of this recent approach.

6.2 Likelihood and Bayesian methods

Estimation of ancestral character states using maximum likelihood proceeds from a straightforward extension of the usual algorithm for calculation of the likelihood function in phylogenetics. Let $f_{ij}(k)$ be the fractional probability of nucleotide k at site i for node j of the phylogeny. That is, $f_{ij}(k)$ is the probability of all the data in site i below node j given that the ancestral state at node j is k. The maximum-likelihood estimate of ancestral state in node j is then obtained by placing the root at node j and maximizing

$$L(k) = f_{ij}(k) \qquad (6.1)$$

with respect to k. Other parameters of the evolutionary model (branch lengths, parameters of the mutational model, etc.) have typically been estimated prior to the analysis and are assumed to be fixed. The method can also be extended to find the joint set of ancestral states for all nodes that maximizes the likelihood (Yang et al., 1995; Koshi and Goldstein, 1996; Pupko et al., 2000). For more details on maximum-likelihood estimation, please see other relevant chapters of this volume. One important thing to notice is that under a uniform prior for all possible ancestral states, the maximum-likelihood estimate is also a Bayesian maximum *a posteriori* probability (MAP) estimate.

However, from a Bayesian perspective it arguably makes little sense to first estimate all parameters of the model (except ancestral states) using maximum likelihood, and then to estimate ancestral states. Instead, it is preferable to estimate ancestral sequences jointly for all sites at the same time while integrating over all other parameters. The advantage is that the phylogenetic uncertainty, and uncertainty regarding the parameters of the evolutionary model, are taken into account in the estimation of ancestral states. This is achievable using Markov chain Monte Carlo (MCMC) methods described in the following section. One advantage of this method is that it directly provides an estimate of an entire evolutionary history; that is, of ancestral states, not only at the nodes of the phylogeny, but also at any point in time along the branches (edges) of the phylogeny.

6.3 MCMC

The basic idea in the MCMC algorithm is to represent mutations directly on the phylogeny. A history of mutations along one or more branches of the phylogeny is called a mutational path. The concept of a mutational path is illustrated in Figure 6.1. Bayesian ancestral reconstruction using MCMC exist in two flavors: (1) a two-step approach where a sample of genealogies and parameter values is first sampled using MCMC, and mutational paths are subsequently simulated given particular sampled phylogenies (Nielsen, 2002; Bollback, 2006), and (2) direct methods where the mutations are represented on the phylogeny while simulating the genealogy (Nielsen, 2001). Both approaches achieve the same goal, but they differ in computational efficiency and in which models they can accommodate. In general, we will assume the model of evolution can be described by a Markov chain on a finite, discrete state space with known generator, such as the set of all possible amino acids, all possible codons, or all possible nucleotides. We will also initially assume that this Markov process is independent among sites, although, as we will show, one of the advantages of these methods is that they can easily be extended to models of correlated evolution among sites.

6.3.1 Sampling mutational paths

In the following we will describe how algorithms of type 1 proceed. Consider first the case of a fixed phylogeny with known branch lengths and evolutionary model. We will first describe how to sample a mutational path stochastically under a particular evolutionary model. From eqn 6.1 we see that,

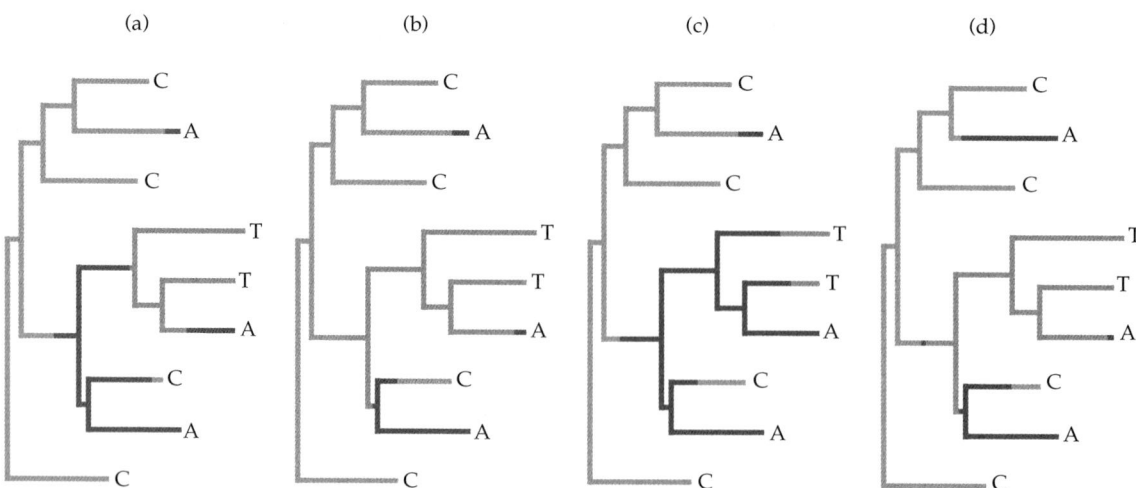

Figure 6.1 Examples of four mutational paths for a single site. Mutational paths were produced using the SIMMAP software package (Bollback, 2006).

$$P(y_{ij} = k | x_i) = \frac{f_{ij}(k)\pi(k)}{\sum_b f_{ij}(b)\pi(b)} \quad (6.2)$$

where π_b is the prior probability of state b (usually assumed to be the stationary frequency of b under the model), x_i is the observed data in site i, and y_{ij} is the unknown ancestral state in site i, node j. The ancestral state at the root of the tree can then be sampled according to these probabilities. At a child node of node j, say node h, the ancestral state, y_{ih}, can then be sampled according to the probabilities

$$P(y_i = k | x_i, y_{ij} = v) = \frac{f_{ih}(k)p_{vk}(h)}{\sum_b f_{ih}(b)p_{vb}(h)} \quad (6.3)$$

This sampling procedure can then be repeated recursively along the branches in the tree until ancestral sequences have been sampled for all nodes. In a sample of n sequences the resulting vector of nucleotides, $y_i = (y_{i1}, y_{i2}, \ldots, y_{i(2n-3)})$, represents a sample from P (y_i, x_i).

For a time-reversible model of substitution, the distribution of y does not depend on where in the tree the root has been placed. An entire mutational path can then be obtained by sampling paths conditional on the ancestral sequences at the nodes. If we let z_{ih} be the mutational path leading to node h from node j for site i, with sampled ancestral states $y_{ih} = k$ and $y_{ij} = v$, a sample from the density $p(z_{ih} | y_{ih} = k, y_{ij} = v)$ can be obtained using standard methods for simulating Markov chains starting at state v. The conditioning can be achieved by simply eliminating paths which do not end in state k, and can be sped up in various ways. Repeating this scheme for all branches of the tree provides a full sample of $z_i | y_i$. The simulation procedure is completed by applying this procedure to all sites, providing a full sample of $z = (z_1, z_2, \ldots)$ and $y = (y_1, y_2, \ldots)$.

6.3.2 Incorporating phylogenetic uncertainty

The preceding description of the simulation procedure assumes that the phylogenetic tree and the parameters where known; that it produces samples from the density $p(z, y | x, \theta)$, where θ is a vector of all the nuisance parameters, including the mutational model and the phylogenetic tree with branch lengths. However, usually θ will not be known. In such cases, we wish to be able to obtain samples from

$$p(z, y | x) = \int p(z, y | x, \theta) p(\theta, x) d\theta \quad (6.4)$$

where the integral is over all supported values of θ. We can think of this integral as a sum over all topologies of the tree and a multiple integral over all possible branch lengths and parameters of the mutational process. The representation in eqn 6.3 suggests the following method for obtaining samples from $p(z, y | x)$:

1. sample $\theta^{(1)}, \theta^{(2)}, \ldots, \theta^{(n)}$ from $p(\theta | x)$
2. sample $z^{(s)}, y^{(s)}$ from $p(z, y | x, \theta^{(s)})$ for $s = 1, 2, \ldots, n$.

The samples from $p(\theta | x)$ can be obtained using any of the well-known MCMC programs, for example MrBayes (Huelsenbeck and Ronquist, 2001). This method provides a method for obtaining samples from $p(z, y | x)$ which takes advantage of existing computational methods for Bayesian inference in phylogenetics and is comparatively easy to implement.

6.3.3 Direct methods

In the direct methods (Nielsen, 2001; Robinson et al., 2003) the MCMC procedure is applied directly to a phylogeny with mutational paths; the state space of the Markov chain is the set of supported values of (y, z, θ). A Markov chain with stationary density $p(y, z, \theta | x)$ can be simulated, for example, by iterating updates of (x, y): for $i = 1, 2, \ldots, B$, where B is the number of sites:

1. simulate a new value of (z_i, y_i), (z_i^*, y_i^*), from a proposal density $q(y_i^*, z_i^* | x_i, \theta)$;
2. accept (z_i^*, y_i^*) with probability $(p(y^*, z^*, x | \theta^*) q(y_i, z_i | x_i, \theta)) / (q(y_i^*, z_i^* | x_i, \theta) p(y, z, x | \theta))$.

and updates of θ:

1. simulate a new value of θ, θ^* from a proposal density $q(\theta^* | x, y, z)$;
2. accept θ^* with probability $(p(x, y, z | \theta) q(\theta | x, y, z) p(\theta^*)) / (q(\theta^* | x, y, z) p(x, y, z | \theta) p(\theta))$.

In the above notation, the current state before an update is simply denoted by (x, y, θ) to simplify the notation and (z^*, y^*) is (z, y), with (z_i^*, y_i^*) replacing (z_i, y_i). Notice that any update kernel $q(\ldots)$ can be used as long as it ensures that all values of (θ, z, y) supported by $p(\theta, z, y|x)$ eventually can be reached. This simulation procedure takes advantage of the fact that $p(c, x|\theta)$ easily can be calculated directly from the generator as the sampling path of a continuous-time Markov chain without the need to calculate time-dependent transition probabilities. However, this MCMC procedure can be slow to converge because of the correlation between (y, z) and θ. One of the advantages of this procedure is that it allows the use of models with correlated evolution among sites (e.g. Robinson et al., 2003; Yu and Thorne, 2006). In models of protein evolution involving tertiary structure, or models involving CpG hypermutations, the state space of the Markov model is the set of 4^B possible sequences of length B. However, the likelihood function can no longer be written as the product of the likelihood in multiple independent sites. This means that the conventional statistical methods for inference are inapplicable. Numerical calculation of the time-dependent transition probabilities of the process are hard or impossible to calculate and methods based on summing over all possible states (e.g. as part of the likelihood calculation) are intractable. However, whereas it is not possible to calculate $p(x|\theta)$ in these models, calculations of $p(x, y, z|\theta)$ are straightforward. This means that MCMC procedures with state space on (y, z, θ) can be implemented relatively easily.

6.4 Statistical inference using sampled mutational paths

After obtaining samples from the posterior distribution of (y, z, θ) using either of the two methods, inference in a Bayesian framework proceeds in a straightforward fashion. The posterior expectation (or summary statistic) of any function of the mutational path can easily be calculated. For example, Nielsen (2001) used this method to calculate the posterior expectation of the ages of non-synonymous and synonymous mutations occurring on a phylogeny. The framework provides a computationally tractable framework for making rigorous statistical inferences based on mutational paths. Bollback (2002) and Nielsen and Huelsenbeck (2001) showed how statistical measures of uncertainty can be obtained in this framework based on the posterior predictive distribution.

Briefly, the posterior predictive distribution measures uncertainty by estimating the "null" or predictive distribution of mutational paths, given the information obtained from the posterior distribution. In this way, a summary statistic can be compared with its predictive distribution to test hypotheses and measures of statistical uncertainty. This approach is appealing because it can be applied and tailored to a wide variety of questions involving only the ability to specify a summary statistic and simulate posterior predictive mutational histories. A typical summary statistic would be calculated by summing over simulated mutational mappings and taking the expectation

$$h(x) = \frac{1}{N}\sum_i h(x, y^{(i)}, z^{(i)}) \qquad (6.5)$$

Using the indirect approach for obtaining samples from the posterior distribution, the estimation of the posterior predictive distribution can be accomplished in the following way.

1 Generate n samples of $\theta, (\theta^{(1)}, \theta^{(2)}, \ldots, \theta^{(n)})$ from the posterior distribution using established MCMC methods.
2 For $j = 1, 2, \ldots, n$
simulate a new set of aligned sequence, $x^{(j)}$, under $\theta^{(j)}$; simulate $k \leq n$ mutational paths, $(z^{(j,i)}, y^{(j,i)})$, $i = 1, 2, \ldots, k$, based on $x^{(j)}$ and $\theta^{(j)}$.
3 The distribution of $h(x_{(j)}) = \sum_i h(x, y^{(j,i)}, z^{(j,i)})$ then approximates the posterior predictive distribution of $h(x)$.

It should be noticed that for each of the simulated predictive data sets the posterior predictive expectation of a site's mutational history is averaged over all of the samples from the posterior. In this way we can effectively integrate out uncertainty in the posterior distribution using a single MCMC run. For additional details on predictive distributions the reader is referred to the literature (e.g. Nielsen and Huelsenbeck, 2001; Bollback, 2002; Suchard et al., 2003).

6.5 Examples

The method of mutational mapping has already found wide use in addressing questions about the age of mutations (Nielsen, 2001), positive selection (Nielsen and Huelsenbeck, 2001), morphological evolution (Huelsenbeck et al., 2003), and modeling non-independence among residues (Dimmic et al., 2005; Yu and Thorne, 2006), among others.

In this section we will illustrate two different uses of mutational mapping. Each of the analyses were performed using the SIMMAP software package (a brief introduction to this package is provided in the next section). We hope that this section will demonstrate the types of question that can be addressed using mutational mapping and motivate its use and further development.

6.5.1 Characterizing transmembrane regions of proteins

Transmembrane proteins span cellular membranes with intra- and extramembrane regions. One feature of these proteins is the highly hydrophobic (water-fearing) nature of the helices spanning the membrane and the hydrophilic (water-loving) nature of the protein domains in the cytoplasmic and periplasmic spaces. An early single-sequence approach evaluated the hydropathy of the primary structure of a protein (Kyte and Doolittle, 1982). This approach plots hydropathy, using a sliding window spanning n residues, as a function of sequence position (residue), which have been called Kyte–Doolittle hydropathy plots.

The Kyte–Doolittle method considers only a single sequence. To demonstrate the use of mutational maps we present an approach that produces a comparative Kyte–Doolittle plot that accommodates multiple sequences, uncertainty in the phylogeny relating the sequences, and the evolutionary model describing changes in those sequences. Whereas more advanced methods exist for discovering transmembrane domains (e.g. Krogh et al., 2001), the simple approach described here offers the ability to summarize data across multiple sequences to initially identify likely transmembrane regions. In addition, other methods may benefit from the inclusion of this type of a multi-sequence approach.

Two data-sets are analyzed using the indirect mutational mapping approach: a primate cytochrome oxidase II data set (Yoder and Yang, 2004) consisting of 52 species of lemur; and a primate chemokine receptor CCR5 data-set (a subset of sequences analyzed by Mummidi et al. (2000)) consisting of the mouse sequence and 11 primate sequences.

Cytochrome oxidase II is one of a number of proteins involved in the electron-transport chain in mitochondria. Cytochrome oxidase II has two transmembrane domains (TM1 and TM2; see Figure 6.2) and a highly aromatic (hydrophilic) region of amino acid residues involved directly in electron transport (Adkins and Honeycutt, 1994). Using the mapping approach the hydropathy of each site, averaged across the probable phylogenies, was evaluated for the primate cytochrome oxidase II. Sampling estimates of the phylogeny and model were first obtained using MrBayes (Huelsenbeck and Ronquist, 2001). SIMMAP was used to map mutational paths for each codon and from these the history of the hydropathy was summarized. This approach weights a site's hydropathy by taking the branch-length-weighted sum of the site's amino acid mutational history. The tree-weighted posterior expectation of hydropathy for each residue was evaluated across 105 trees and was plotted in a Kyte–Doolittle plot with a sliding window of size 11 (see Figure 6.2). The method clearly identifies the two transmembrane regions of cytochrome oxidase II protein and identifies the conserved hydrophilic region with amino acid residues involved directly in electron transport.

The second data-set consisted of 11 primate sequences and the mouse sequence (Mummidi et al., 2000), which were retrieved and aligned from the GenBank Nucleotide Sequence Database; mouse, human, chimpanzee, gorilla, green monkey, spider monkey, squirrel monkey, golden lion tamarin, golden-rumped tamarin, a marmoset, and two species of lemur. The chemokine receptor (CCR5) protein is a coreceptor target of HIV and SIV, and possibly other related viruses (Mummidi et al., 2000; Paterlini, 2002). CCR5 has seven

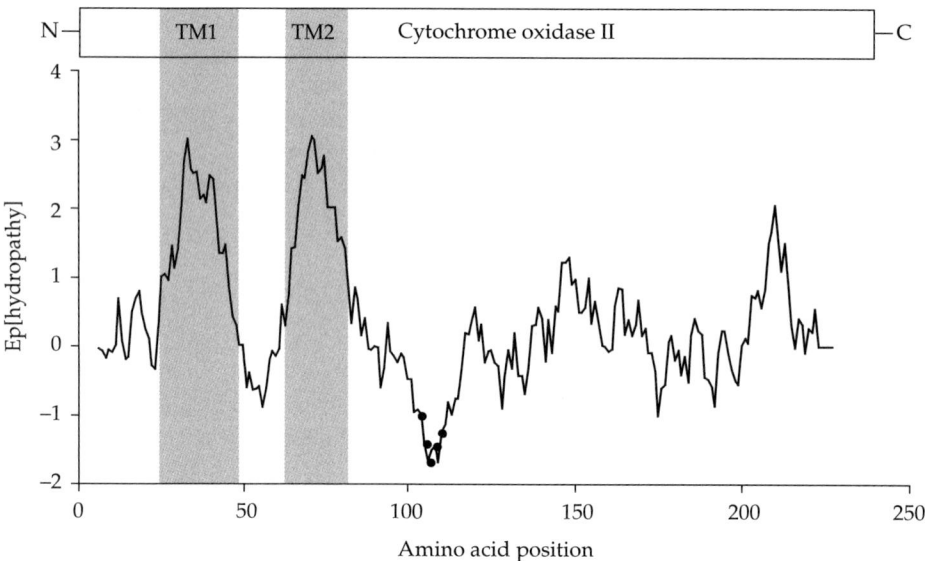

Figure 6.2 A comparative Kyte–Doolittle hydropathy plot of the hydopathic character of the primate cytochrome oxidase II protein using mutational mapping to summarize multiple sequences. The location of the transmembrane (TM) domains are shown at the top and the circles show the predicted amino acids involved in electron transport.

transmembrane domains, labeled TM1–TM7 (Paterlini, 2002). Using mutational histories we analyzed CCR5 for the hydropathic signature of the transmembrane regions (Figure 6.3) and the association between hydropathic change and positive selection (Figure 6.4). In addition, for each site we calculated the posterior expectation of the change in hydropathy.

Of the seven transmembrane domains in CCR5 TM1–TM6 are clearly identified as having had a highly hydrophobic history, whereas TM7 does not show a large deviation in its hydropathy. The cytoplasmic/periplasmic regions show a considerable bias towards being hydrophilic. Inspection of the mean change in hydropathy indicates that there is a general pattern of transmembrane regions showing values close to 0 or slightly negative, whereas hydrophilic regions exhibit larger changes with a tendency towards hydrophobicity (Figure 6.4). This latter observation at first seems surprising but may reflect tertiary packing in these regions in which evolution is occurring at mostly buried residues or at residues that are under selection for interaction with other proteins; sites that show evidence for positive selection occur, predominantly, in cytoplasmic/periplasmic domains (74% of sites; Figure 6.4). In addition, sites in transmembrane regions that show evidence for positive selection show a large tendency for change to more hydrophobic residues. Nevertheless the approach provides comparative information indicating domains that are likely to have had a highly hydrophobic mutational history and, thus, likely to be transmembrane regions.

6.5.2 Nucleotide covariation in mitochondrial tRNAs

One powerful use of the mutational mapping approach is in detecting non-independence, or covariation, among nucleotide residues. Mutational histories have been successfully applied to detecting compensatory changes in amino acid residues (Dimmic et al., 2005) and correlation between morphological characters (Huelsenbeck et al., 2003).

To illustrate the use of mutational histories for detecting nucleotide covariation we analyzed all 21 mitochondrial tRNAs for 106 species with the hope

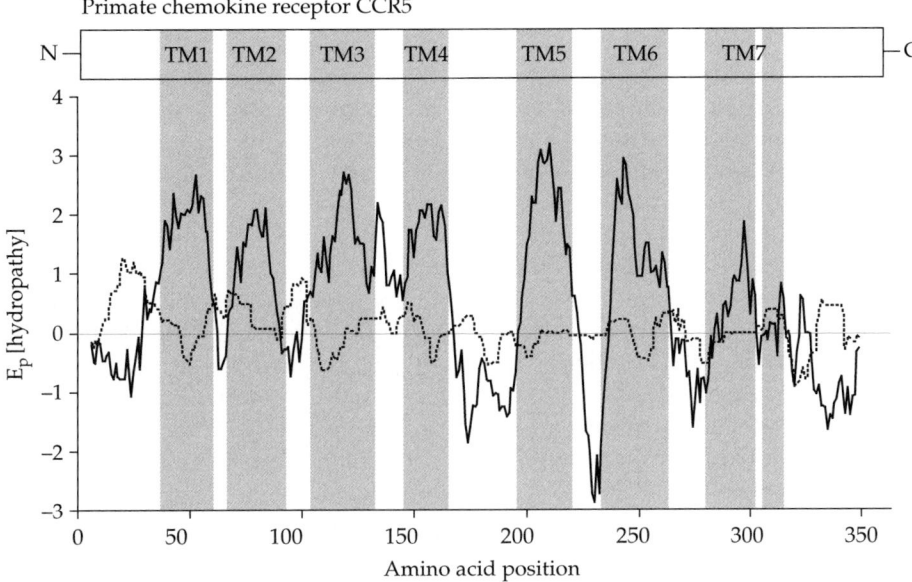

Figure 6.3 A comparative Kyte–Doolittle hydropathy plot of the hydropathic features of the primate CCR5 protein using mutational mapping to summarize multiple sequences. The posterior expectation of hydropathy along the gene is shown by the black line while the dashed line is the posterior expectation of the direction and magnitude of change in hydropathy (see Figure 6.4 for a plot of the hydropathic magnitude and direction of changes on a residue-by-residue basis). The plots were generated using a window size of 11 residues. Locations of the transmembrane (TM) regions are shown at the top.

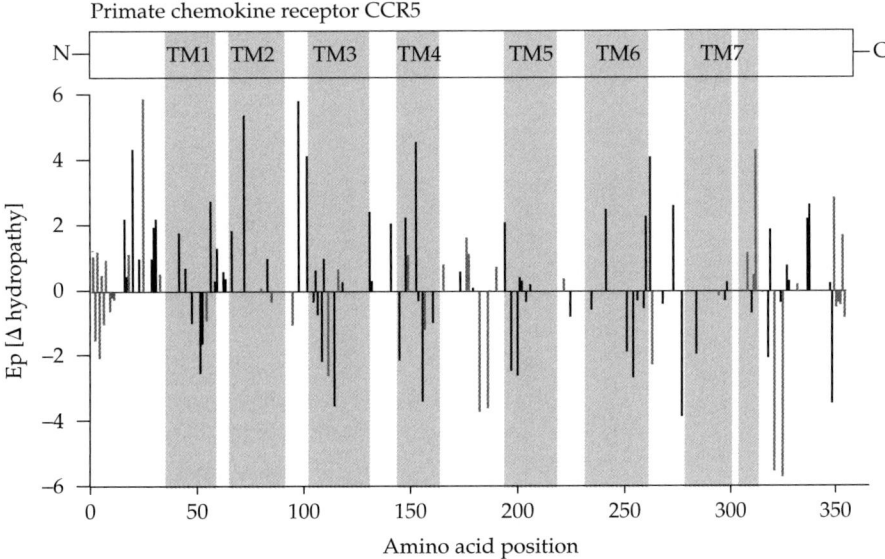

Figure 6.4 Patterns of change in hydropathy and positive selection. The magnitude and direction of change in hydropathy are plotted along the sequence. Sites that show at least two substitutions and with a non-synonymous/synonymous ratio >1 are shown in gray.

of detecting positive evidence for covariation among base-pairing residues. As above, we adopted the indirect approach using MrBayes (Huelsenbeck and Ronquist, 2001) to provide a sampling approximation of the phylogeny and substitution model parameters. All 21 tRNA genes were concatenated and the GTR + gamma substitution model was used (Lanavé et al., 1984; Yang, 1994) to obtain samples from the posterior distribution of the phylogeny and substitution model parameters.

But how can we measure covariation among nucleotides using mutational histories? The answer is straightforward. For each site we sample a mutational path and then compare the covariation (coincidence of states) between each site's history. Any number of statistics can be used. To evaluate the degree of covariation we calculated the association between different states along each branch of a phylogeny using the following statistic:

$$m_{ij} = f_{ij} \log_2 \frac{f_{ij}}{f_i f_j} \quad (6.6)$$

where f_{ij} is the fraction of time state i is associated with j in a character history, and f_i is the fraction time in a particular state independent of associations (i.e. the sum of time spent in a particular state on the phylogeny). We refer to this statistic as the mutual historical information content, or MHIC, because of its relationship in form to the classical mutual information content statistic (Chiu and Kolodziejczak, 1991; Gutell et al., 1992). By averaging over all state associations we get the following statistic for the overall character correlation:

$$M = \sum_{i=1}^{x} \sum_{j=1}^{y} m_{ij} \quad (6.7)$$

In addressing whether we can detect covariation in RNA molecules, which is the result of base-pairing constraints, we perform the summation over only Watson–Crick (AU and GC) and wobble pairs (GU). In this way we focus on covariation that is due to base-pairing. To accommodate uncertainty in the phylogeny and substitution model parameters we calculated the posterior expectation of the MHIC statistic for each pair and then compared the known pairs with the unknown pairs for each tRNA. A comparison of the MHIC reveals that paired sites show a strong signature of covariation relative to unpaired sites (Figure 6.5) in 20 of the 21 tRNAs; in the single case of the tRNA$_{Met}$ the difference between paired and unpaired is indistinguishable. This is likely the result of very little variation in the tRNA$_{Met}$ at paired sites. These results clearly indicate that the signature of covariation can be easily identified using the mutational history approach.

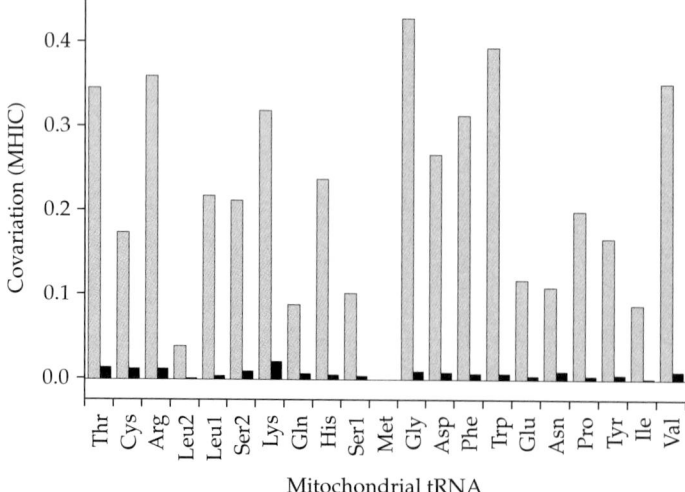

Figure 6.5 Nucleotide covariation measures for the 21 mammalian tRNAs. Gray bars represent the mean mutual historical information content (MHIC) for known pairs and black bars show the means of unpaired sites. A comparison of the standard deviations (not shown) indicates that in all but the tRNA$_{Met}$ these differences are significant.

Whereas the previous results indicate a strong difference in the signature of covariation between known pairs and unpaired sites we might wish to determine how well the method predicts true pairs (true positives) relative to mis-assigning pairs (false positives). One commonly used measure is Mathew's correlation coefficient (MCC):

$$\text{MCC} = \frac{TP \times TN - FP \times FN}{\sqrt{(TP+FP)(TP+FN)(TN+FP)(TN+FN)}} \quad (6.8)$$

where TN is the number of true negatives, TP is the number of true positives, FN is the number of false negatives, and FP is the number of false positives. In using this measure the first step is to rank each comparison MHIC value and then to break this into discrete intervals or thresholds. Values above the threshold are considered to be *paired* and values below to be *unpaired*. At each threshold the MCC is calculated and is a measure of the method's ability to correctly identify true pairs while minimizing false positives. Figure 6.6 shows the MCC values for the the tRNA$_{Gly}$ and tRNA$_{His}$.

We evaluated overall performance by calculating the mean and 95% confidence intervals of MCC at each threshold value for all 21 tRNAs (Figure 6.7). The performance of the method, as measured by the MCC, is high, with a maximum mean MCC value of 0.42 averaged across the 21 tRNAs, and with 11 out of 21 (52%) of the tRNAs with individual maximum MCC values above 50%. The performance of this approach does decline with divergence as expected (data not shown).

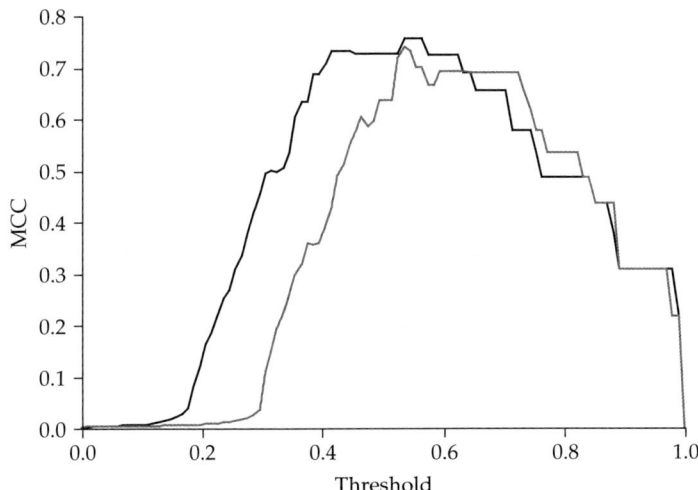

Figure 6.6 Mathew's correlation coefficient (MCC) for the mammalian tRNA$_{Gly}$ (black) and tRNA$_{His}$ (gray).

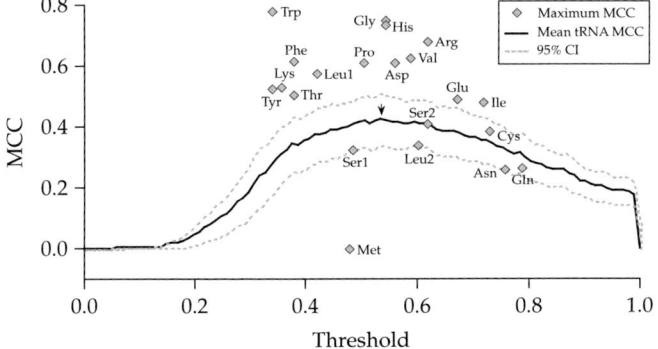

Figure 6.7 Mathew's correlation coefficient (MCC) for the 21 mammalian tRNAs. The mean value for all 21 tRNAs is shown as the black line with the 95% confidence interval (CI) is shown by dashed lines. The arrow indicates the point at which the mean MCC is maximized.

6.6 Software

Few user-friendly software packages have been written that generalize the mutational mapping method to different kinds of data (molecular and morphological) and a wide variety of biological questions. We introduce one such package, SIMMAP, which provides researchers with the ability to address a wide variety of questions using either molecular or morphological data. Briefly, SIMMAP is a software implementation of the indirect method of mapping mutational histories described in the first sections of this chapter. For example, SIMMAP can be used to sample character histories on phylogenies to address questions about general evolutionary patterns of change (trends), positive selection at focal sites or across a gene region, and correlation among characters (as described above), and will generate the raw mutational histories for custom analyses not available in the software package. In addition, SIMMAP can be used to calculate the posterior distribution of ancestral states in either a full hierarchical or empirical Bayesian framework. The software accommodates a wide variety of substitution models and priors. SIMMAP is free for academic use and a download can be obtained at www.simmap.com.

6.7 Acknowledgments

We would like to acknowledge the following funding agencies for support during the writing: a Danish FSS grant (271050599) to J.P.B and R.N, and a Carlsberg Foundation grant (21000680) to P.P.G.

References

Adkins, R.M. and Honeycutt, R.L. (1994) Evolution of the primate cytochrome c oxidase subunit II gene. *J. Mol. Evol.* **38**: 215–231.

Bishop, J.G., Dean, A.M., and Mitchell-Olds, T. (2000) Rapid evolution of plant chitinases: molecular targets of selection in plant-pathogen coevolution. *Proc. Natl. Acad. Sci. USA* **97**: 5322–5327.

Bollback, J.P. (2002) Bayesian model adequacy and choice in phylogenetics. *Mol. Biol. Evol.* **19**: 1171–1180.

Bollback, J.P. (2006) SIMMAP: stochastic character mapping of discrete traits on phylogenies. *BMC Bioinformatics* **7**: 88.

Brooks, D.R. and McLennan, D.A. (1991) *Phylogeny, Ecology, and Behavior: a Research Program in Comparative Biology*. University of Chicago Press, Chicago.

Chiu, D.K. and Kolodziejczak, T. (1991) Inferring consensus structure from nucleic acid sequences. *Comput. Appl. Biosci.* **7**: 347–352.

Dimmic, M.W., Hubisz, M.J., Bustamante, C.D., and Nielsen, R. (2005) Detecting coevolving amino acid sites using bayesian mutational mapping. *Bioinformatics* **21** (suppl. 1): i126–i135.

Felsenstein, J. (1985) Phylogenies and the comparative method. *Am. Nat.* **125**: 1–15.

Gillespie, J. (1991) *The Causes of Molecular Evolution*. Oxford University Press, Oxford.

Gutell, R.R., Power, A., Hertz, G.Z., Putz, E.J., and Stormo, G.D. (1992) Identifying constraints on the highe order structure of RNA: continued development and application of comparative sequence analysis methods. *Nucleic Acids Res.* **20**: 5785–5795.

Harvey, P.H. and Pagel, M.D. (1991) *The Comparative Method in Evolutionary Biology*. Oxford University Press, Oxford.

Huelsenbeck, J.P. and Ronquist, F. (2001) MRBAYES: Bayesian inference of phylogenetic trees. *Bioinformatics Applications Note* **17**: 754–755.

Huelsenbeck, J.P., Nielsen, R., and Bollback, J.P. (2003) Stochastic mapping of morphological characters. *Syst. Biol.* **52**: 131–158.

Koshi, J.M. and Goldstein, R.A. (1996) Probabilistic reconstruction of ancestral protein sequences. *J. Mol. Evol.* **42**: 313–320.

Krogh, A., Larsson, B., von Heijne, G., and Sonnhammer, E.L. (2001) Predicting transmembrane protein topology with a hidden Markov model: application to complete genomes. *J. Mol. Biol.* **305**: 567–580.

Kyte, J. and Doolittle, R.F. (1982) A simple method for displaying the hydropathic character of a protein. *J. Mol. Biol.* **157**: 105–132.

Lanavé, C., Preparata, G., Saccone, C., and Serio, G. (1984) A new method for calculating evolutionary substitution rates. *J. Mol. Evol.* **20**: 86–93.

Langley, C.H. and Fitch, W.M. (1974) An estimation of the constancy of the rate of molecular evolution. *J. Mol. Evol.* **3**: 161–177.

Messier, W. and Stewart, C.-B. (1997) Episodic adaptive evolution of primate lysomzymes. *Nature* **385**: 151–154.

Mummidi, S., Bamshad, M., Ahuja, S.S., Gonzalez, E., Feuillet, P.M., Begum, K. *et al.* (2000) Evolution of human and non-human primate cc chemokine receptor 5 gene and mRNA. Potential roles for haplotype and mRNA diversity, differential haplotype specific transcriptional activity, and altered transcription factor binding to polymorphic nucleotides in the pathogenesis of HIV-1 and simian immunodeficiency virus. *J. Biol. Chem.* **275**: 18946–18961.

Nielsen, R. (2001) Mutations as missing data: inferences on the ages and distributions of nonsynonymous and synonymous mutations. *Genetics* **159**: 401–411.

Nielsen, R. (2002) Mapping mutations on phylogenies. *Syst. Biol.* **51**: 729–732.

Nielsen, R. and Huelsenbeck, J.P. (2001) Detecting positively selected amino acid sites using posterior predictive p-values. In *Pacific Symposium on Biocomputing Proceedings*, pp. 576–588. World Scientific, Singapore.

Pagel, M.D. (1999) Inferring the historical patterns of biological evolution. *Nature* **401**: 877–884.

Paterlini, M.G. (2002) Structure modeling of the chemokine receptor CCR5: implications for ligand binding and selectivity. *Biophys. J.* **83**: 3012–3031.

Pupko, T., Pe'er, I., Shamir, R., and Graur, D. (2000) A fast algorithm for joint reconstruction of ancestral amino acid sequences. *Mol. Biol. Evol.* **17**: 890–896.

Robinson, D.M., Jones, D.T., Kishino, H., Goldman, N., and Thorne, J.L. (2003) Protein evolution with dependence among codons due to tertiary structure. *Mol. Biol. Evol.* **20**: 1692–1704.

Suchard, M.A., Weiss, R.E., Sinsheimer, J.S., Dorman, K.S., Patel, P., and McCabe, E.R.B. (2003) Evolutionary similarity among genes. *J. Am. Stat. Assoc. Theory Methods* **98**: 653–662.

Templeton, A.R. (1996) Contingency tests of neutrality using intra/interspecific gene trees: the rejection of neutrality for the evolution of the cytochrome oxidase II gene in the hominoid primates. *Genetics* **144**: 1263–1270.

Yang, Z. (1994) Maximum likelihood phylogenetic estimation from DNA sequences with variable rates over sites: approximate methods. *J. Mol. Evol.* **39**: 306–314.

Yang, Z., Kumar, S., and Nei, M. (1995) A new method of inference of ancestral nucleotide and amino acid sequences. *Genetics* **141**: 1641–1650.

Yoder, A.D. and Yang, Z. (2004) Divergence dates for Malagasy lemurs estimated from multiple gene loci: geological and evolutionary context. *Mol. Ecol.* **13**: 757–773.

Yu, J. and Thorne, J.L. (2006) Dependence among sites in RNA evolution. *Mol. Biol. Evol.* **23**: 1525–1537.

CHAPTER 7

Coarse projections of the protein-mutational fitness landscape

F. Nicholas Braun

As described in any textbook on evolution, at the core of the neo-Darwinian synthesis (natural selection, populations, genetics) lies the concept of a fitness landscape. Although in broad outline credited to Wright and Fisher, the *de facto* first application of the landscape idea specifically to protein sequence seems to have come much later, in a paper by Maynard-Smith (1970), where he sets out an attractive analogy between evolutionary paths in sequence space and a simple word game. The object of the game is to pass from one word to another by changing a single letter at a time, always observing the rule that intermediate words must be meaningful in the same language. For example,

LIFE LIFT LIST LUST DUST

Maynard-Smith's concern here was to stress the pivotal role in protein evolution of continuity of function. A meaningful word plucked from the sea of largely nonsensical four-letter combinations corresponds to a biologically fit (i.e. functionally viable) protein sequence from among many deleterious alternatives. Hence, a connected thread of meaningful words corresponds to a run of fitness-preserving mutational steps in sequence space; in effect, to a ridge or plateau in the figurative fitness landscape.

In ancestral sequence reconstruction, a modified version of this game is played where, by changing one letter at a time, the aim is to spell out some particularly interesting ancient word long since forgotten. The challenge of this version derives in part from the player's general inability to judge from looking at a word whether it has any meaning; that is, whether a sequence maps to a viable ridge in the landscape or to an unviable hole.

We have seen from the chapters of this book that in fact it is feasible in practice to get around this handicap by playing fast and loose with the rule about continuity of meaning/function. The strategy is to essentially ignore whether intermediate sequences extrapolated back to the ancestor are at all functionally viable, relying instead on the option to "phone a friend" at the end of the day, in the form of an experimental assay. It is reasonably clear, nevertheless, that there exists scope for an improved strategy built around knowledge of the fitness landscape, even if only to the extent that this would allow us to immediately identify and weed out unviable intermediates of the reconstruction. But there are two obvious hurdles which seem initially to obstruct its practical implementation: (1) we do not know the fitness landscape and (2) even if we did, arguably a main scenario of interest concerns specifically those situations where the landscape has changed since the ancestor.

Toward negotiating the first and largest of these hurdles, this chapter explores the premise that proteins tend generically to become marginally fit with respect to certain coarse-grained traits of protein sequence. Given phylogenetic data for such traits, I argue, this property may be exploited to infer crude outlines of the fitness landscape. At this crude level, the second point above may become moot.

To begin with, what is marginal fitness and how does it come about in protein evolution? Several theoretical studies have explored the ways in which functional proteins disperse evolutionarily across the space of sequence and structural phenotype, assigning fitness according to various statistical mechanical models of folding stability

(Bornberg-Bauer and Chan, 1999; Van Nimwegen et al., 1999; Bastolla et al., 2003; Bloom et al., 2005), or other functional criteria such as solubility (Braun, 2004; Ito et al., 2004; Sear, 2004). A key insight which emerges is that, even in the absence of selective gradients and peaks, protein evolution nevertheless remains highly directed; toward sequences and phenotypes which are maximally *designable*, in the sense that they may be reached by a large number of viable mutational pathways. Thus, Taverna and Goldstein (2001) are able to argue in particular that a design pressure of this nature, as distinct from a positive selection pressure, is responsible for the frequent evolutionary incidence among globular proteins of marginal folding stability. Neutral mutational events in this interpretation tend to reduce rather than enhance folding stability, hence pushing the phenotype toward some marginal edge region of the folding fitness landscape. Several instructive related points concerning "weak" and "strong" designability are made by England and Shakhnovich (2003).

A generic association between design pressure and marginal protein fitness can be extended, beyond folding, to other kinds of functional criteria as well. As a template for illustration, consider the sequence architecture sketched in Figure 7.1 for major ampullate spidroin 1 (MaSp1), one of the principal fibrous components of spider silk. This protein comprises a large number of tandem repeats featuring a polyalanine stretch and several glycine-based motifs (GGX/GX). The polyalanine stretches form crystalline (β-sheet) bands which, according to current structural hypotheses (Hayashi et al., 1999), hold the assembled silk fiber together. The more structurally flexible GGX/GX motifs, on the other hand, afford elasticity. MaSp1 genes have been identified in a dozen or so different spider species, essentially retaining this overall architecture, but with differences in the number of elastic motifs in the block (let us denote this Λ) and also in the density of motif deletions (n), depicted as empty squares in Figure 7.1.

These parameters characterizing the MaSp1 sequence constitute specific examples, against the more general backdrop of this discussion, of what I mean by coarse-grained traits likely to be implicated in fitness. I suspect that both n and Λ are intimately connected to mechanical performance of the assembled silk fiber. Consider in this respect a one-dimensional projection of silk mechanical fitness over the space of one of these traits. If we postulate that design pressure tends to push MaSp1 either to the left or to the right across

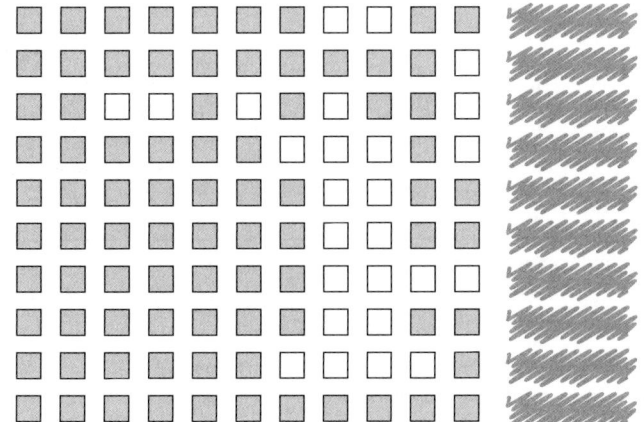

Figure 7.1 Fragments of the silk protein MaSp1 can be found on the SwissProt database for a number of phylogenetically close spider species. Self-alignment of these fragments in a tandemly repeated block pattern suggests several novel examples of coarse-grained traits of sequence likely to influence the mechanical fitness of the assembled silk. In the text we have focused on the total number, Λ of elastic motifs (squares) in the repeated unit, and also on the deleted proportion, n, of such motifs (empty squares).

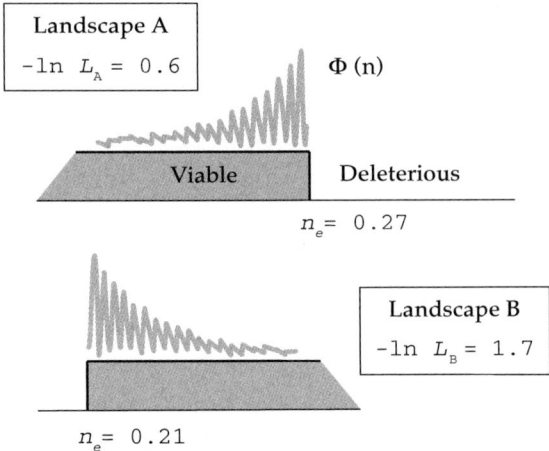

Figure 7.2 There are two possible mirror-image projections of a marginal fitness equilibrium in one dimension. To determine which of these pertains to a given coarse-grained trait, we look for their respective footprints in phylogenetic data. Phylogenetic variability $\Phi(n)$ of the MaSp1 trait n (Table 7.1) is left-skewed, a situation which supports model A. Details of the statistical analysis are given in the text.

this space, then eventually a deleterious impasse must inevitably by reached corresponding to one or other of the alternative scenarios sketched in Figure 7.2. Having reached such an impasse, the mutational dynamics subsequently relax into a marginally fit steady-state equilibrium.

The marginally fit mutational equilibrium, if it exists, ought to be reflected in the phylogenetic variability of the protein. It should therefore be possible in principle to ascertain from phylogenetic data which landscape topography, A or B, pertains to the given coarse-grained trait. To do this, we describe the equilibrium qualitatively according to some plausible normalized distribution function in mutational distance d from the hypothesized fitness impasse. An expedient choice is

$$\Phi(d) = \langle d \rangle^{-1} \exp[-d/\langle d \rangle] \qquad (7.1)$$

which has the useful property that the mean also trivially defines the second moment, $\langle d^2 \rangle = 2\langle d \rangle^2$. To a given landscape hypothesis, we can thus assign a likelihood score derived from the observed values of d in a phylogenetic sample:

$$-\ln L = \left[2\langle d \rangle^2 - \langle d^2 \rangle\right]^2 / 2\langle d \rangle^2 \qquad (7.2)$$

where Poisson-like error is assumed.

Figure 7.2 presents an analysis of the density n of MaSp1 elastic defects by this method. With mutational distance defined as $d = |n - n_e|$, where n_e is the location of the deleterious edge under the respective models, we find that the phylogenetic data of Table 7.1 support model A. This result seems also physically plausible. Design pressure in this case acts to generate the defects, rather than to mend them. When their density exceeds $n_e = 0.27$, mechanical performance of the silk becomes adversely affected. Block period Λ may be analyzed in similar fashion, setting $d = |\Lambda - \Lambda_e|$. Here we find that the table instead supports landscape B, indicating a design pressure to reduce Λ rather than increase it ($-\ln L_B = 0.03$ for $\Lambda_e = 11$, compared with $-\ln L_A = 0.90$ for $\Lambda_e = 14$).

Figure 7.3 gives an example of how these projections might usefully be applied in conjunction with a traditional approach to ancestral reconstruction. In the example, a putative reconstruction is revealed to be evolutionarily deleterious in the Λ projection. It also fails to pass muster in the n projection, in so far as it lies below a lower critical value n_c for the density of elastic defects likely to be observed in the marginal equilibrium. For the probability distribution function of eqn 7.1, we have

$$d_c = -\langle d \rangle \ln P \qquad (7.3)$$

where P is the significance level. Hence, substituting $d_c = |n_c - n_e|$ with, from the analysis above, $n_e = 0.27$ and $\langle d \rangle = \langle |n - n_e| \rangle = 0.03$ we obtain for a $P = 0.01$ level of significance,

$$n_c = n_e + \langle |n - n_e| \rangle \ln P \cong 0.13 \qquad (7.4)$$

It is possible to think of further statistical manipulations in this spirit. However, let us

Table 7.1 Phylogenetic variation of the coarse-grained spidroin sequence traits defined in Figure 7.1

Species (SwissProt ref.)	Block period, Λ	Defect density, n
Nephila clavipes (Q692G2)	11	0.24
Nephila madagascariensis (Q9BIT6)	13	0.21
Nephila senegalensis (Q9BIT4)	11	0.27
Latrodectus geometricus (Q9BIUO)	12	0.24
Latrodectus hesperus (Q4G1Y2)	12	0.27
Argiope aurantia (Q9BIV1)	12/14	0.23

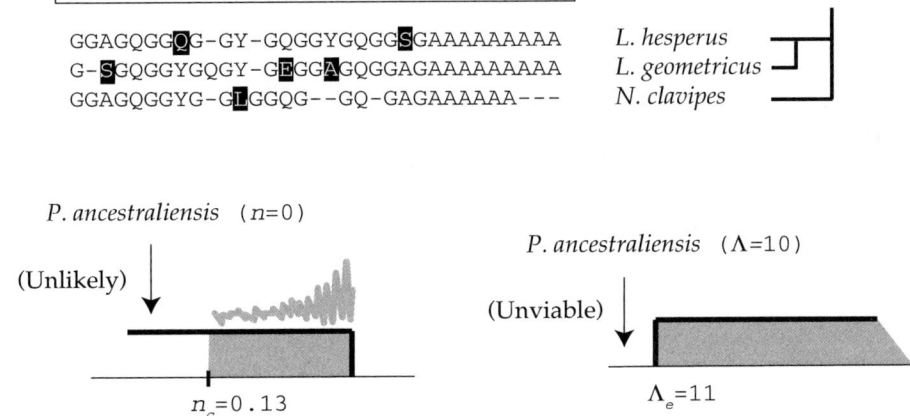

Figure 7.3 Coarse-grained projections of the fitness landscape may in certain cases eliminate an ancestral reconstruction which looks plausible otherwise. In this example, single repeats arbitrarily drawn from different MaSp1 sequences are aligned to infer a putative ancestor, *Parsimonius ancestraliensis*. In both the n and Λ projections of the fitness landscape, however, a MaSp1 sequence uniquely comprised of such repeats is seen to be untenable.

conclude here. The main point is to demonstrate that the principle of a marginal fitness equilibrium is extremely powerful. From a few almost trivially simple calculations exploiting this principle, we have been able to derive quite striking semi-quantitative insights into the way in which coarse-grained traits of the MaSp1 sequence evolve and can be extrapolated back to the ancestor.

Of course, a criticism which might be leveled at this framework from a general perspective is that it relies on a rather haphazard identification of relevant coarse-grained traits, that the examples given here for silk proteins are obscure one-offs. Against this, it can be countered on the one hand that a degree of *ad hoc* resourcefulness with regard to physicochemical context is necessary to avoid the study of ancestral sequence and function becoming merely a sub-discipline of the phylogenetic tree-building paradigm of sequence evolution. There are also a number of coarse-grained parameters which have across-the-board relevance to proteins very generally, for example length, net charge, and hydrophobic content (Cheung and Truskett, 2005).

With regard to the opening remark (2), above, concerning changes to the landscape since the ancestor, it is appropriate to stress finally that "the map is not the territory", borrowing from the semanticist Alfred Korzybski his well-known phrase. Considerable changes to the fitness landscape are possible without affecting the low-resolution map of our approach. In the exploration of ancestral sequence and function, the map remains useful nevertheless.

References

Bastolla, U., Proto, M., Roman, H.E., and Vendruscolo, M. (2003) Statistical properties of neutral evolution. *J. Mol. Evol.* **57**: S103–S119.

Bloom, J.D., Silberg, J.J., Wilke, C.O., Drummond, D.A., Adami, C., and Arnold, F.H. (2005) Thermodynamic prediction of protein neutrality. *Proc. Natl. Acad. Sci. USA* **102**: 606–611.

Bornberg-Bauer, E. and Chan, H.S. (1999) Modelling evolutionary landscapes: mutational stability, topology and superfunnels in sequence space. *Proc. Natl. Acad. Sci. USA* **96**: 10689–10694.

Braun, F.N. (2004) Sequence variability of proteins evolutionarily constrained by solution-thermodynamic function. *Phys. Rev. E* **69**: 011903.

Cheung, J.K. and Truskett, T.M. (2005) Coarse-grained strategy for modelling protein stability in concentrated solutions. *Biophys. J.* **89**: 2372–2384.

England, J.L. and Shakhnovich, E.I. (2003) Structural determinant of protein designability. *Phys. Rev. Lett.* **90**: 218101.

Hayashi, C.Y., Shipley, N.H., and Lewis, R.V. (1999) Hypotheses that correlate the sequence, structure and mechanical properties of spider silk proteins. *Int. J. Biol. Macromol.* **24**: 271–275.

Ito, Y., Kawama, T., Urabe, I., and Yomo, T. (2004) Evolution of an arbitrary sequence in solubility. *J. Mol. Evol.* **58**: 196–202.

Maynard Smith, J. (1970) Natural selection and the concept of a protein space. *Nature* **225**: 563–564.

Sear, R.P. (2004) Solution stability and variability in a simple model of globular proteins. *J. Chem. Phys.* **120**: 998–1005.

Taverna, D.M. and Goldstein, R.A. (2001) Why are proteins marginally stable? *Proteins* **46**: 105–109.

Van Nimwegen, E., Crutchfield, J.P., and Huynen, M. (1999) Neutral evolution of mutational robustness. *Proc. Natl. Acad. Sci. USA* **96**: 9716–9720.

CHAPTER 8

Dealing with uncertainty in ancestral sequence reconstruction: sampling from the posterior distribution

David D. Pollock and Belinda S.W. Chang

8.1 Introduction

Whereas the evolution of morphology, particularly bone morphology, can be studied by digging up fossilized remains, DNA and proteins unfortunately do not survive the ravages of time as well. Nevertheless, the evolution of ancient protein function can be studied by inferring ancestral sequences using phylogenetic techniques (Fitch, 1971; Yang *et al.*, 1995; Koshi and Goldstein, 1996) and applying gene-synthesis methods to resurrect them and assay their function *in vitro* (Geman and Geman, 1984; Malcolm *et al.*, 1990; Jermann *et al.*, 1995; Krawczak *et al.*, 1996; Ivics *et al.*, 1997; Messier and Stewart, 1997; Benner *et al.*, 2000; Benner, 2002; Chang *et al.*, 2002a, 2002b; Zhang and Rosenberg, 2002; Gaucher *et al.*, 2003a, 2003b; Thornton *et al.*, 2003; Thornton, 2004). The power of this approach lies in the opportunity to directly test hypotheses concerning the evolution of ancestral protein structure and function. Molecular evolution is a field dominated by inferences about the past based primarily on examination of present-day protein function, making this hypothesis-testing ability particularly important. Since an experimentally recreated ancestral sequence is inferred rather than observed, however, the question remains as to whether the functional features of the reconstructed protein are a true approximation of the functional features of the ancestral protein. A thorough statistical assessment and justification of the method used is therefore critical to success.

Parsimony methods have been used for ancestral reconstruction due to their ease of implementation; however, certain limitations such as the lack of an explicit model of evolution, and bias towards more frequent amino acids (or nucleotides, in the case of DNA or codon reconstructions; Collins *et al.*, 1994; Zhang and Nei, 1997; Eyre-Walker, 1998; Sanderson *et al.*, 2000; Krishnan *et al.*, 2004), have rendered the use of model-based maximum likelihood in an empirical Bayesian approach more prevalent (Chang *et al.*, 2002a, 2002b; Gaucher *et al.*, 2003b; Thornton *et al.*, 2003; Thornton, 2004). More recently, full Bayesian methods have also been implemented (Huelsenbeck *et al.*, 2003; Ronquist and Huelsenbeck, 2003; see also Chapter 16 in this volume).

Depending on the levels of sequence divergence at which ancestral nodes are being reconstructed, there can be substantial variation at certain amino acid sites in the ancestral sequences inferred under different models of evolution. This issue of model variability in reconstructed ancestral sequences has been addressed experimentally using a variety of methods. These methods have ranged from using site-directed mutagenesis techniques to generate variants at sites differing among maximum-likelihood results from different models (Chang *et al.*, 2002a), to the incorporation of degenerate oligonucleotides into the synthesis of the most likely ancestral gene, allowing for random sampling of sites that vary under different models (Ugalde *et al.*, 2004). At the levels of divergence investigated thus far, these experiments have not demonstrated significant functional differences among reconstructed ancestral protein variants. However, this issue of model variability in experimental recreation of ancestral proteins is distinct from the uncertainty that can arise under a single model of evolution, which is the main concern of this chapter. Model variability, and how best to address it experimentally, is the subject of Chapter 15 in this volume, and will not be discussed at length here.

Due to the constraints imposed by the effort and resources required to reconstruct ancestral proteins in the laboratory, most studies of ancestral protein function have by necessity tended to focus on a single optimal ancestral sequence. This is either the most parsimonious ancestor for the maximum-parsimony approaches or the most probable ancestor (MPA) for the Bayesian and empirical Bayesian approaches (Krishnan et al., 2004). If time has permitted, variants based on that initial sequence have also been synthesized. This was assumed to make sense, given the high costs of synthesizing and expressing proteins in vitro, but the potential pitfalls of focusing on the MPA were not thoroughly considered. Unfortunately, just as with maximum-parsimony reconstructions, optimality-based MPAs can be biased toward more frequent amino acid states, even when the model is correct (Krishnan et al., 2004). This bias in amino acid frequencies may in turn lead to biases in the inferred function of the reconstructed ancestors (Krishnan et al., 2004; Williams et al., 2006).

The goal of ancestral inference, of course, is to have as accurate a picture of ancestral function as possible, so it is worthwhile to try to understand the nature and cause of the sequence and functional bias, and how to overcome this bias. The principle source of bias in the MPA and maximum-parsimony reconstructions is (as their names imply) the choice of a "best" reconstruction for every site in the protein. Although this sounds intuitively preferable, the result of repeatedly making the optimal or best choice at every site is that you will preferentially choose amino acids that are more frequent at every site. If these more frequent amino acids are preferentially associated with some aspect of function, such as thermodynamic stability, then the cumulative effect will be an error in reconstruction of that aspect of function. Although an optimist might prefer to assume that such association is rare, in nearly-neutral population genetics models and in thermodynamic-based population genetics simulations, slightly deleterious variants tend to be incorporated into substitutions less often than the more fit alternatives. Simulations show that this can lead to reconstructed ancestors that are *more* stable than the true ancestral sequences (Williams et al., 2006).

It is important to note that the bias toward more frequent amino acids in ancestral reconstruction is due to the choice of the most probable amino acid residue at each site, from the posterior probability distribution of all amino acids at that site. To our knowledge, it has not yet been shown that there are important differences in ancestral reconstruction depending on the whether the posterior sampling comes from a full Bayesian analysis of topology and other parameters, or from an empirical Bayesian analysis that produces a marginal posterior distribution of amino acid frequencies at each node and site. In this paper, we will refer to the bias in ancestral reconstruction as optimization bias, regardless of the method of generating the posterior distribution.

In considering how ancestral amino acid frequency optimization biases might lead to ancestral functional biases, it might be assumed that biases are only a problem if the frequency of a particular amino acid residue is biased across the entire protein under consideration. This is not the case, however, since the bias arises according to the frequency of the particular amino acid residues at each site. Thus, if slightly deleterious variants are the less frequent variant at every site, it does not matter which residue is slightly deleterious, and it does not matter if these slightly deleterious residues are consistently one or a few particular amino acids, or not.

8.2 A case study

Consider the case of the ancestral archosaur visual pigment, rhodopsin (Chang et al., 2002a). This protein was chosen for ancestral reconstruction analysis partly because there is very little divergence (no more than 16%) among all vertebrates, and indeed the posterior probability for the most likely amino acid reconstructions at almost all sites is in the range of 0.9–1.0 (see Figure 2 from Chang *et al.*, 2002a), with only six sites having values of less than 0.9. However, in total there are, on average, nearly three (2.44) sites that would have been sampled differently if sampled from the posterior (or marginal posterior) rather than choosing the MPA at each site (on average, two of these differently sampled sites would have come from the six sites with the least likely maxima). In a random sample of 10 sequences from the posterior, the MPA sequence was never sampled, and the sequences sampled differed from the MPA sequence by up to five sites (Figure 8.1a).

We can also consider whether the amino acid frequencies are different in the most likely ancestor than in the extant vertebrates, but this turns out to be somewhat difficult to answer with any certainty. The average amino acid frequencies are different between the sites that are conserved among all vertebrates and those that are variable (Figure 8.2a), particularly for the amino acids alanine, cysteine, glycine, isoleucine, proline, arginine, serine, valine, and tyrosine. The rare variants (those observed only once) also have a notably different profile: for the amino acids noted above, the rare variants match the conserved frequencies for alanine, glycine, proline, and tyrosine, match the variable frequencies for isoleucine and arginine, and are uniquely different for cysteine, serine, and valine. Furthermore, they are notably different from both conserved and variable site averages for aspartic acid, histidine, methionine, and serine, for which the frequencies are greater, and glutamic acid, lysine, leucine, and valine, for which the frequencies are less.

Using the marginal posterior probabilities calculated in PAML (from the analysis in Chang *et al.* 2002; see also Chapter 15 in this volume), it can be seen that there are also frequency differences between the sites that are certain (posterior probability, 1.0) and uncertain at the archosaur ancestral node (Figure 8.2b). These are moderate for cysteine, phenylalanine, histidine, isoleucine, lysine, and tryptophan, but quite large for

glutamic acid, glycine, proline, valine, and tyrosine. In contrast, the frequencies of amino acids in

(a)

37	54	107	112	137	173	189	213	308
F	I	A	I	V	V	V	T	V
F	V	A	I	V	V	V	T	V
F	I	A	I	V	V	I	A	M
Y	I	A	I	V	V	V	T	V
Y	V	I	I	V	V	V	A	V
F	I	I	I	I	V	V	T	V
Y	I	A	V	V	V	I	T	V
Y	V	I	I	V	V	I	T	V
F	I	A	I	V	I	I	T	V
F	V	I	I	V	V	I	A	V

(b)

```
AEFLLLIIPYVATYIKAVGEIFIVVLQTEAQMSVKT
GEFLLLVIAYVATYIWVVGEIFIVVLPTEAHFSVKT
AEFLLLIIPYVATSIWAVGEICIVILQADAQFSMKT
AEYLLFIIPYVATYIWVVGEIFIVVLPTEAQFSVKT
AEYVLLVIPYMITYIWVVGEIFIVVLPAEAQFCVKT
AEFLLLIIPYVITYIWVIGEIFIVVLPTEAQFSVKM
AEYLLLILPYVATYVWVVGEIFIVIRPTERQFSVKT
ADYLMLVIPYVIIYPWVVTNIFAVILPTEAQFSVKT
AEFLLLIIPLVATYIWVVGELFIIILPTEAQFSVST
AEFLLLVIPYVIAYIWVVGEIFIVILPAEAQFSVKT
```

Figure 8.1 Amino acid variation among 10 sampled ancestors. In (a) the sequences are random samples from the posterior distribution under the general time-reversible (GTR) model. Only the sites that were variable among the sampled sequences are shown, and residues that differed from the MPA sequence are highlighted. The MPA sequence was never sampled; three sequences differed from the MPA at one site, four sequences differed at two sites, two differed at three sites, and one sequence differed from the MPA sequence at five sites. The alignment number of each variable site is shown above the site column. In (b) a sampling of rare variants (residues observed only once at a site) is added to each random sequence. The rare variant is highlighted in dark gray, and these new sequences differ from the MPA sequence by three (two sequences), four (two sequences), five (three sequences), seven (two sequences), or 10 (one sequence) sites. The number of rare variants per sequence was determined by random sampling from a Poisson distribution with a mean of 3.43 (see legend for Figure 8.3). Each rare variant was chosen randomly from among the 103 residues observed only once at a site. For aesthetic reasons, we did not display the rare variant sites selected, but in order these are 32, 33, 40, 49, 50, 63, 71, 74, 81, 108, 111, 112, 126, 130, 150, 154, 159, 162, 194, 196, 235, 237, 273, 281, 311, 232, 319, and 349. Note that, as expected, we have occasionally sampled the same rare variant (130, 196), different rare variants at the same site (108), or rare variants at sites that were already sampled differently based on the posterior (112).

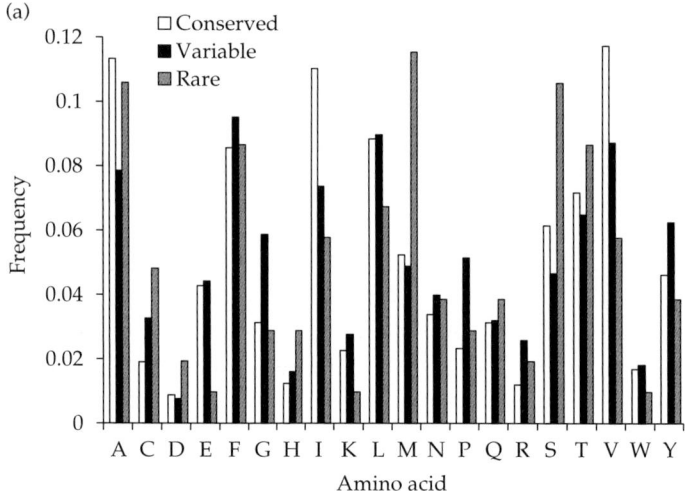

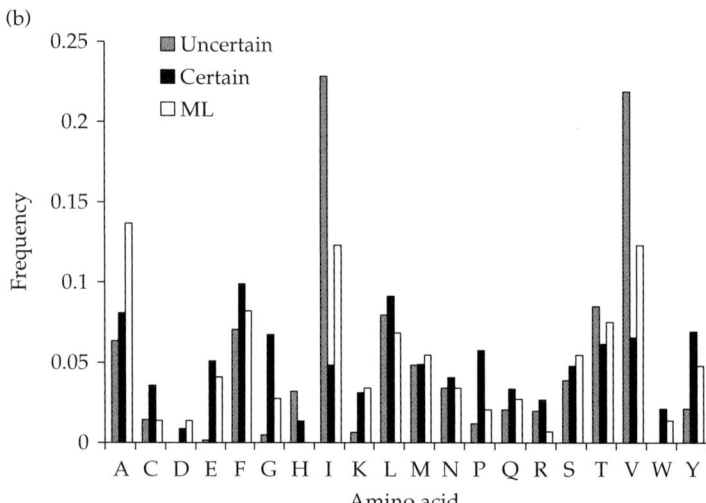

Figure 8.2 Amino acid frequencies for extant and ancestral sequences. Frequencies in extant organisms were divided into those sites that were variable or conserved among all sequences from extant organisms (a), and were also divided into those sites that were certain (posterior probability = 1.0) or uncertain (posterior probability <1.0) in the archosaur ancestor under the general time-reversible (GTR) model (b). Amino acid frequencies in the rare variants (those residues observed only once in the extant sequences at a site) and in the MPA sequence are also compared in a and b, respectively. For the uncertain sites, the frequencies are calculated based on the extant sequences, not the posterior probability at that site. ML, maximum likelihood.

the most likely archosaur reconstruction at variable sites are often midway between the certain and uncertain frequencies, but there are a number of notable exceptions, including alanine, histidine, and arginine.

The question of the effect of rare variants on ancestral reconstruction is a thorny one, mostly because it is difficult to assess their prevalence in ancestral sequences with any degree of accuracy (precisely because of their rarity). Models determined by averaging over many sites will tend to obscure variants that are rare at some sites but not at others, while the best conceivable site-specific models will not accurately assess the true frequencies of rare variants because there are not enough data at an individual site. Nevertheless, we wanted to provide an example to illustrate the possibility for rare variants to be under-sampled in this data-set, regardless of *which* amino acid is the rare variant. With over 100 variants observed only once at a site, it is reasonable to wonder whether a substantial number of low-frequency or rare amino acid variants are missing from the reconstructed ancestor. If such variants tend to affect function, then their absence could bias results.

If the number of times that a variant is observed among all vertebrates is used to estimate its expected frequency, then rare variants clearly tend to be under-represented in the reconstructed ancestor. The distribution of variant counts on a logarithmic scale (Figure 8.3a) is somewhat U-shaped, meaning that at many sites it is common to have a single dominant variant and one or more rare variants. A rough estimate of the expected number of times that a variant should be sampled in the ancestor can be calculated by weighting the variant count by the number of times each variant is observed in the vertebrates. We did this by first counting over all L sites the number of times, N_x, that any residue is observed x times at each site; that is, $N_x = \sum_i^L C_x^i$, where C_x^i is the count of variants observed x times at site i. Assuming that there is no bias in ancestral reconstruction, the expected representation of these variants in the maximum-likelihood ancestor is just the frequency of each variant count among all sequences in the alignment, $E_x = xN_x/S$, where S is the number of sequences (in this case, 30).

For example, there were 103 rare variants that were observed only once in the 30 extant vertebrates sampled, and we expect that on average $1 \times 103/30 = 3.43$ of these should have been sampled in the maximum-likelihood archosaur ancestor (whereas in practice, no rare variants were sampled in the maximum-likelihood ancestor). When expectations are compared with the number of times that variants were sampled in the

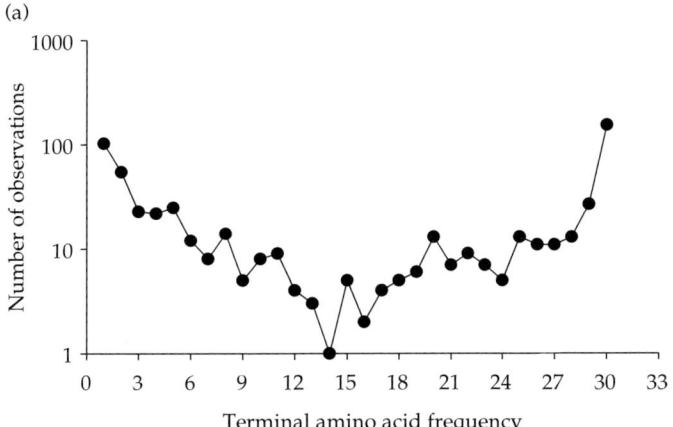

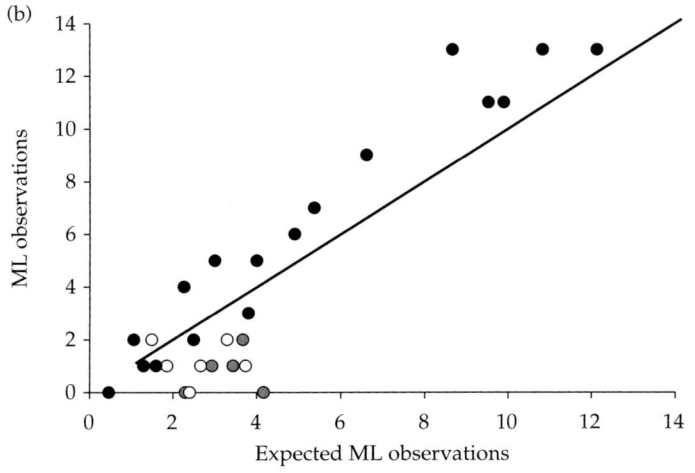

Figure 8.3 Distribution of variant counts among extant rhodopsins and predicted under-sampling of rare variants in the ancestral archosaur. The distribution of variant count observations in the extant archosaurs was assessed (a) and used to determine whether variants of a specific count were under- or over-represented in the MPA archosaur compared to expectation (b). The observed versus expected points for variants observed five or fewer times are shown as gray circles, and the points for variants observed between six and 11 times are shown as white circles. Cumulatively, variants at frequencies of 11 or fewer are observed almost 20 times less than expected. ML, maximum likelihood.

maximum-likelihood reconstruction at variable sites, it can be seen that most of the variants that were under-sampled relative to expectation were observed 11 or fewer times in the vertebrate sequences sampled (19.7 variants). An even greater proportion of under-sampling was found among those variants observed five or fewer times (13.5 variants; Figure 8.3b).

These estimates seem high, and this may be due partly to our use of a rough non-phylogenetic estimator of variant frequencies. The underestimate of rare variants at a particular node will of course depend upon the phylogenetic structure around that node, the frequencies of rare variants at each site in the alignment, and the rates of substitution to and from those rare variants at each site. Nevertheless, the estimate serves as an illustration of the potential for a substantial number of rare variants to be omitted from the MPA. Whereas it is true that the degree that rare variants are absent will depend somewhat on the phylogenetic position of a node, and the assumption that site-specific frequencies are well estimated by this (or any other method) is almost certainly wrong, it is also true that less-frequent variants will be systematically missing from MPA or maximum-parsimony reconstructions, regardless of phylogenetic position.

We took the frequency of rare variants observed only once (3.44) as a much more conservative estimate of the possible number of missing rare variants, and added a sample of those rare variants to the 10 previously sampled sequences from the posterior (Figure 8.1b). These new sequences differ from the MPA sequence at up to 10 sites. As with the bias of MPA sampling, missing rare variants may or may not contribute to functional bias in ancestral reconstruction, and it would be worthwhile to test their effect empirically. We suggest that although the number of missing rare variants is difficult to estimate, and precise knowledge of which rare variant is missing at each site may be nearly impossible to acquire, empirical testing of the cumulative effect of substituting variants that are rare in the extant species would help to gauge the potential severity of the problem.

In addition to MPA reconstruction bias, the inability of a global model to accurately account for rare variants at each site could contribute heavily to reconstruction errors. Mixture models that incorporate sharply different models tend to reconstruct function fairly accurately in thermodynamics-based population simulations (Williams et al., 2006), and these may in general be preferable so long as the taxonomic sampling density is high enough that the posterior probability of the particular model at each site is relatively high (Pollock and Bruno, 2000). Whereas Bayesian methods are preferable, their accuracy may still depend on the accuracy of the substitution model used in the reconstruction, and a detailed understanding of the conditions under which model inaccuracy can lead to biased reconstruction is largely unknown. In a situation such as the example given here with so many rare variants out of 30 sequences, it is expected that due to lack of data even the best phylogeny-based site-specific models would not be able to estimate the frequency of specific rare variants accurately.

8.3 Discussion and practical recommendations

It is apparent that even for archosaur rhodopsin, our relatively ideal case study, the bias inherent in choosing to reconstruct the ancestral sequence with the highest posterior probability, along with the optimization bias due to site-specific model inaccuracy, may have biased the frequencies with which certain amino acids are inferred. Amino acids that tend to have consistently low posterior probabilities are most probably undersampled. The lack of evidence linking aspects of rhodopsin function such as absorption spectrum to rare substitutions, along with the paucity of uncertain sites and the small difference between the predicted ancestral function and the range of extant functions, lead us to expect that this amino acid sampling bias did not strongly affect the functional inference in this case. Nevertheless, whether or not this bias may in fact have affected the functional assessment of the ancestor remains to be determined. Here we discuss some of the theoretical and practical considerations with regards to this problem, and present a simple strategy for addressing it when the goal is to reconstruct ancestral proteins in the laboratory.

8.3.1 Theoretical considerations

Determination of functional bias
This can be determined by sampling at least one ancestral sequence from the posterior distribution, in addition to the most likely (MPA) sequence. If there is no important difference in the function assayed, then the bias inherent in reconstructing the most likely ancestral sequence may not be a problem for that particular aspect of protein function. If functional differences are found between the most likely sequence and ancestral sequences sampled from the posterior, then further sampling from the posterior may be required. The number of sequences needing to be sampled will depend on the nature of the variability contained in the data-set, but in this case at least a few sequences from the posterior should be sampled; and more if resources allow. This will provide an estimate of the variability of the functional inference, rather than accepting a point estimate with unknown error. It can then be determined whether the differences between the most likely sequence and the posterior samples are significant. Relevant questions can then be asked, including whether the uncertainty in ancestral function is less or greater than any notable differences between the ancestor and some of its descendants, and whether the magnitude of uncertainty is greater or less than the difference between the most likely and average posterior functions.

Model variability
It is important to explore different models of evolution, and use as detailed a substitution probability mixture model as is justified based on the data sampled. Although this is an important issue, it is addressed in detail in another Chapter 15, and will not be discussed extensively here. It is also important to try to detect any unusual situations, such as coevolution (Pollock *et al.*, 1999; Wang and Pollock, 2005), or adaptive bursts (Stewart *et al.*, 1987; Messier and Stewart, 1997; Bishop *et al.*, 2000; Liberles *et al.*, 2001), that might add uncertainty to the posterior inference at particular sites. If the three-dimensional structure and the structural basis of the function of the protein are known, it may be useful to sample as many plausible combinations of amino acid residues as possible at sites proximal to ligand binding or enzymatic function (Chang *et al.*, 2002a, 2005).

Rare variants
As rare variants may be likely to affect function, it is important to experimentally investigate their effects if possible, and to include sampling of rare variants according to their occurrence among all extant sequences. These sequences will not represent a reconstruction of the exact ancestor, but they will represent a sampling of rare variants and their effect on function. The particular rare variants that were missed in the ancestor of interest may be unknowable.

8.3.2 Experimental considerations

Methods of site-directed mutagenesis
If the number of variable sites is not large, then this may be a feasible way of generating ancestral protein variants based on the posterior distribution. The idea would be to first synthesize the MPA, and then use site-directed mutagenic methods to synthesize variants. This approach is a good one when there are only a few variable sites, but it is difficult to scale it up when there are large numbers of variable sites involved.

Degenerate oligonucleotides
Rather than introducing variability after synthesis of an ancestor, variability can instead be incorporated directly in the initial gene-synthesis steps. This is particularly easy if the gene synthesis uses relatively short oligonucleotide fragments of no longer than 50 bases in length. The use of shorter oligonucleotide fragments has been found to improve the speed and efficiency of gene synthesis over longer fragments, while retaining a relatively low error rate (Chang *et al.*, 2005). Additionally, the use of shorter oligonucleotides makes it easy to introduce variability at sites in the gene-synthesis step by including oligonucleotides that are degenerate at those sites. The advantage of this approach is that even a large amount of variability

can be incorporated in a single synthesis step. It is limited, however, in that it is costly and not particularly accurate to incorporate nucleotides at unequal frequencies, such as would be necessary to sample variants proportional to their frequencies in the posterior distribution.

Variable oligonucleotide frequencies
Another means of incorporating variability during gene synthesis would be to synthesize different reasonably likely ancestral oligonucleotides, and then to mix them at the appropriate frequencies to sample from the posterior distribution. Although this method offers a significant advantage by directly incorporating ancestral variants at the frequencies in which they occur in the posterior distribution, it suffers from a major drawback as well. If more than a few variants with substantial posterior probabilities occur on the same oligonucleotide fragment, then many different oligonucleotides will need to be synthesized to encode all possible variant combinations. For a long and variable gene, this could be quite expensive. Note that this is not as much of a problem for rare variants since we can simply sample the rare variants only rarely (for example, a 5% variant in the posterior might be included at 50% frequency in an oligonucleotide mix only 10% of the time).

Sampling in silico
In contrast to the above methods, in which sampling from the posterior distribution is accomplished by experimental incorporation of proportionally sampled variants, sampling can also be separated from the gene-synthesis procedures and done instead beforehand, on a computer. In this approach, a small number of ancestral variants are sampled *in silico*. These variants can then be synthesized in the laboratory by site-directed mutagenic methods from an initial (or MPA) variant.

8.3.3 A proposed strategy

In consideration of the factors discussed above, we propose that an optimal strategy would be a hybrid approach incorporating both sampling *in silico* and a single-step gene-synthesis strategy using degenerate oligonucleotides. This approach would use sampling *in silico* to produce a reduced set of ancestral sequence variants drawn from the posterior distribution. Degenerate oligonucletides containing these restricted variants would then be mixed randomly in the synthesis step, in the proportions in which they occur in the *in silico* sampling. Such an approach would have notable advantages over creating a full library of sequences drawn from the posterior, as it would be composed of a greatly reduced number of variable amino acid sites relative to the full posterior distribution. Moreover, the number of different oligonucleotides required in the synthesis step would be greatly reduced, and degeneracy at any site could be limited to 50:50 mixes, thus greatly reducing the expense of the procedure. Note that this approach also takes advantage of the fact that sites in the protein are assumed to be independent, and can therefore be sampled independently when creating these libraries, as long as the frequencies of the states at variable sites are maintained.

It is worth noting that with a single synthesis step, the pre-sampled oligonucleotides are mixed randomly across oligonucleotide positions. In other words, although the recoverable nucleotide variants are pre-specified by the sampling *in silico*, and thus limited in number, the oligonucleotide combinations are not, and a very large number of gene sequences may be recovered. Increasing the sample size of the computer-generated sequences and thus increasing the number of variable oligonucleotides synthesized might be statistically preferable (to reduce sampling variance), but will only be worthwhile if a very large number of overall sequences are recovered and tested, which is unlikely to be feasible in many experimental systems. Using this synthesis strategy, the final result would be a set of ancestral variants that contained variable amino acid sites in the proportion that occur in the computer-sampled sequences from the posterior. Note that this synthesis strategy does not necessarily result in synthesis of the overall MPA sequence, which would need to be synthesized separately.

Although choosing to synthesize the most likely sequence (MPA), rather than sampling from the posterior distribution, can lead to biased results if the functional consequences are also biased, these are issues that are easily addressed, and should in no way hinder the use of ancestral reconstruction as a fundamentally useful and powerful technique in comparative functional genomics. The strategy recommended above does not require any more work than is now standard in the field (for example, recent studies have tested single mutation variants from the MPA or maximum-parsimony construct, or have tested a variety of models). This is particularly true if the variants are incorporated in the initial steps of gene synthesis (instead of using mutagenesis methods afterwards). Our suggested strategy is technically feasible and will immediately provide more reliable results than synthesizing the MPA alone. While we do not yet know, in practice, how often the MPA is functionally biased compared to a sample from the posterior, it appears preferable to us to sample from the posterior in the first place.

In addition to providing a potentially unbiased reconstruction of ancestral function, incorporation of the above recommendations would allow ancestral reconstruction to provide additional fundamental information about protein biochemistry, about the sequence/structure/function relationship, and about the context-dependent effect of variants that have been accepted at some time in the evolutionary process. The distribution of effects among evolutionarily accepted variants are different from the distribution of effects from random mutations, and a better understanding of such distributions is essential to a realistic theoretical model of protein evolution.

8.4 Acknowledgments

We thank Zhengyuan Wang, Matthew Reynolds, and Judith Beekman for technical assistance in creating practical web-based perl scripts for implementing our proposed strategy (www.EvolutionaryGenomics.com). This work was supported by grants to D.D.P. from the National Institutes of Health (GM065612-01 and GM065580-01) and the National Science Foundation through Louisiana EPSCOR and the Center for Biomodular Multi-scale Systems, and from startup funds from the University of Colorardo Health Sciences Center. This work was also supported by a grant to B.S.W.C. from the Natural Sciences and Engineering Research Council.

References

Benner, S.A. (2002) The past as the key to the present: resurrection of ancient proteins from eosinophils. *Proc. Natl. Acad. Sci. USA* **99**: 4760–4761.

Benner, S.A., Chamberlin, S.G., Liberles, D.A., Govindarajan, S., and Knecht, L. (2000) Functional inferences from reconstructed evolutionary biology involving rectified databases–an evolutionarily grounded approach to functional genomics. *Res. Microbiol.* **151**: 97–106.

Bishop, J.G., Dean, A.M., and Mitchell-Olds, T. (2000) Rapid evolution in plant chitinases: molecular targets of selection in plant-pathogen coevolution. *Proc. Natl. Acad. Sci. USA* **97**: 5322–5327.

Chang, B.S., Jonsson, K., Kazmi, M.A., Donoghue, M.J., and Sakmar, T.P. (2002a) Recreating a functional ancestral archosaur visual pigment. *Mol. Biol. Evol.* **19**: 1483–1489.

Chang, B.S., Kazmi, M.A., and Sakmar, T.P. (2002b) Synthetic gene technology: applications to ancestral gene reconstruction and structure-function studies of receptors. *Methods Enzymol.* **343**: 274–294.

Chang, B.S., Ugalde, J.A., and Matz, M.V. (2005) Applications of ancestral protein reconstruction in understanding protein function: GFP-like proteins. *Methods Enzymol.* **395**: 652–670.

Collins, T.M., Wimberger, P.H., and Naylor, G.J.P. (1994) Compositional bias, character-state bias, and character-state reconstruction using parsimony. *Syst. Biol.* **43**: 482–496.

Eyre-Walker, A. (1998) Problems with parsimony in sequences of biased base composition. *J. Mol. Evol.* **47**: 686–690.

Fitch, W.M. (1971) Toward defining the course of evolution: minimum change for a specific tree topology. *Syst. Zool.* **20**: 406–416.

Gaucher, E.A., Miyamoto, M.M., and Benner, S.A. (2003a) Evolutionary, structural and biochemical evidence for a new interaction site of the leptin obesity protein. *Genetics* **163**: 1549–1553.

Gaucher, E.A., Thomson, J.M., Burgan, M.F., and Benner, S.A. (2003b) Inferring the palaeoenvironment of ancient bacteria on the basis of resurrected proteins. *Nature* **425**: 285–288.

Geman, S. and Geman, D. (1984) Stochastic relaxation, Gibbs distributions, and the Bayesian restoration of images. *IEEE Trans Pattern Anal Machine Intelligence* **6**: 721–741.

Huelsenbeck, J.P., Nielsen, R., and Bollback, J.P. (2003) Stochastic mapping of morphological characters. *Syst. Biol.* **52**: 131–158.

Ivics, Z., Hackett, P.B., Plasterk, R.H., and Izsvak, Z. (1997) Molecular reconstruction of Sleeping Beauty, a Tc1-like transposon from fish, and its transposition in human cells. *Cell* **91**: 501–510.

Jermann, T.M., Opitz, J.G., Stackhouse, J., and Benner, S.A. (1995) Reconstructing the evolutionary history of the artiodactyl ribonuclease superfamily. *Nature* **374**: 57–59.

Koshi, J.M. and Goldstein, R.A. (1996) Probabilistic reconstruction of ancestral protein sequences. *J. Mol. Evol.* **42**: 313–320.

Krawczak, M., Wacey, A., and Cooper, D.N. (1996) Molecular reconstruction and homology modelling of the catalytic domain of the common ancestor of the haemostatic vitamin-K-dependent serine proteinases. *Hum. Genet.* **98**: 351–370.

Krishnan, N.M., Seligmann, H., Stewart, C.B., De Koning, A.P., and Pollock, D.D. (2004) Ancestral sequence reconstruction in primate mitochondrial DNA: compositional bias and effect on functional inference. *Mol. Biol. Evol.* **21**: 1871–1883.

Liberles, D.A., Schreiber, D.R., Govindarajan, S., Chamberlin, S.G., and Benner, S.A. (2001) The adaptive evolution database (TAED). *Genome Biol.* **2**: RESEARCH0028.

Malcolm, B.A., Wilson, K.P., Matthews, B.W., Kirsch, J.F., and Wilson, A.C. (1990) Ancestral lysozymes reconstructed, neutrality tested, and thermostability linked to hydrocarbon packing. *Nature* **345**: 86–89.

Messier, W. and Stewart, C.B. (1997) Episodic adaptive evolution of primate lysozymes. *Nature* **385**: 151–154.

Pollock, D.D. and Bruno, W.J. (2000) Assessing an unknown evolutionary process: effect of increasing site-specific knowledge through taxon addition. *Mol. Biol. Evol.* **17**: 1854–1858.

Pollock, D.D., Taylor, W.R., and Goldman, N. (1999) Coevolving protein residues: maximum likelihood identification and relationship to structure. *J. Mol. Biol.* **287**: 187–198.

Ronquist, F. and Huelsenbeck, J.P. (2003) MrBayes 3: Bayesian phylogenetic inference under mixed models. *Bioinformatics* **19**: 1572–1574.

Sanderson, M.J., Wojciechowski, M.F., Hu, J.M., Khan, T.S., and Brady, S.G. (2000) Error, bias, and long-branch attraction in data for two chloroplast photosystem genes in seed plants. *Mol. Biol. Evol.* **17**: 782–797.

Stewart, C.B., Schilling, J.W., and Wilson, A.C. (1987) Adaptive evolution in the stomach lysozymes of foregut fermenters. *Nature* **330**: 401–404.

Thornton, J.W. (2004) Resurrecting ancient genes: experimental analysis of extinct molecules. *Nat. Rev. Genet.* **5**: 366–375.

Thornton, J.W., Need, E., and Crews, D. (2003) Resurrecting the ancestral steroid receptor: ancient origin of estrogen signaling. *Science* **301**: 1714–1717.

Ugalde, J.A., Chang, B.S., and Matz, M.V. (2004) Evolution of coral pigments recreated. *Science* **305**: 1433.

Wang, Z.O. and Pollock, D.D. (2005) Context dependence and coevolution among amino acid residues in proteins. *Methods Enzymol.* **395**: 779–790.

Williams, P.D., Pollock, D.D., Blackburne, B.P., and Goldstein, R.A. (2006) Assessing the accuracy of ancestral protein reconstruction methods. *PLoS Computat. Biol.* **2**: 598–605.

Yang, Z., Kumar, S., and Nei, M. (1995) A new method of inference of ancestral nucleotide and amino acid sequences. *Genetics* **141**: 1641–1650.

Zhang, J. and Nei, M. (1997) Accuracies of ancestral amino acid sequences inferred by the parsimony, likelihood, and distance methods. *J. Mol. Evol.* **44**: (suppl. 1): S139–S146.

Zhang, J. and Rosenberg, H.F. (2002) Complementary advantageous substitutions in the evolution of an antiviral RNase of higher primates. *Proc. Natl. Acad. Sci. USA* **99**: 5486–5491.

CHAPTER 9

Evolutionary properties of sequences and ancestral state reconstruction

Lesley J. Collins and Peter J. Lockhart

9.1 The nature of sequence evolution

Ancestral sequences have recently been reconstructed to test protein function for a number of proteins (e.g. Bapteste *et al.*, 2005; Thomson *et al.*, 2005) as well as for enhancing searches of homologous sequences (Collins *et al.*, 2003; Cai *et al.*, 2004). The reliability of these phylogenetic inferences is dependent on understanding the evolutionary properties of sequences. Current understanding is still developing. One of the earliest theoretical models suggested that whereas some sequence positions evolved more slowly than others, properties of substitution were nevertheless universal across evolutionary lineages (Uzzell and Corbin, 1971). This model with discrete rate classes is the framework most commonly used today for reconstructing evolutionary trees from DNA and protein sequences (Waddell, 1995; Yang, 1996; Posada and Crandal, 2001).

However, it was not the only model proposed early in the development of methods of comparative sequence analysis. Some authors emphasized observations suggesting that the process of evolution at individual sites of homologous sequences may differ in different evolutionary lineages. Fitch and Markowitz (1970) first suggested that whereas only 10% of the sites in cytochrome oxidase were free to accept substitutions at any point in their evolutionary history, animal and plant orthologs had only partially overlapping distributions for these variable sites. Their explanation was that sequence evolution involved covariation among different sequence positions (i.e. if a site became free to vary in one part of a molecule it became invariable in another). This phenomenon was first called covarion evolution in the case of protein sequences, or covariotide evolution in the case of nucleotide sequences. For over 20 years, this model and the phenomenon it sought to describe received little attention. However, in recent years, numerous authors have reported observations and inferences on the properties of sequence evolution that are poorly explained by discrete-rate-class models of evolution (e.g. Miyamoto and Fitch, 1995; Simon *et al.*, 1996; Lockhart *et al.*, 2000, 2006; Philippe and Germot, 2000; Steel *et al.*, 2000; Lopez *et al.*, 2002; Misof *et al.*, 2002; Susko *et al.*, 2002; Ané *et al.*, 2005; Guindon *et al.*, 2004; Inagaki *et al.*, 2004; Brown, 2005; Guo and Stiller, 2005; Baele *et al.*, 2006).

Some of these properties, although covarion-like, are nevertheless not strictly those described by the model of covariation envisaged by Fitch and Markowitz (1970). This point was first noted by Philippe and colleagues, who invented the term heterotachy to describe a site property of multiple sequence alignments (Lopez *et al.*, 2002) wherein, at a particular site in a multiple sequence alignment, the rate of evolution is inferred to be significantly different between evolutionary lineages. This would be the case under the earlier covarion model of Fitch and Markowitch (1970). However, this difference in evolutionary rate could also occur if the proportion of variable sites (p_{var}) in different lineages also changed (Philippe and Germot, 2000; Lockhart *et al.*, 2006). Constancy of p_{var} is central in the covarion model of Fitch and Markowitch (1970), and also in subsequent and more formal mathematical covarion models

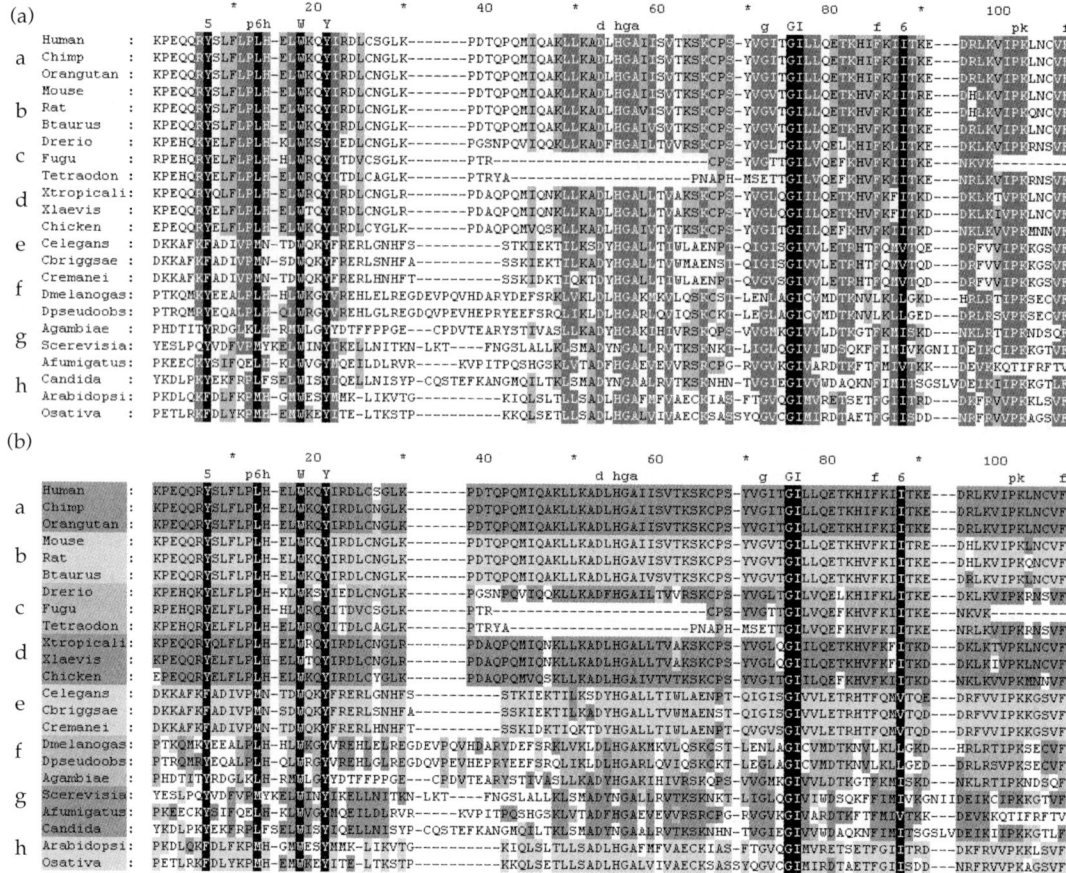

Figure 9.1 Segment of the alignment of Pop4 proteins. (A) Conservation of amino acid sites across all species. The darker the shading, the higher the conservation. (B) The same alignment but shaded according to lineage conservation. Lineages: a, primate; b, other mammal; c, fish; d, frog/bird; e, nematode; f, insect; g, fungi; and h, plants. A feature of covarion evolution is a greater than expected occurrence of substitution patterns constant in some lineages but variable in others (Lockhart et al., 1998; Ané et al., 2005). Tests for covarion evolution typically ask the question: is there a greater occurrence of N3 and N4 substitution patterns (constant within one lineage but variable in other) than expected under a rates-across-sites model (e.g. Lockhart et al., 1998, 2000; Ané et al., 2005). Protein accession numbers (from the top): CAG38806, XP_512983, Q5R7B0, NP_079666, AAH88183, AAX46326, XP_701926, SINFRUP00000058608, CAF95496, NP_001017252, AAH74186, XP_414118, CAB02730, CAE56215, AAGD01000954, NP_648168, EAL30008, EAA00065, CAA85220, XP_753533, CAG60581, NP_973678, and AAT94020. Btaurus, *Bos taurus*; Drerio, *Danio rerio*; Fugu, *Fugu rubripes*; Xtropicali, *Xenopus tropicalis*; Xlaevis, *Xenopus laevis*; Celegans, *Caenorhabditis elegans*; Cbriggsae, *Caenorhabditis briggsae*; Cremanei, *Caenorhabditis remanei*; Dmelanogas, *Drosophila melanogaster*; Dpseudoobs, *Drosophila pseudoobscura*; Agambiae, *Anopheles gambiae*; Scerevisia, *Saccharomyces cerevisiae*; Afumigatus, *Aspergillus fumigatus*; Arabidopsi, *Arabidopsis thaliana*; Osativa, *Oryza sativa*.

(Tuffley and Steel, 1998; Galtier, 2001; Huelsenbeck, 2002; Ané et al., 2005; Beale et al., 2006). However, some authors have discussed observations suggesting that p_{var} may change across the underlying tree (e.g. Philippe and Germot, 2000; Steel et al., 2000; Lockhart et al., 2006).

A further point that has caused some confusion arises from the lack of recognition that the concept of heterotachy also includes the scenario of Felsenstein (1978). In this case, p_{var} is assumed to be constant in all evolutionary lineages, but the rate of substitution at some sites would be faster in some lineages than in others. This scenario can be modeled as a reversible stationary process (Felsenstein, 1981). Distinguishing between the possible underlying processes that result in observed patterns of sequence variation is a difficult problem (Lockhart and Steel, 2005), and one that has been identified as

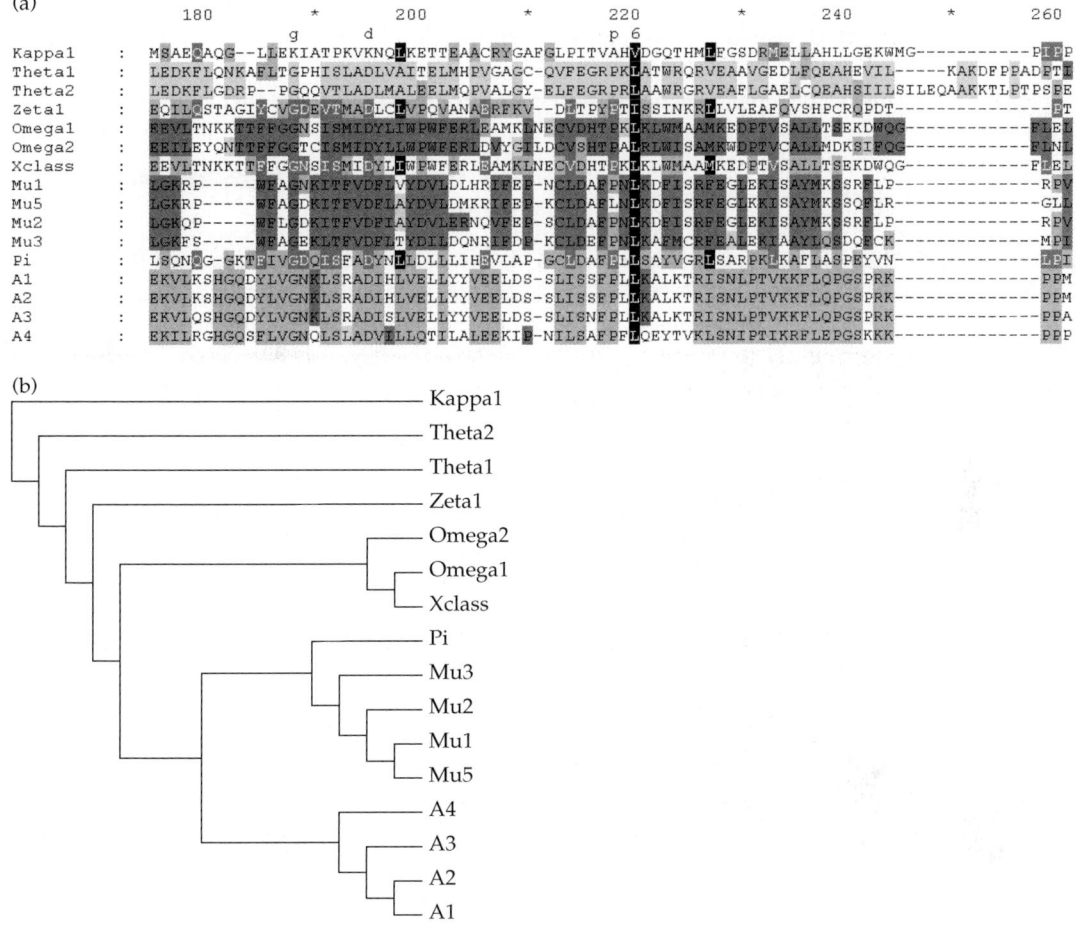

Figure 9.2 Heterotachy can also be observed in paralogous sequences. (a) Section of an alignment of human glutathione transferase proteins (Retief et al., 1999) showing lineage conservation. (b) Phylogenetic tree showing how these sequences are related. Protein accession numbers (from the top): Q9Y2Q3, NP_000844, AAC13317, NP_665877, AAV68046, CAC16040, U90313 (Retief et al., 1999), NP_000552, AAH58881, NP_000839, AAH08790, AAV38753, NP_665683, NP_000837, NP_000838, and NP_001503.

suggesting important directions for future research (Thornton and Kolaczkowski, 2005).

In this respect, a number of authors (e.g. Buck and Atchley, 2005; Guo and Stiller, 2005; Lopez et al., 2002; Lockhart et al., 2006) have recently stressed the importance of understanding the evolution of protein–protein interactions as an underlying cause of covarion-like patterns of evolution in sequences. Coevolution of interacting partners is seen as important, first in directing the evolution of proteins, and second in explaining why structural constraints may become lineage-specific while function remains the same. The importance of protein–protein interactions for understanding both relative rates of protein evolution and patterns of sequence divergence is currently of much research interest (e.g. Fraser et al., 2002; Fraser and Hirsh, 2004; Hahn and Kern, 2004; Mintseris and Weng, 2005; Rekha et al., 2005; Makino and Gojobori, 2006). The extent to which interactions of ancestral proteins differ from those of modern relatives has received little attention to date.

Although the concept of heterotachy and covarion evolution in the literature is most often restricted to describing the evolution of orthologs (sequences related by descent and having the same function; an example is shown in Figure 9.1), the same or similar principles of sequence divergence apply to proteins related by gene duplication (paralogs), and which have different functions and functional constraints (Gu, 1999; Gaucher et al., 2001; Gribaldo et al., 2003; an example is shown in Figure 9.2). In both cases, homologs may differ from each other in respect of exhibiting lineage-specific (1) distributions and proportions of variable sites (Lockhart et al., 1996; Gaucher et al., 2002), (2) patterns of non-conservative substitution, and (3) patterns of insertion and deletion. Awareness of these properties of sequence evolution may be helpful when gene finding since such evolutionary properties directly impact upon the extent of sequence conservation across the length of homologs.

9.2 Heterotachy and gene finding

It is well known that the use of profiles and hidden Markov models (HMMs) can improve efficiency of gene finding in genome-database searches. However, it is also known that to find distantly related homologs it is sometimes necessary to search genomes using distantly related sequences (Retief et al., 1999). A likely explanation for this is that as sequences diverge, evolution of their functional and/or structural constraints means that sequence conservation strong enough to detect homology may not be a universal property of all evolutionary lineages for any single region of a molecule. This concept is illustrated in Figure 9.1, which shows an alignment of an RNase P protein component Pop4. Pop4 is the most conserved protein in the RNase P complex (van Eenennaam et al., 1999; Welting et al., 2004) and may bind to both the RNA component and other proteins within the complex (Houser-Scott et al., 2002; Welting et al., 2004). Figure 9.1 shows that sites that are not highly conserved across all eukaryotes may be conserved or invariant within the individual lineages (N3 and N4 patterns *sensu* Lockhart et al., 1998), a reflection of possible lineage-specific protein and RNA binding. Presumably regions that are evolving very rapidly in some lineages are under relaxed evolutionary constraint and are likely to be highly diverged or even absent in some lineages. Including these sequence regions in HMM profiles will be inefficient for identifying homologs in those lineages where the evolutionary constraints differ from that of the profile.

9.3 Gene finding with ancestral sequences

When comparing sequences from distantly related genomes, it is often difficult to identify homologous or paralogous protein-coding regions that are highly diverged. However, searching with a calculated ancestral sequence can aid in gene finding (Collins et al., 2003). When BLAST (Altschul et al., 1997) fails to detect a distant homolog, another solution is to use a HMM-based program such as HMMER (Eddy, 1998). Instead of searching with an individual sequence, HMM-based programs use a training data-set (i.e. a multiple sequence alignment), and can therefore be more powerful than pairwise BLAST. However, the relationship between the sequences within the alignment is not taken into account. This can cause problems if in the alignment there are many sequences that are closely related and only a few distantly related sequences (i.e. a tree representing these sequences will have long and short branches). This can have the effect that a resulting HMM profile is well trained for sequences closely related to the majority in the initial alignment and not so well trained for under-represented or unrepresented sequences that are more divergent. HMM-based tools can also be difficult in the absence of information from newly sequenced genomes.

The ancestral sequence reconstruction (ASR) technique (Collins et al., 2003) uses an alignment and a tree indicating the relationship between the sequences in that alignment to produce potential ancestral sequences for the nodes of the represented tree. The idea is that the predicted ancestral sequence reduces the evolutionary distance for the search and thus provides a greater chance of detecting a distant homolog. Because the output is a sequence it can be used in combination with any sequence-based tools, such as BLAST or HMMER.

An example can be seen in Figure 9.3. Pop1 is another protein component of the eukaryotic RNase P (Lygerou et al., 1996; van Eenennaam et al., 2000; Welting et al., 2004). Unlike the Pop4

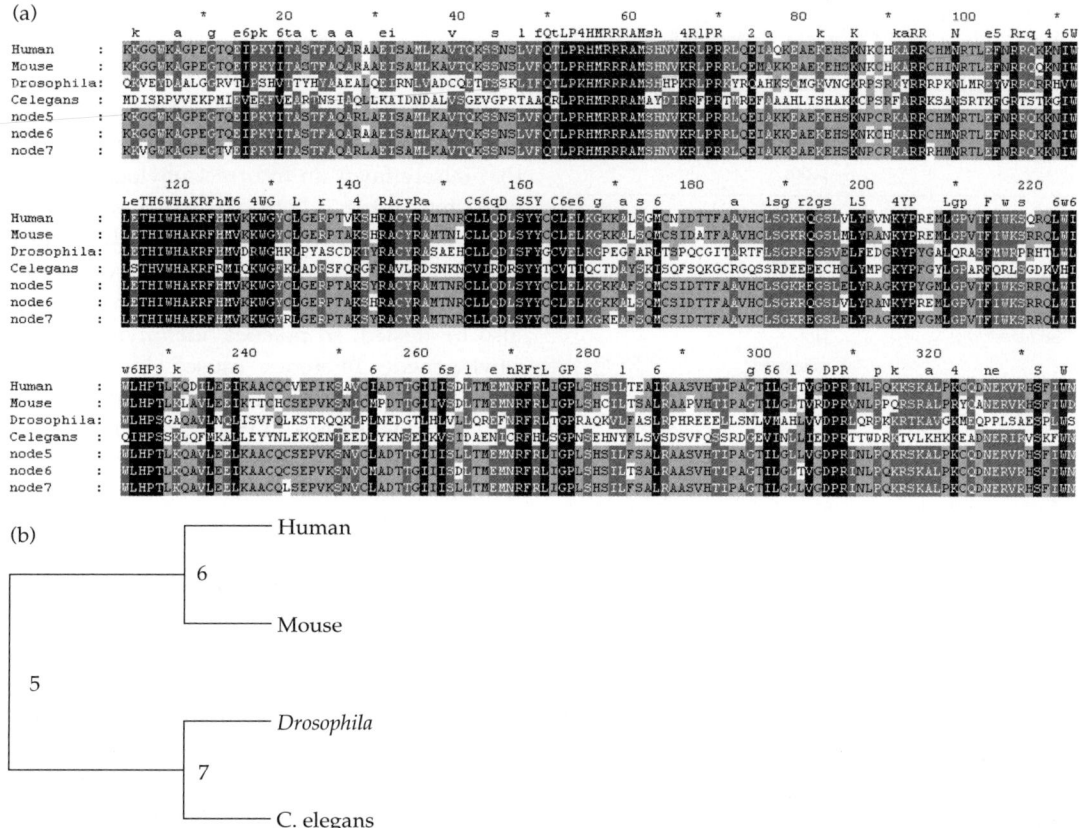

Figure 9.3 Ancestral sequences calculated from four Pop1 proteins. (a) Section of a Pop1 protein alignment using four modern sequences to produce three ancestral nodal sequences. (b) Phylogenetic tree representing the relationship of the sequences and the predicted nodes. Protein accession numbers (from the top): X99302, BAB30296, U00048, and AAF46049.

protein, used in our previous example, it is not highly conserved and functions as a scaffolding-type protein (van Eenennaam et al., 2000). Pop1 sequences do share some homology but only in small motif regions which may not be significant enough to be detected in a BLAST search of the complete protein (Table 9.1). Ancestral nodes can be constructed using just four sequences, in this case sequences from human, mouse, the fruit fly *Drosophila melanogaster*, and the nematode worm *Caenorhabditis elegans*. Any of these four sequences on their own fail to detect any Pop1 homolog above reasonable cut-off levels (e.g. E value = 0.001; the E value (Expect value) is a parameter that describes the number of hits one can expect to see just by chance when searching a database of a particular size) from a database of *Giardia lamblia*

(a unicellular protist) proteins (www.mbl.edu/Giardia). However, some of the ancestral sequences reconstructed from these four proteins are able to detect the Pop1 homolog from the *G. lamblia* database. In a way, the ancestral sequence is shortening the evolutionary distance that needs to be crossed from the query sequences to the subject database.

As expected, the accuracy of the reconstruction of an ancestral sequence is reduced sharply with increasing evolutionary distance (Koshi and Goldstein, 1996). However, often the most interesting evolutionary changes do not occur at the low levels of sequence divergence where the ancestral rates are more easily inferred (Chang and Donoghue, 2000). However, the application of ASR as a gene-finding tool means that the accuracy of the

Table 9.1 Results of BLAST searches with modern and ancestral Pop1 sequences against the *Giardia lamblia* genome (see Figure 9.3 for phylogenetic tree and nodal placements)

Query sequence: *Giardia* proteins	Resulting BLAST bit score	Resulting E value[a]
Human	35	0.06
Mouse	32	0.5
Drosophila melanogaster	32	0.54
Caenorhabditis elegans	39	0.004
Schizosaccharomyces pombe	Not detected	–
Node 5	41	6e-04
Node 6	37	0.01
Node 7	46	2e-05

[a]The E value (Expect value) is a parameter that describes the number of hits one can expect to see just by chance when searching a database of a particular size.

sequence does not have to be as stringent as for use in function-based experimentation, only close enough to advance the construction of the next node or to find a candidate homolog. An advantage with this technique is that one can use their favourite method of ancestral sequence prediction. The above example used maximum-likelihood methods (the PAML package) but there are a number of other methods available (including parsimony). A possible weakness of the approach, as currently implemented, relates to recent issues raised over the relative performance of different tree-building methods.

9.4 Heterotachy and ancestral state reconstruction

Considerable discussion has appeared recently in the literature over the relative reliability of parsimony and likelihood tree-building methods when sequences evolve with the property of heterotachy (e.g. Kolaczkowski and Thornton, 2004; Gaucher and Miyamoto, 2005; Philippe *et al.*, 2005; Spencer *et al.*, 2005; Steel, 2005; Thornton and Kolaczkowski, 2005). Much of this discussion is limited to the restrictive case for variations of the scenario studied by Felsenstein (1978). In the case of changing distributions and proportions of variable sites, the reliability of tree building and ancestral sequence reconstruction may also be affected in two ways. It can contribute to a form of substitution model misspecification, which is expected to reduce the accuracy of both parsimony and likelihood methods. In the particular case that p_{var} changes it may also induce topological biases that falsely favor an incorrect phylogeny. Such an effect may mislead both parsimony and likelihood methods (Lockhart *et al.*, 2006). In the context of ancestral state reconstruction, this is significant since inference of ancestral nodal sequences may also be misled. In practice, identifying erroneous phylogenetic inference caused by model misspecification and topological bias is made more difficult due to our current and limited understanding of the evolutionary properties of sequences. Recent work has suggested that ancestral sequence reconstruction is more accurate when phylogenetic reconstructions implement corrections for positional rate heterogeneity with a discrete gamma distribution of rate classes (Cai *et al.*, 2004). It is well known that under some conditions this correction approximates the effect of a covarion model (Miyamoto and Fitch, 1995). However, in cases where p_{var} changes, such a correction is unlikely to be sufficient to correct for the effects of model misspecification (Lockhart and Steel, 2005; Lockhart *et al.*, 2006).

9.5 Conclusions

As a technique ASR is perhaps unlikely to replace other more popular gene-finding tools such as BLAST and HMMER. However, improving our understanding of the evolutionary properties of sequences will help us to more fully recognize the potential of ASR and understand its limitations better. We are still in the early stages of successfully implementing the approach and automating the process. It is another tool for the toolbox, and one that will have greatest impact where historical information rather than criteria of similarity are important for recognizing homologies.

References

Altschul, S.F., Madden, T.L., Schaffer, A.A., Zhang, J., Zhang, Z., Miller, W., and Lipman, D.J. (1997) Gapped BLAST and PSI-BLAST: a new generation of protein database search programs. *Nucleic Acids Res.* **25**: 3389–3402.

Ané, C., Burleigh, J.G., McMahon, M.M., and Sanderson, M.J. (2005) Covarion structure in plastid genome evolution: a new statistical test. *Mol. Biol. Evol.* **22**: 914–924.

Baele, G., Raes, J., Van de Peer, Y., and Vansteelandt, S. (2006) A powerful multiple testing method to detect heterotachy in nucleotide sequences. *Mol. Biol. Evol.* **23**: 1397–1405.

Bapteste, E., Charlebois, R.L., MacLeod, D., and Brochier, C. (2005) The two tempos of nuclear pore complex evolution: highly adapting proteins in an ancient frozen structure. *Genome Biol.* **6**: R85.

Brown, R.P. (2005) Large subunit mitochondrial rRNA secondary structures and site-specific rate variation in two lizard lineages. *J. Mol. Evol.* **60**: 45–56.

Buck, M.J. and Atchley, W.R. (2005) Networks of coevolving sites in structural and functional domains of Serpin proteins. *Mol. Biol. Evol.* **22**: 1627–1634.

Cai, W., Pei, J., and Grishin, N.V. (2004) Reconstruction of ancestral protein sequences and its applications. *BMC Evol. Biol.* **4**: 33.

Chang, B.S. and Donoghue, M.J. (2000) Recreating ancestral proteins. *Trends Ecol. Evol.* **15**: 109–114.

Collins, L.J., Poole, A.M., and Penny, D (2003) Using ancestral sequences to uncover potential gene homologues. *Appl. Bioinformat.* **2**: S85–S95.

Eddy, S.R. (1998) Profile hidden Markov models. *Bioinformatics* **14**: 755–763.

Felsenstein, J. (1978) Number of evolutionary trees. *Syst. Zool.* **27**: 27–33.

Felsenstein, J. (1981) Evolutionary trees from DNA sequences: a maximum likelihood approach. *J. Mol. Evol.* **17**: 368–376.

Fitch, W.M. and Markowitz, E. (1970) An improved method for determining codon variability in a gene and its application to rate of fixation of mutations in evolution. *Biochem. Genet.* **4**: 579–593.

Fraser, H.B. and Hirsh, A.E. (2004) Evolutionary rate depends on number of protein-protein interactions independently of gene expression level. *BMC Evol. Biol.* **4**: 13.

Fraser, H.B., Hirsh, A.E., Steinmetz, L.M., Scharfe, C., and Feldman, M.W. (2002) Evolutionary rate in the protein interaction network. *Science* **296**: 750–752.

Galtier, N. (2001) Maximum-likelihood phylogenetic analysis under a covarion-like model. *Mol. Biol. Evol.* **18**: 866–873.

Gaucher, E.A. and Miyamoto, M.M. (2005) A call for likelihood phylogenetics even when the process of sequence evolution is heterogeneous. *Mol. Phylogenet. Evol.* **37**: 928–931.

Gaucher, E.A., Miyamoto, M.M., and Benner, S.A. (2001) Function-structure analysis of proteins using covarion-based evolutionary approaches: elongation factors. *Proc. Natl. Acad. Sci. USA* **98**: 548–552.

Gaucher, E.A., Gu, X., Miyamoto, M.M., and Benner, S.A. (2002) Predicting functional divergence in protein evolution by site-specific rate shifts. *Trends Biochem. Sci.* **27**: 315–321.

Gribaldo, S., Casane, D., Lopez, P., and Philippe, H. (2003) Functional divergence from evolutionary analysis: a case study of vertebrate haemoglobin. *Mol. Biol. Evol.* **20**: 1754–1759.

Gu, X. (1999) Statistical methods for testing functional divergence after gene duplication. *Mol. Biol. Evol.* **16**: 1664–1674.

Guindon, S., Rodrigo, A.G., Dyer, K.A., and Huelsenbeck, J.P. (2004) Modeling the site-specific variation of selection patterns along lineages. *Proc. Natl. Acad. Sci. USA* **101**: 12957–12962.

Guo, Z. and Stiller, J.W. (2005) Comparative genomics and evolution of proteins associated with RNA polymerase II C-terminal domain. *Mol. Biol. Evol.* **22**: 2166–2178.

Hahn, M.W. and Kern, A.D. (2005) Comparative genomics of centrality and essentiality in three eukaryotic protein-interaction networks. *Mol. Biol. Evol.* **22**: 803–806.

Houser-Scott, F., Xiao, S., Millikin, C.E., Zengel, J.M., Lindahl, L., and Engelke, D.R. (2002) Interactions among the protein and RNA subunits of Saccharomyces cerevisiae nuclear RNase P. *Proc. Natl. Acad. Sci. USA* **99**: 2684–2689.

Huelsenbeck, J.P. (2002) Testing a covariotide model of DNA substitution. *Mol. Biol. Evol.* **19**: 698–707.

Inagaki, Y., Susko, E., Fast, N.M., and Roger, A.J. (2004) Covarion shifts cause a long-branch attraction artifact that unites microsporidia and archaebacteria in EF-1 alpha phylogenies. *Mol. Biol. Evol.* **21**: 1340–1349.

Kolaczkowski, B. and Thornton, J.W. (2004) Performance of maximum parsimony and likelihood phylogenetics when evolution is heterogeneous. *Nature* **431**: 980–984.

Koshi, J.M. and Goldstein, R.A. (1996) Probabilistic reconstruction of ancestral protein sequences. *J. Mol. Evol.* **42**: 313–320.

Lockhart, P.J., Larkum, A.W.D., Steel, M.A., Waddell, P.J., and Penny, D. (1996) Evolution of chlorophyll and bacteriochlorophyll: the problem of invariant sites in sequence analysis. *Proc. Natl. Acad. Sci. USA* **93**: 1930–1934.

Lockhart, P.J., Steel, M.A., Barbrook, A.C., Huson, D.H., Charleston, M.A., and Howe, C.J. (1998) A covariotide model explains apparent phylogenetic structure of oxygenic photosynthetic lineages. *Mol. Biol. Evol.* **15**: 1183–1188.

Lockhart, P.J., Huson, D., Maier, U., Fraunholz, M.J., Van de Peer, Y., Barbrook, A.C. et al. (2000) How molecules evolve in eubacteria. *Mol. Biol. Evol.* **17**: 835–838.

Lockhart, P.J. and Steel, M.A. (2005) A tale of two processes. *Syst. Biol.* **54**: 948–951.

Lockhart, P.J., Novis, P., Milligan, B.G., Riden, J., Rambaut, A., and Larkum, A.W.D. (2006) Heterotachy and tree building: a case study with plastids and eubacteria. *Mol. Biol. Evol.* **23**: 40–45.

Lopez, P., Casane, D., and Philippe, H. (2002) Heterotachy, an important process of protein evolution. *Mol. Biol. Evol.* **19**: 1–7.

Lygerou, Z., Pluk, H., van Venrooij, W.J., and Seraphin, B. (1996) hPop1: an autoantigenic protein subunit shared by the human RNase P and RNase MRP ribonucleoproteins. *EMBO J.* **15**: 5936–5948.

Makino, T. and Gojobori, T. (2006) The evolutionary rate of a protein is influenced by features of the interacting partners. *Mol. Biol. Evol.* **23**: 784–789.

Mintseris, J. and Weng, Z. (2005) Structure, function, and evolution of transient and obligate protein-protein interactions. *Proc. Natl. Acad. Sci. USA* **102**: 10930–10935.

Misof, B., Anderson, C.L., Buckley, T.R., Erpenbeck, D., Rickert, A., and Misof, K. (2002) An empirical analysis of mt 16S rRNA covarion-like evolution in insects: site-specific rate variation is clustered and frequently detected. *J. Mol. Evol.* **55**: 460–469.

Miyamoto, M.M. and Fitch, W.M. (1995) Testing the covarion hypothesis of molecular evolution. *Mol. Biol. Evol.* **12**: 503–513.

Philippe, H. and Germot, A. (2000) Phylogeny of eukaryotes based on ribosomal RNA: long-branch attraction and models of sequence evolution. *Mol. Biol. Evol.* **17**: 830–834.

Philippe, H., Zhou, Y., Brinkman, H., Rodrigue, N., and Delsuc, F. (2005) Heterotachy and long branch attraction in phylogenetics. *BMC Evol. Biol.* **5**: 50.

Posada, D. and Crandal, K.A. (2001) Selecting the best-fit model of nucleotide substitiution. *Syst. Biol.* **50**: 580–601.

Rekha, N., Machado, S.M., Narayanan, C., Krupa, A., and Srinivasan, N. (2005) Interaction interfaces of protein domains are not topologically equivalent across families within superfamilies: implications for metabolic and signaling pathways. *Proteins* **58**: 339–353.

Retief, J.D., Lynch, K.R., and Pearson, W.R. (1999) Panning for genes--a visual strategy for identifying novel gene orthologs and paralogs. *Genome Res.* **9**: 373–382.

Simon, C., Nigro, L., Sullivan, J., Holsinger, K., Martin, A., Grapputo, A. et al. (1996) Large differences in substitutional pattern and evolutionary rate of 12S ribosomal RNA genes. *Mol. Biol. Evol.* **13**: 923–932.

Spencer, M., Susko, E., and Roger, A.J. (2005) Likelihood, parsimony and Heterogeneous evolution. *Mol. Biol. Evol.* **22**: 1161–1164.

Steel, M.A. (2005) Should phylogenetic models be trying to fit an elephant? *Trends Genet.* **21**: 307–309.

Steel, M., Huson, D., and Lockhart, P.J. (2000) Invariable sites models and their use in phylogeny reconstruction. *Syst. Biol.* **49**: 225–232.

Susko, E., Inagaki, Y., Field, C., Holder, M.E., and Roger, A.J. (2002) Testing for differences in rates-across-sites distributions in phylogenetic subtrees. *Mol. Biol. Evol.* **19**: 1514–1523.

Thomson, J.M., Gaucher, E.A., Burgan, M.F., De Kee, D.W., Li, T., Aris, J.P., and Benner, S.A. (2005) Resurrecting ancestral alcohol dehydrogenases from yeast. *Nat. Genet.* **37**: 630–635.

Thornton, J.W. and Kolaczkowski, B. (2005) No magic pill for phylogenetic error. *Trends Genet.* **21**: 310–311.

Tuffley, C. and Steel, M. (1998) Modeling the covarion hypothesis of nucleotide substitution. *Math. Biosci.* **147**: 63–91.

Uzzell, T. and Corbin, K.W. (1971) Fitting discrete probability distributions to evolutionary events. *Science* **172**: 1089–1096.

van Eenennaam, H., Pruijn, G.J., and van Venrooij, W.J. (1999) hPop4: a new protein subunit of the human RNase MRP and RNase P ribonucleoprotein complexes. *Nucleic Acids Res.* **27**: 2465–2472.

van Eenennaam, H., Jarrous, N., van Venrooij, W.J., and Pruijn, G.J. (2000) Architecture and function of the human endonucleases RNase P and RNase MRP. *IUBMB Life* **49**: 265–272.

Waddell, P.J (1995) *Statistical Methods of Phylogenetic Analysis, Including Hadamard Conjugations, LogDEt Transforms and Maximum Likelihood*. Massey University of Palmerston North, New Zealand.

Welting, T.J., van Venrooij, W.J., and Pruijn, G.J. (2004) Mutual interactions between subunits of the human RNase MRP ribonucleoprotein complex. *Nucleic Acids Res.* **32**: 2138–2146.

Yang, Z.H. (1996) Among-site rate variation and its impact on phylogenetic analyses. *Trends Ecol. Evol.* **11**: 367–372.

10.2.2 Combined protein data-sets conserved across all domains of life

An approach that has been designed to reconstruct the tree of life rather than the eukaryotic clade specifically is that of conserved protein analysis (Brown et al., 2001). The idea is to use the most ancient proteins to reconstruct the most ancient branches. On comparing open reading frames from complete (or almost complete) Bacteria, Archea and Eukaryota, 23 orthologous proteins were found to be present across all domains using a sample size of 45 species. To verify that these were single gene orthologs, homology searches and individual gene phylogenies were constructed. In some cases there were both cytoplasmic and mitochondrial copies present in eukaryote species: only the cytoplasmic copies were retained for analysis as they represent the more closely related copy. Poorly conserved regions of alignment were removed, leaving 6591 positions, which were concatenated. Using a number of phylogenetic reconstruction methods including maximum-likelihood quartet puzzling, maximum parsimony, minimum evolution, and neighbor joining, support was found for the Coelomata clade rather than the Ecdysozoa (Brown et al., 2001), with support values of 100, 100, and 96% for maximum parsimony, neighbor joining, and quartet puzzling respectively.

10.2.3 Introns

Besides the use of protein-coding sequences to reconstruct the phylogeny of the animals, other characters such as the pattern of spliceosomal intron conservation have been employed. Introns have a very slow rate of insertion and loss, with intron-turnover estimates ranging from around 10^{-9}/year for flies and worms to 10^{-11}/year for mammals (Lynch and Richardson, 2002; Roy et al., 2003). A high proportion of introns should therefore persist for very long periods, giving them a desirable slow rate of evolution. It seemed improbable that an intron, once lost, would be regained in exactly the same position; this gave the added benefit of irreversibility to this approach. Possibly the most significant advantage to using introns is that they were believed to be immune to rate variation between branches (Roy and Gilbert, 2005).

A method for analyzing the pattern of shared intron positions for an unresolved tree consisting of molecular data (from complete genomes) for arthropods, nematodes, and deuterostomes, and a plant outgroup, was developed (Roy and Gilbert, 2005). Using 684 identified eukaryotic orthologs and measuring the pattern of intron conservation across all species, support was found for the Ecdysozoa hypothesis with a significance score of $P < 10^{-6}$. The Coelomata grouping received no more support in this analysis than the universally rejected grouping of nematodes with vertebrates. The method described in Roy and Gilbert (2005) takes into account variation in rates of intron loss in a specific lineage but does not incorporate possible differences in rates of loss between introns within a single lineage. The Dollo parsimony approach used in combination with intron data should place species with similarly high or low rates of character loss together to resolve a highly supported and uncontroversial phylogeny. This is not the case with intron data (Wolf et al., 2004), suggesting that this is an unsuitable data source. It is conceivable that intron-position data are suitable but that the Dollo parsimony treatment of this new data form (though suitable for nucleotide or protein sequences) is not appropriate. It was necessary to test these data independently; such analysis has shown that the same intron has been lost independently on multiple branches of the *Caenorhabditis* clade (Coghlan and Wolfe, 2004). In the genomes of *Caenorhabditis elegans* and *Caenorhabditis briggsae* there are more than 6000 introns that are lineage-specific, with a very high rate of intron turnover (at least 0.005 intron gains or losses on average over a gene per million years; Coghlan and Wolfe, 2004). Also, intron–exon structure is not retained even within closely related nematode species. Introns therefore, are unlikely to be valid phylogenetic characters (Cho et al., 2004).

10.2.4 Analysis of protein domains

Rather than comparing aligned sequences it is possible to adopt a molecular cladistic approach by examining the presence or absence of specific domains, any shared/derived higher-order character would therefore be indicative of a close relationship. Using this tactic it is possible to hypothesize unique or shared patterns/combinations of protein domains. Applying this approach to the Metazoa and the unicellular choanoflagellates, the pattern of domains that are present clusters these groups of organisms into a single clade (King and Carroll, 2001). A derivative of this approach has also been applied to bacterial phylogeny reconstruction, where the presence of specific gene orthologs and families are used as the heritable characters (House and Fitz-Gibbon, 2002). This approach held promise for the Coelomata/Ecdysozoa topology. To date the largest data-set assembled in this fashion consisted of 1712 orthologous genes and 2906 protein domains from completed metazoan genomes, the result of which was greater support for the Coelomata (Copley et al., 2004). However, it is known that the nematode lineage has a higher rate of character loss than the arthropods and vertebrates. An increased rate of secondary loss in one taxon produces the same effect on phylogeny reconstruction as rapidly evolving species, and therefore is subject to long-branch attraction (LBA; see section 10.3). An attempt to correct these data for LBA was proposed (Copley et al., 2004) that calculates a coefficient of secondary loss for the nematodes, arthropods, and vertebrates. This is done by examining characters that are present in the yeast outgroup and in at least one of the animal ingroups. For each ingroup the tendency to lose characters is calculated as a ratio of the number of characters lost in that lineage compared to the total number of characters that existed in the metazoan common ancestor. On correcting these data for LBA in this way, the opposing phylogeny (Ecdysozoa) is supported, therefore highlighting the importance of LBA and also showing that protein-domain combination data and orthologous genes, regardless of the large data-sets available, are not exempt from this issue.

10.2.5 rRNA secondary structure

The secondary structure for RNAs transcribed from rRNA is complex, driven by the base pairing between regions of the rRNA molecule, and forming the well-known structure of loops and stems. The selective pressure to retain the secondary structure of this molecule results in different evolutionary rates in the stem and loop regions, with stem mutations having a different probability of fixation than their equivalent mutations in loops. Not all mutations between base pairs are equally likely. A novel method has been developed that uses a 16-state model rather than a simpler four-state single-nucleotide model. This model takes the single substitution rate, double substitution rate, double transversion rate, and the substitutions between paired and mismatched states into account for calculating the phylogeny (Telford et al., 2005). Using this more sophisticated approach, small-subunit rRNA data from bilateria were tested. The resultant topology supported the ecdysozoan hypothesis (Telford et al., 2005).

10.2.6 Insertion/deletion/fusion events

Possibly one of the most important observations supporting the Ecdysozoa hypothesis is the presence of a mulitmeric form of the β-thymosin gene in the genomes of *Drosophila melanogaster* and *C. elegans*, whereas other metazoans used in the analysis showed the presence of a monomeric form (Manuel et al., 2000). This gene was therefore taken to be a molecular synapomorphy that consists of a change from the primitive monomeric character to the derived multimeric form. This represents a rare event and was provided as a convincing line of evidence linking the arthropods and nematodes (Manuel et al., 2000). However, the multimeric form was also found to be present in a deuterosome, *Ciona intestinalis*, and a lophotrochozoan, *Hemissenda crassicornis*, and also exists outside the Metazoa in a fungus, thereby demoting the multimeric form from being an ecdysozoan-specific state (Telford, 2004b). The β-thymosin gene is therefore not a valid character to show support for the ecdysozoan clade, and it is advisable to take a more comprehensive taxon sample size when considering molecular synapomorphies.

The identification of a rare genomic event, such as the insertion of a 12-amino acid sequence in a primarily highly conserved region of the EF-1α protein, seemed to add serious weight to the argument for the grouping of the animal and fungal clades (Baldauf and Palmer, 1993). Later studies showed that this insertion, while conserved in position and strikingly conserved among fungi, varies extensively in both length and sequence (Baldauf, 1999) and is not present at all in some platyhelminths (Littlewood et al., 2001). The EF-1α protein is also present in multiple copies in most genomes, and in the case of some flatworms phylogenetic reconstruction using this protein does not produce uncontroversial monophyletic groups (Littlewood et al., 2001).

10.3 Methodological biases

10.3.1 Gene sampling and taxon sampling

The choice to use large data-sets or indeed completed genomes in phylogenetic reconstructions is generally made at the expense of taxon sampling. The number of characters and taxa required for accurate phylogenetic reconstruction is debatable (Rokas and Carroll, 2005). Clearly the sample of taxa used had a profound effect on the analyses that lead to the reconstruction of the moulting clade. More slowly evolving nematode species produce the ecdysozoa grouping while faster-evolving species produce the Coelomata relationship. A number of studies suggested that gene or character sampling may have an even greater effect on phylogeny reconstruction than taxon sampling (Mitchell et al., 2000; Rosenberg and Kumar, 2001). This finding spurred on a plethora of phylogenetic studies using large amounts of sequence data supporting the coelomata hypothesis (Mushegian et al., 1998; Blair et al., 2002). Complete eukaryotic genomes and clusters of orthologous groups of proteins permitted large-scale analysis of over 500 eukaryotic orthologous genes (known as KOGs) using a variety of phylogenetic methods, and support was found for the Coelomata hypothesis (Koonin et al., 2004). Using complete genomes of 11 eukaryotic species, homologous sequences derived from 18 human chromosomes (25 000 amino acid sequences). Following adjustment for unequal evolutionary rates among lineages, the Coelomata grouping was favoured using distance, maximum parimony, and Bayesian phylogeny-reconstruction methods (Dopazo et al., 2004). This study highlighted the large number of exons/characters required to reliably reconstruct the animal phylogeny, stating that those analyses supporting the Ecdysozoa hypothesis did not reach the sample-size requirement.

Over the following years, the analyses became larger, with data-sets growing from 100 (Blair et al., 2002) to 500 genes (Wolf et al., 2004) and with the most recent boasting more than 800 genes (Philip et al., 2005). These large-scale gene analyses support, without exception, the grouping of coelomate arthropods and vertebrates to the exclusion of the pseudocoelomate nematodes. So it is clear that smaller numbers of genes from a wide taxon sampling support the Ecdysozoa and large scale or genome wide analyses from a small number of taxa support the Coelomata.

We might be tempted at this stage to retire the debate, giving victory over to the Coelomate followers, and dismissing the Ecdysozoa claim as a result of poor gene sample size. It seems that this might be a little premature, however, as recent studies suggest that the phylogenies from large-scale analyses might be the result of LBA (Felsenstein, 1978).

One of the first molecular studies into the structure of the three kingdoms of eukaryotic life involved sequence data from both large- and small-subunit rRNA, 10 isoaccepetor tRNA families, and six highly conserved proteins from all three kingdoms (Gouy and Li, 1989). Applying a transformed distance method and a maximum-parsimony method, using these three distinct data-sets, a single phylogenetic tree was obtained that placed the fungi at the base, with plant and animal kingdoms as closest neighbors. This was the first sequence analysis to provide statistically robust reconstructions of the base of the eukaryotic tree.

This traditional phylogeny has been challenged with data from ultrastructural characters such as the presence of a uniflagellate reproductive stage grouping the fungi with the animals (Cavalier-Smith, 1987b). These ultrastructural characters are not unique or consistent synapomorphies for animals and fungi; the uniflagellate condition and flattened mitochondrial cristae have been detected outside of the opisthokonts (Steenkamp *et al.*, 2006), calling for the need for molecular synapomorphies. Sequence data used include studies of four protein-encoding genes (α-tubulin, β-tubulin, EF-1α, and enolase; Baldauf and Palmer, 1993; Keeling and Doolittle, 1996). Support for a topology that grouped animals and fungi together to the exclusion of plants was found, although the support varied considerably depending on the protein. On closer analysis of the tubulin data, it was found that these proteins are members of highly paralogous families, with α-tubulin for example consisting of 23 members. In addition the enolase protein has a history of duplication, loss, and horizontal gene transfer, therefore making identification of true orthologs for these proteins difficult (Harper and Keeling, 2004). The use of these characters for phylogeny reconstruction is therefore dubious.

As the analysis of paralogs is difficult due to the lack of sequence data or indeed complete genomes a different approach is to completely disgard multigene families and focus instead on only single gene families from completed genomes. This way we are sure that we are comparing like with like (Blair *et al.*, 2002; Philip *et al.*, 2005). This method of single-gene ortholog identification and analysis ensures that hidden paralogy is minimized. The conservative approach for single-gene ortholog identification involves a two-tier process (see Figure 10.1). Supertree construction, consensus-tree construction, and total evidence methods have all been applied to the data (Creevey and McInerney, 2005; Philip *et al.*, 2005).

10.3.2 Running out of steam

When considering the analysis of molecular data it is necessary for any given data-set to ask whether or not it is capable of reconstructing the phylogeny, a key facet of this being the number of

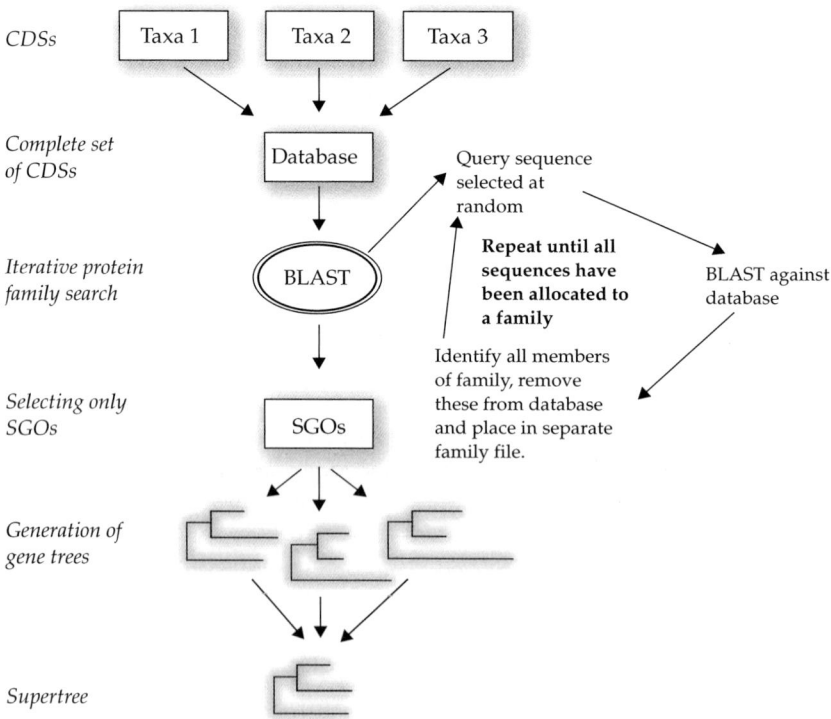

Figure 10.1 Schematic showing the steps involved in identifying single-gene orthologous protein families using coding DNA sequences (CDSs) from completed (or other) genomes. This method is believed to remove many of the biases that can be encountered in generating data-sets for the reconstruction of ancestral relationships. SGO, single-gene ortholog.

characters required to reliably reconstruct a phylogeny. Analysis has shown that molecular data run quickly out of steam when the process of evolution is rapid at the edges of the tree. It is estimated that in these circumstances a polynomial number of samples is required for the phylogeny to be resolved (Mossel and Steel, 2004).

10.3.3 Markov chain Monte Carlo (MCMC) methods of phylogenetic reconstruction and data concatenation

MCMC algorithms play a critical role in Bayesian inference of phylogeny. The rate of convergence of many of the widely used Markov chains has been tested. The practical application of this is that Bayesian MCMC methods can be misleading when the data are generated from a mixture of trees. Thus, in cases of data containing conflicting/potentially conflicting phylogenetic signals, phylogenetic reconstruction should be performed separately on each signal (Mossel and Vigoda, 2005).

10.3.4 LBA

Site-stripping has been applied to analyse sites with different mutation rates in the data-set to reduce the effect of LBA (Philip *et al.*, 2005). Site-stripping is a method of dividing the data-set into different categories of site depending on mutation rate. The purpose of treating the data in this way is to reduce the effect of LBA. In the case of Philip *et al.* (2005), eight different rate categories were defined (see Figure 10.2). The progress of the phylogeny was then followed as the faster-evolving sites are methodologically removed and using different combinations of these rate categories the emerging phylogeny with absence of (or at least reduced) LBA is generated (see Figure 10.2a and b; this method has also been applied to

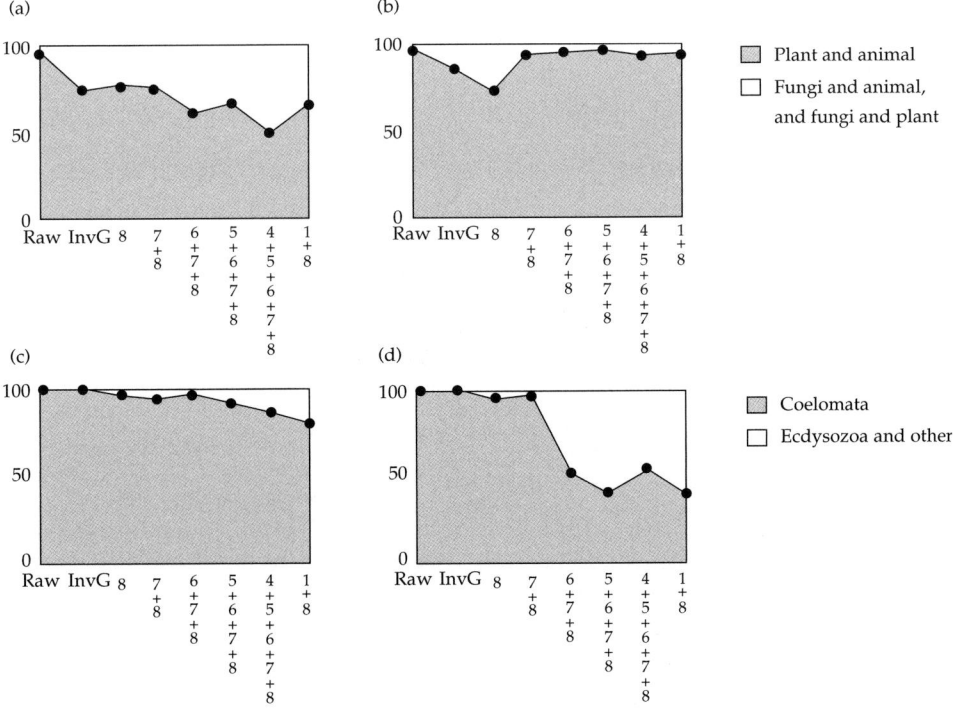

Figure 10.2 (a, b) Data from completed plant, animal, and fungal genomes. (c, d) Data from completed animal genomes. (b, d) Data controlled for sequence length and ability to recover uncontroversial parts of the tree. The *x* axis indicates rate categories removed from data, going from the most slowly to the most rapidly evolving. Raw, represents the data in its original form and InvG indicates data analysed using an invariable gamma model of rate variation. Modified from Philip *et al.* (2005).

the Coelomata/Ecdysozoa data-set and the results are shown in Figure 10.2c and d). A number of statistical and randomization tests were carried out on the data which consisted of 1452 aligned positions and in no case using this taxon selection was support found for the Opisthokont grouping; instead the animals were grouped with the plants; see Figure 10.2a and b (Philip et al., 2005). The issue still remains, however, that we do not have a large-enough taxa sample size. Indeed, the need for complete genome sequences of lower plants, animals, and protists is evident from all the literature to date. The inclusion of data from sister protista has the effect of producing the Opisthokonta grouping (Steenkamp et al., 2006).

It is most interesting that in a recent analysis of 500 genes, the least support for the Coelomata group comes from the maximum-likelihood analyses (Wolf et al., 2004). Maximum-likelihood methods are expected to be the most robust when dealing with rate heterogeneity, and therefore we would expect that, if the phylogenetic signal is strong enough, the use of maximum-likelihood methods would retrieve the most probable tree. The largest data-set applied includes 780 single-gene orthologous protein families (representing some 436 450 amino acid positions) and the site-stripping method described above (see Figure 10.2) and in the vast majority of analyses (24 out of 26) the Coelomata topology was favored over the Ecdysozoa (Philip et al., 2005). This analysis, while sensitively and thoroughly examining available eukaryotic complete genomes, has a limitation in that the available completed genomes are biased towards the higher eukaryotes. A major concern therefore is that the phylogeny supported by this and other large-scale analyses is not due to true phylogenetic signal but rather the rapid evolution of the nematode, causing the nematode to be dragged to the root of the phylogeny. It has been shown that multiple gene analyses

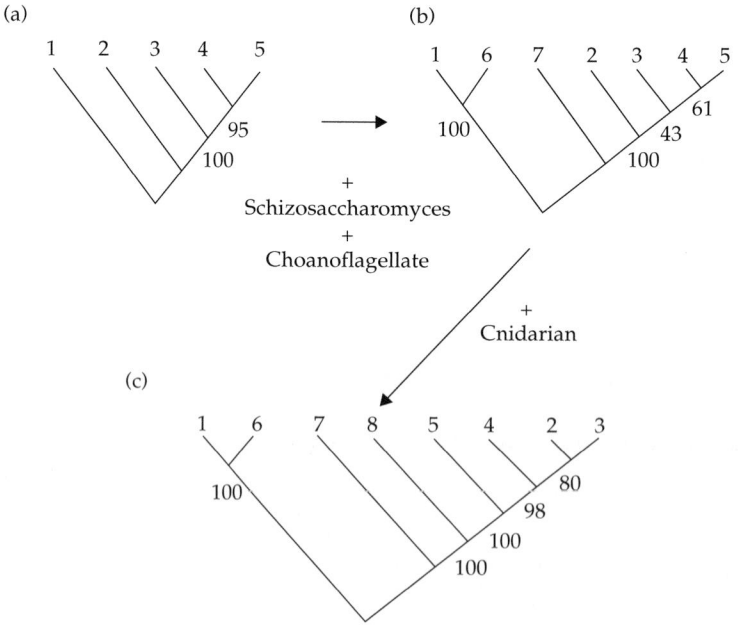

Figure 10.3 This figure shows the importance of taxa sampling on the resulting animal phylogeny. The numbers (1–8) correspond to the following species respectively: *Saccharomyces cerevisiae*, *Caenorhabditis elegans*, *Schistosoma mansoni*, *Drosophila melanogaster*, *Homo sapiens*, *Schizosaccharomyces cerevisiae*, *Monosiga brevicolis*, *Hydra* sp. Using distant outgroups such as yeast in the analysis draws the nematode towards the base of the tree, Coelomata grouping (a). The addition of a second yeast species (*Schizosaccharomyces* sp.) and the choanoflagellate (*Monosiga* sp.) results in a decrease in the support for the grouping of *Drosophila* sp., *H. sapiens*, and *C. elegans* (b). The addition of the Cnidarian (*Hydra* sp.) causes a rearrangement of the species, placing the nematodes as a sister taxa to *Schistosoma* sp. with very high confidence/support (c). Modified from Philippe et al. (2005).

(a)

```
Ecdysozoa    LCYGLPALMAKGQITIVFSPLIALIQDQIDHLMKLKVPWNS
Coelomata    LCYQLPAVMDEGQITIVFSPLIALMKDQIYYLMKKEIPCDS
```

(b)

```
Ecdysozoa    NDFSVSQAEMSGSQQAMLENAMDIKIEKFSISAQGKELKVN
Coelomata    DHFTVSQVAKTGTQQAMMENSMDIKIENFNISAQGKQLFDK
```

(c)

```
Ecdysozoa    YDVTNKASFDNIQAWLTEIHEYAQHDVALMLLGNKVDSSAH
Coelomata    YDITNKASFENCRDWLSQIKEYGQEDVQIMLIGNKCDSSAN
```

Figure 10.4 Ancestral reconstructions of each of the following proteins from the KOG database were generated using the PAML software package for the Ecdysozoan ancestor (top row of each alignment) and for the Coelomata ancestor (bottom row of each alignment). (a) ATP-dependent DNA helicase (KOG0352), positions 210–250 of the reconstructed amino acid sequences; (b) eIF2-interacting protein ABC50 (KOG0066), positions 330–370 of the reconstructed amino acid sequence; and (c) GTPase Rab26/Rab37, small G-protein superfamily (KOG0083); positions 550–590 of the amino acid sequence.

reinforce LBA if it is present in the data (Phillips et al., 2004).

To address the issue of LBA, Wolf et al. (2004) assumed that Ecdysozoa was in fact the correct phylogeny (null hypothesis). Then, by differing the degree of branch-length inequality, they determined the frequency of recovering the Coelomata. By simulating a large number of sequences evolving according to the Ecdysozoan phylogeny, giving the nematode ever-increasing branch lengths relative to the fly and vertebrate lineages, it was found that with longer nematode branch lengths the Coelomata tree was recovered. However, for any given nematode branch length, and comparing the simulated and real data-sets, it was found that the frequency of support for Coelomata was significantly lower. Therefore these results show that the real data-set contains phylogenetic signal that is not fully due to LBA (Telford, 2004a).

It is clear that a larger sampling of taxa is needed, as we have too few completed genomes. The analysis of 129 orthologous proteins from 35 animal species (many of which are expressed sequence tags), including the use of an early-branching and slowly evolving animal (*Hydra magnipapillata*) rather than a fungal species such as yeast to root the phylogeny, was an important advance (Philippe et al., 2005). By increasing the number of taxa sampled and removing rapidly evolving species and sites the Ecdysozoa phylogeny is supported most strongly. It is highly probable that the use of very-long-branched species such as yeast to root the phylogeny has attracted nematodes to the base of the phylogeny in previous analyses (see Figure 10.3; taken from Philippe et al., 2005).

What are the consequences of the two alternative phylogenies with regard to the ancestral sequences that they infer? We analysed three KOG protein families using PAML and produced an ancestral sequence for each based on both the Coelomata phylogeny and the Ecdysozoa phylogeny independently (see Figure 10.4). The resultant ancestral sequences were similar at many positions; this is reflected in the pairwise distances between the Ecdysozoa and Coelomata reconstructed ancestors for genes a, b and c (relating to panels of Figure 10.4) being 0.221, 0.243, and 0.105 respectively. However, there are a number of positions which differ between the two reconstructed ancestral proteins, a sample of which is shown in Figure 10.4.

In examining controversial issues in eukaryote phylogeny reconstruction, it must be borne in mind that the reasons why these issues have remained controversial is that it has been possible to recover contradictory signals. The reasons why these signals exist have been discussed and include simple stochastic error and sampling, systematic biases, and the selection of genes for analysis that are not orthologs.

References

Aguinaldo, A.M., Turbeville, J.M., Linford, L.S., Rivera, M.C., Garey, J.R., Raff, R.A., and Lake, J.A. (1997) Evidence for a clade of nematodes, arthropods and other moulting animals. *Nature* 387: 489–493.

Baldauf, S.L. (1999) A search for the origins of animals and fungi: comparing and combining molecular data. *Am. Nat.* 154: S178–S188.

Baldauf, S.L. and Palmer, J.D. (1993) Animals and fungi are each other's closest relatives: congruent evidence from multiple proteins. *Proc. Natl. Acad. Sci. USA* 90: 11558–11562.

Baldauf, S.L., Roger, A.J., Wenk-Siefert, I., and Doolittle, W.F. (2000) A kingdom-level phylogeny of eukaryotes based on combined protein data. *Science* 290: 972–927.

Blair, J.E., Ikeo, K., Gojobori, T., and Hedges, S.B. (2002) The evolutionary position of nematodes. *BMC Evol. Biol.* 2: 7.

Brown, J.R., Douady, C.J., Italia, M.J., Marshall, W.E., and Stanhope, M.J. (2001) Universal trees based on large combined protein sequence data sets. *Nat. Genet.* 28: 281–285.

Cavalier-Smith, T. (1987a) The simultaneous symbiotic origin of mitochondria, chloroplasts, and microbodies. *Ann. NY Acad. Sci.* 503: 55–71.

Cavalier-Smith, T. (1987b) *The Origin of Fungi and Pseudofungi*. Cambridge University Press, Cambridge.

Cho, S., Jin, S., Cohen, A., and Ellis, R.E. (2004) A phylogeny of *Caenorhabditis* reveals frequent loss of introns during nematode evolution. *Genome Res.* 14: 1207–1220.

Coghlan, A. and Wolfe, K.H. (2004) Origins of recently gained introns in Caenorhabditis. *Proc. Natl. Acad. Sci. USA* 101: 11362–11367.

Copley, R.R., Aloy, P., Russell, R.B. and Telford, M.J. (2004) Systematic searches for molecular synapomorphies in model metazoan genomes give some support for Ecdysozoa after accounting for the idiosyncrasies of *Caenorhabditis elegans*. *Evol. Dev.* 6: 164–169.

Creevey, C.J. and McInerney, J.O. (2005) Clann: investigating phylogenetic information through supertree analyses. *Bioinformatics* 21: 390–392.

Darwin, C. (1859) *On the Origin of the Species by Means of Natural Selection or the Preservation of Favoured Races in the Struggle for Life*. John Murray, London.

Dopazo, H., Santoyo, J., and Dopazo, J. (2004) Phylogenomics and the number of characters required for obtaining an accurate phylogeny of eukaryote model species. *Bioinformatics* 20 (suppl. 1): I116–I121.

Felsenstein, J. (1978) Cases in which parsimony or compatibility methods will be positively misleading. *Syst. Zool.* 27: 401–410.

Gouy, M. and Li, W.H. (1989) Molecular phylogeny of the kingdoms Animalia, Plantae, and Fungi. *Mol. Biol. Evol.* 6: 109–122.

Harper, J.T. and Keeling, P.J. (2004) Lateral gene transfer and the complex distribution of insertions in eukaryotic enolase. *Gene* 340: 227–235.

House, C.H. and Fitz-Gibbon, S.T. (2002) Using homolog groups to create a whole-genomic tree of free-living organisms: an update. *J. Mol. Evol.* 54: 539–547.

Hyman, L.H. (1940) *The Invertebrates*. McGraw-Hill, New York.

Keeling, P.J. and Doolittle, W.F. (1996) Alpha-tubulin from early-diverging eukaryotic lineages and the evolution of the tubulin family. *Mol. Biol. Evol.* 13: 1297–1305.

King, N. and Carroll, S.B. (2001) A receptor tyrosine kinase from choanoflagellates: molecular insights into early animal evolution. *Proc. Natl. Acad. Sci. USA* 98: 15032–15037.

Koonin, E.V., Fedorova, N.D., Jackson, J.D., Jacobs, A.R., Krylov, D.M., Makarova, K.S. *et al.* (2004) A comprehensive evolutionary classification of proteins encoded in complete eukaryotic genomes. *Genome Biol.* 5: R7.

Littlewood, D.T., Olson, P.D., Telford, M.J., Herniou, E.A., and Riutort, M. (2001) Elongation factor 1-alpha sequences alone do not assist in resolving the position of the acoela within the metazoa. *Mol. Biol. Evol.* 437–442.

Lynch, M. and Richardson, A.O. (2002) The evolution of spliceosomal introns. *Curr. Opin. Genet. Dev.* 12: 701–710.

Mader, S.S. (1993) *Biology*. Wm. C. Brown, Dubuque.

Manuel, M., Kruse, M., Muller, W.E., and Le Parco, Y. (2000) The comparison of beta-thymosin homologues among metazoa supports an arthropod-nematode clade. *J. Mol. Evol.* 51: 378–381.

Mitchell, A., Mitter, C., and Regier, J.C. (2000) More taxa or more characters revisited: combining data from nuclear protein-encoding genes for phylogenetic analyses of Noctuoidea (Insecta: Lepidoptera). *Syst. Biol.* 49: 202–224.

Mossel, E. and Steel, M. (2004) A phase transition for a random cluster model on phylogenetic trees. *Math. Biosci.* 187: 189–203.

Mossel, E. and Vigoda, E. (2005) Phylogenetic MCMC algorithms are misleading on mixtures of trees. *Science* 309: 2207–2209.

Mushegian, A.R., Garey, J.R., Martin, J., and Liu, L.X. (1998) Large-scale taxonomic profiling of eukaryotic model organisms: a comparison of orthologous proteins encoded by the human, fly, nematode, and yeast genomes. *Genome Res.* **8**: 590–598.

Philip, G.K., Creevey, C.J., and McInerney, J.O. (2005) The Opisthokonta and the Ecdysozoa may not be clades: stronger support for the grouping of plant and animal than for animal and fungi and stronger support for the Coelomata than Ecdysozoa. *Mol. Biol. Evol.* **22**: 1175–1184.

Philippe, H., Lartillot, N., and Brinkmann, H. (2005) Multigene analyses of bilaterian animals corroborate the monophyly of Ecdysozoa, Lophotrochozoa, and Protostomia. *Mol. Biol. Evol.* **22**: 1246–1253.

Phillips, M.J., Delsuc, F., and Penny, D. (2004) Genome-scale phylogeny and the detection of systematic biases. *Mol. Biol. Evol.* **21**: 1455–1458.

Rokas, A. and Carroll, S.B. (2005) More genes or more taxa? The relative contribution of gene number and taxon number to phylogenetic accuracy. *Mol. Biol. Evol.* **22**: 1337–1344.

Rokas, A., Williams, B.L., King, N., and Carroll, S.B. (2003) Genome-scale approaches to resolving incongruence in molecular phylogenies. *Nature* **425**: 798–804.

Rosenberg, M.S. and Kumar, S. (2001) Traditional phylogenetic reconstruction methods reconstruct shallow and deep evolutionary relationships equally well. *Mol. Biol. Evol.* **18**: 1823–1827.

Roy, S.W. and Gilbert, W. (2005) Resolution of a deep animal divergence by the pattern of intron conservation. *Proc. Natl. Acad. Sci. USA* **102**: 4403–4408.

Roy, S.W., Fedorov, A., and Gilbert, W. (2003) Large-scale comparison of intron positions in mammalian genes shows intron loss but no gain. *Proc. Natl. Acad. Sci. USA* **100**: 7158–7162.

Steenkamp, E.T., Wright, J., and Baldauf, S.L. (2006) The protistan origins of animals and fungi. *Mol. Biol. Evol.* **23**: 93–106.

Telford, M.J. (2004a) Animal phylogeny: back to the coelomata? *Curr. Biol.* **14**: R274–R276.

Telford, M.J. (2004b) The multimeric beta-thymosin found in nematodes and arthropods is not a synapomorphy of the Ecdysozoa. *Evol. Dev.* **6**: 90–94.

Telford, M.J., Wise, M.J., and Gowri-Shankar, V. (2005) Consideration of RNA secondary structure significantly improves likelihood-based estimates of phylogeny: examples from the bilateria. *Mol. Biol. Evol.* **22**: 1129–1136.

Woese, C.R. (1987) Bacterial evolution. *Microbiol. Rev.* **51**: 221–271.

Wolf, Y.I., Rogozin, I.B., and Koonin, E.V. (2004) Coelomata and not Ecdysozoa: evidence from genome-wide phylogenetic analysis. *Genome Res.* **14**: 29–36.

III

Computational applications of ancestral sequence reconstruction

CHAPTER 11

Using ancestral sequence inference to determine the trend of functional divergence after gene duplication

Xun Gu, Ying Zheng, Yong Huang and Dongping Xu

11.1 Introduction

Under the framework of phylogenomic annotation of gene function (Golding and Dean, 1998; Eisen and Fraser, 2003), the importance of gene function can be measured quantitatively in terms of the functional constraints of the protein sequence. In other words, the evolutionary function is defined as the parameter that measures their contributions to the fitness of the organism. Following this idea, many research groups including ours have developed statistical methods for testing and predicting functional divergence after the gene duplication (Lichtarge et al., 1996; Gu, 1999, 2001, 2006; Landgraf et al., 1999, 2001; Knudsen and Miyamoto, 2001; Gaucher et al., 2002; Bielawski and Yang, 2004; Gao et al., 2005; Rastogi and Liberles, 2005). Recently, this notion of functional divergence was challenged by Lopez et al. (1999). They claimed that change of sequence conservation during protein evolution may largely result in functional equivalence, predicting no association between site-specific rate shifts and experimental evidence of functional divergence. However, several case studies (Landgraf et al., 1999, 2001; Jordan et al., 2001; Wang and Gu, 2001; Gaucher et al., 2002; Gu, 2003) indeed showed the association between sequence and function/structure divergence, supporting the notion of functional divergence.

Ancestral sequence inference under a given phylogeny is becoming an important approach in molecular biology and functional comparative genomics (Golding and Dean, 1998). This is partly because evolution has selected proteins for function over hundreds or even thousands of millions of years, keeping those that carried out critical functions, and eliminating deleterious mutations. Thus, it provides an alternative—shaped by evolution—to understanding protein function through random mutagenesis.

We have recognized that ancestral sequence reconstruction is a powerful technique for linking sequence to function. In this chapter, we attempt to carry out a new approach to functional divergence analysis with the combination of ancestral sequence inference, using the family of animal G-protein α subunits as an example. Consequently, we are not only able to identify amino acid residues that are crucial for functional divergence after gene duplication, but also infer the trend of evolutionary changes at these residues.

11.2 Types of functional divergence in protein sequence evolution

From the view of molecular evolution, an amino acid residue is said to be functionally or structurally important if it is evolutionarily conserved (Kimura, 1983). Therefore, change in the evolutionary conservation of a particular residue may indicate the involvement of functional divergence during the evolution of a gene family (Gu, 1999). Furthermore, Gu (2001) made a distinction between type I and type II functional divergences. Note that these two types of functional divergence may have other names. For instance, the basic

Evolutionary Trace approach (Lichtarge et al., 1996; Madabushi et al., 2004) has mainly focused on cluster-specific residues related to type II functional divergence. Gribaldo et al. (2003) also looked at type II functional divergence but called it 'constant-but-different'. Meanwhile, the weighted Evolutionary Trace approach proposed by Landgraf et al. (1999, 2001) was similar to type I functional divergence (Gu, 1999). A discussion of these and additional methods can be found in the following sections of this chapter and is also extended in Chapter 19 of this volume.

11.3 Type I functional divergence (site-specific rate shift)

This type of functional divergence refers to the evolutionary process that results in site-specific rate shifts after gene duplication (Gu, 1999; Landgraf et al., 1999, 2001; Knudsen and Miyamoto, 2001; Gaucher et al., 2002). Typically, an amino acid residue is highly conserved in one duplicate gene, but highly variable in the other one. Gu (1999) has developed a statistical method to test the significance of type I functional divergence between duplicate genes. Briefly, the two-state model proposed by Gu (1999) assumed that an amino acid residue (site) is in either one of two states: *related to functional divergence* if its evolutionary rate is shifted (up or down) after the gene duplication or *unrelated to functional divergence* otherwise. The coefficient of (type I) functional divergence between duplicate genes, denoted by θ_I, is defined as the probability of being related to functional divergence. Apparently, a large value for θ_I indicates a high level of type I functional divergence, and vice versa. In a typical case when two gene clusters are generated by a gene-duplication event, the coefficient of (type I) functional divergence between them can be estimated (Gu, 1999; Gu and Vander Velden, 2002). Rejection of the null hypothesis of $\theta_I = 0$ means that the evolutionary rate has become different between the duplicate genes at some sites. Using this method, many case studies have demonstrated the structural basis of functional divergence; for example, the caspase family (Wang and Gu, 2001) and the Jak family of protein kinases (Gu et al., 2002).

Moreover, a site-specific profile based on the empirical posterior analysis is useful for predicting amino acid residues that are crucial for functional divergence.

11.4 Type II functional divergence

As opposed to a site-specific shift of evolutionary rate (type I functional divergence), type II functional divergence results in a site-specific property shift. A typical case is when, at a position that is homologous between members of a gene family, a radical shift of amino acid property—for example, a change from positive to negative charge—occurs between two duplicate genes; the contrasting case is when the position is evolutionarily conserved within each of the orthologous gene sets. Whereas several methods for assessing type I functional divergence are available, the implementation of statistical testing for type II functional divergence of gene families has not been well resolved. We have recently solved this problem (Gu, 2006) using the software package DIVERGE2. In the following, we give a brief introduction to the methodology.

In principle, the evolution of protein sequences of duplicate genes can be divided into two stages, the early (E) stage after gene duplication, and the late (L) stage. We assume that (type II) functional divergence between duplicate genes has occurred in the E stage, whereas in the L stage purifying selection plays a major role to maintain related but distinct functions of the two duplicate genes (Ohno, 1970; Kimura, 1983; Force et al., 1999). Gu (2006) thus extended the two-state model (Gu, 1999, 2001) to type II (cluster-specific) functional divergence, as follows. (1) In the E stage, an amino acid residue can be in either of two states: type II unrelated and type II related. The probability of a residue being under the type II related state is denoted by θ_{II}, the coefficient of type II functional divergence. (2) In the L stage, there is no further type II functional divergence for any amino acid residue, so amino acid substitutions in this stage are mainly under purifying selection.

Under the functional divergence-unrelated status, the substitution model largely reflects the conserved evolution of protein sequences, which can be determined empirically by the Dayhoff model (Dayhoff *et al.*, 1978) or the JTT model (Jones *et al.*, 1992). In contrast, under the functional divergence-related status, radical amino acid substitutions may occur more frequently (Lichtarge *et al.*, 1996). To avoid over-parameterization, Gu (2006) proposed a simple model that can distinguish between the radical and conserved amino acid substitutions. First, we tentatively classify 20 amino acids into four groups: charge positive (K, R, H), charge negative (D, E), hydrophilic (S, T, N, Q, C, G, P), and hydrophobic (A, I, L, M, F, W, V, Y). An amino acid substitution is called radical (denoted by R) if it changes from one group to another; otherwise it is called conservative—within the group—denoted by C. The status of no substitution is denoted by N.

Based on these assumptions, Gu (2006) developed an evolutionary model for this type of functional divergence. In a typical case when two gene clusters are generated by a gene-duplication event, the coefficient of (type II) functional divergence between them can be estimated. Moreover, a site-specific profile based on the empirical posterior analysis is useful to predict amino acid residues that are crucial for type II functional divergence.

11.5 DIVERGE2: an analytical pipeline for comprehensive functional divergence analysis

The software DIVERGE (Gu and Vander Velden, 2002) is a software system for studying functional divergence between member genes of a protein family based on (site-specific) shifted evolutionary rates (type I) after gene duplication. Posterior analysis results in a site-specific profile for predicting important amino acid residues that are responsible for this type of functional divergence. Moreover, when the three-dimensional protein structure is available, these predicted amino acid residues can be mapped to the three-dimensional-structure viewer to explore their structural basis.

The updated version, named DIVERGE2, has provided more options (e.g. type II functional divergence) to explore functional evolution of protein family sequences. One can use the site-specific profiles to detect amino acid residues that are crucial for this type (I or II) of functional divergence. In practice, one may use the site-(k)-specific scores $Q_I(k)$ or $Q_{II}(k)$, the posterior probability that site k is related to type I or type II functional divergence. Another commonly used measure is based on the posterior ratio: in our case, it is given by

$$R_I(k) = Q_I(k)/[1 - Q_I(k)] \text{ or }$$
$$R_{II}(k) = Q_{II}(k)/[1 - Q_{II}(k)]$$

When a cutoff is given, important residues for two types of functional divergence are predicted.

An important development of DIVERGE2 is to provide an analytical pipeline for combining functional divergence and ancestral sequence inference, which can be used to infer the trend of functional divergence. Currently, DIVERGE2 adopts the Bayesian algorithm of Zhang and Nei (1997) to infer the ancestral sequences under a known phylogeny of a gene family. It is a simplified version of Yang *et al.* (1995) in which the branch lengths of the phylogenetic tree are estimated using a least-squares method rather than the maximum-likelihood method. Each site in the inferred ancestral sequence receives the assignment of amino acid with the highest posterior probability. Using this approach one may determine whether an amino acid residue that was highly conserved in the ancestral protein sequence now becomes highly variable, or vice versa.

11.6 Case study: animal G-protein α subunits

G-protein receptor activation of intracellular targets is probably the most ancient of the metazoan signal-transduction pathways. There are over 16 G-protein α subunits in animals, which can be further divided into four major classes, G_s, G_i, G_q, and G12 (Simon *et al.*, 1991; Neer, 1995), depending on their actions upon the effectors. We have updated the multiple alignment of 81 amino acid sequences of animal G-protein α subunits using a standard BLAST search followed by the program ClustalX. The phylogenetic tree of the G-protein α gene family was inferred by the neighbor-joining method (Saitou and Nei, 1987). The parsimony (PAUP4.0) and likelihood (PHYLIP) methods give virtually the same topology. Consistent with previous results, the four major classes of G-protein α subunit are monophyletic.

As an example, we choose G_s and G_q, the two major classes of G-protein α subunits, to demonstrate the ancestral sequence-based analysis of functional divergence. See Figure 11.1 for the phylogenetic tree of the G_s and G_q classes. The G_s class, which consists of G_s and Golf subtypes, is involved in hormonal stimulation of adenylate cyclase and the opening of Ca^{2+} channels. Whereas the G_s subtype is expressed in almost all tissue types, the Golf subtype is expressed exclusively in olfactory cells and is thought to be involved specifically in odorant signal transduction (Kaziro et al., 1991). On the other hand, class G_q has the function of simulating phospholipase C,

which has four subtypes, G_q, G11, G14, and G15. Despite G_q and G11 being widely distributed and often found in the same cell, they may have different receptors and effectors or act at different developmental stages. The other two subtypes, G14 and G15, are tissue-specific, and may interact with different members of the phospholipase family.

11.7 Functional divergence between G_s and G_q proteins

11.7.1 Type I functional divergence

We first tested the site-specific shift of evolutionary rate (type I functional divergence) after the gene-duplication event leading to the G_s and G_q subtypes. The coefficient of type I functional

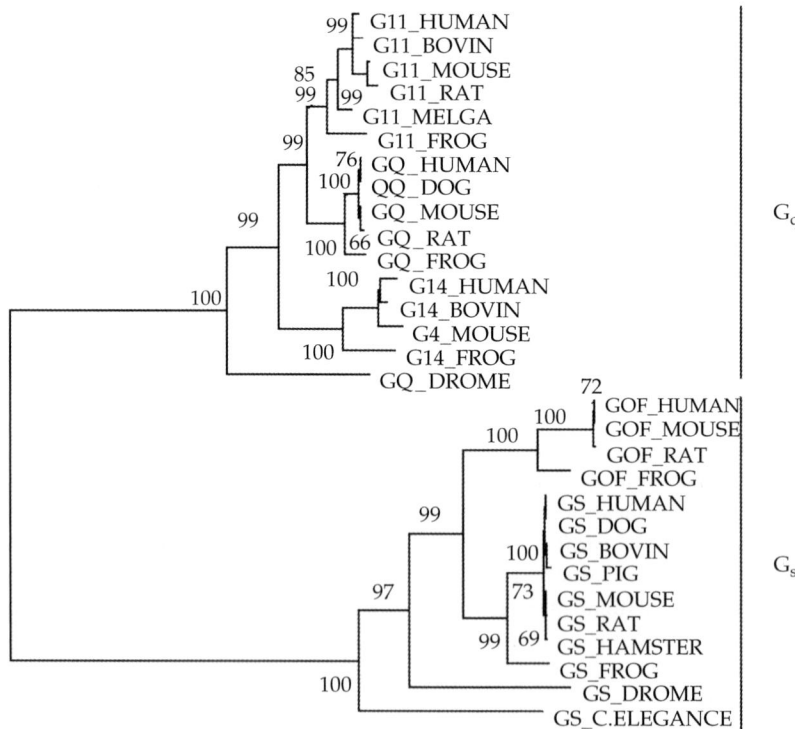

Figure 11.1 The neighbor-joining tree of G_q and G_s based on the multiple sequence alignment with Poisson distance. Bootstrap scores of >50% are presented. Accession numbers: P50148 (GQ_HUMAN), Q28294 (GQ_DOG), P21279 (GQ_MOUSE), P82471 (GQ_RAT), P38410 (GQ_FROG), P29992 (G11_HUMAN), P38409 (G11_BOVIN), P21278 (G11_MOUSE), Q9JID2 (G11_RAT), P45645 (G11_MELGA), P43444 (G11_FROG), O95837 (G14_HUMAN), P38408 (G14_BOVIN), P30677 (G14_MOUSE), O73819 (G14_FROG), JN0115 (GQ_DROME), P04895 (GS_HUMAN), CAA78161 (GS_DOG), P04894 (GS_MOUSE), AAA40827 (GS_RAT), CAA35516 (GS_HAMSTER), P04896 (GS_BOVIN), P29797 (GS_PIG), CAA39571 (GS_FROG), Q8CGK7 (GOF_MOUSE), P38406 (GOF_RAT), P38405 (GOF_HUMAN), CAC82735 (GOF_FROG), NP_477506 (GS_DROME), and NP_490817 (GS_C. ELEGANCE).

divergence between G_s and G_q is $\theta_I = 0.53 \pm 0.08$, which is significantly larger than 0. Hence, site-specific rate differences may occur at some amino acid residues after the gene-duplication event. Figure 11.2 shows the site-specific profile of the posterior ratio, $R_I(k)$; apparently, most sites are unlikely to be involved in type I functional divergence. We used the cutoff $R_I > 2$ to identify the (type I) functional divergence-related residues between G_s and G_q, and obtained 25 amino acid residues (Figure 11.3). These sites clearly show a typical pattern of type I functional divergence; that is, conserved amino acids in one cluster, and diverse amino acids in the other. Moreover, these predicted sites can be divided into two groups: group a in Figure 11.3 includes 15 sites that are conserved in G_q but not conserved in G_s, whereas group b includes 10 sites that are conserved in G_s but not conserved in G_q. Consequently, G_q proteins become more conserved than G_s proteins.

11.7.2 Type II functional divergence

Based on the same multiple alignment of protein sequences, we obtained the estimate of the coefficient of type II functional divergence, $\theta_{II} = 0.325 \pm 0.055$, between the G_s and G_q α proteins, which is significantly larger than 0. It suggests that, after the gene-duplication event, some amino acid positions that were evolutionarily conserved in both G_s and G_q proteins may have undergone radical changes in their amino acid properties. Figure 11.4 shows the site-specific profile based on the posterior ratio, $R_{II}(k)$, for type II functional divergence between G_s and G_q proteins. Apparently, most residues received very low scores, indicating that only a small portion of amino acid residues were involved in this type of functional divergence. Noticeably, 29 amino acid positions with the highest scores (posterior ratio $R_{II} > 17$) show a typical shift of amino acid properties at conserved sites (Figure 11.5), as shown in Table 11.1.

11.7.3 The evolutionary trend of functional divergence: ancestral inference analysis

Using the Bayesian ancestral sequence inference implemented in the DIVERGE2 software, we inferred the ancestral sequences of all internal nodes on the phylogeny of G-protein α subunits, which will provide some new information about the evolutionary trend of functional divergence. In particular, the inferred ancestral amino acid residues that are related to type I or type II functional divergence are presented in Figures 11.3 and 11.5, respectively. The ancestral inference points are summarized in Figure 11.6.

As shown in Figure 11.3, there are two groups of type I functional divergence. Among 15 residues in group a—conserved in G_q but variable in G_s—five residues (e.g. position 088) show the conserved G_q-type amino acid in all four major internal nodes (X, Y, Z, and z), whereas six residues (e.g. position 016) show the conserved G_q-type amino acid in three internal nodes but not at Y (the common ancestor of G_s). Putting these two ancestral patterns together, it appears that the ancestral

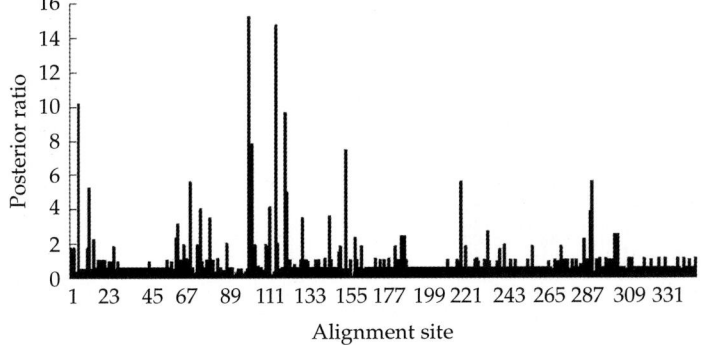

Figure 11.2 Type I divergence site-specific profiles for the G_q and G_s clusters.

122 ANCESTRAL SEQUENCE RECONSTRUCTION

	(a)		(b)	
	Position(k)	000001111111222 134881124778078 692780615565937	Position(k)	1112223332 4570448991 7700014684
G_q	G11_HUMAN	EIEEEKIQAIKQRVN	G11_HUMAN	LTQAGENAVL
	G11_BOVIN	EIEEEKIQAIKQRVN	G11_BOVIN	LTRAGENAVL
	G11_MOUSE	EIEEEKIQAIKQRVN	G11_MOUSE	LTQAGENAVL
	G11_RAT	EIEEEKIQAIKQRVN	G11_RAT	LTQAGENAVL
	G11_MELGA	EIEEEKIQAIKQRVN	G11_MELGA	LMPAGENAVL
	G11_FROG	EIEEEKIQAIKQRVN	G11_FROG	VCPTGENAVL
	8GQ_HUMAN	EIEDEKIQAIKQRVN	GQ_HUMAN	LSPTAQSAVL
	GQ_DOG	EIEDEKIQAIKQRVN	GQ_DOG	LSPTAQSAVL
	GQ_MOUSE	EIEDEKIQAIKQRVN	GQ_MOUSE	LSPTSQSAVL
	GQ_RAT	EIEDEKIQAIKQRVN	GQ_RAT	LSPTSQSAVL
	GQ_FROG	EIEDEKIQAIKQRVN	GQ_FROG	LAPTGQSAVL
	G14_HUMAN	EIEDEKIQAIKQRVN	G14_HUMAN	ISEASENVQQ
	G14_BOVIN	EIEDEKIQAIKQRVN	G14_BOVIN	LSDAAENVQQ
	G14_MOUSE	EIEDEKIQAIKQRVN	G14_MOUSE	ITDASENVQQ
	G14_XENLA	EIEDEKIQAIKQRVN	G14_XENLA	VSKTGENSQQ
	GQ_DROME	EIEDEKIQAIKQRVN	GQ_DROME	LTPADDGHLC
G_s	GOF_HUMAN	VAKPELVVSVKKSAN	GOF_HUMAN	YFEADDKVLI
	GOF_MOUSE	VAKPELVVSVKKSAN	GOF_MOUSE	YFEADDKVLI
	GOF_RAT	VAKPELVVSVKKSAN	GOF_RAT	YFEADDKVLI
	GOF_FROG	IAKSEQVVIAQKHVN	GOF_FROG	YFEADDKVLI
	GS_HUMAN	NAKGEQLEVAKRVVQ	GS_HUMAN	YFEADDKVLI
	GS_DOG	NAKGEQLEVAKRVVQ	GS_DOG	YFEADDKVLI
	GS_BOVIN	NAKGEQLEVAKRVVQ	GS_BOVIN	YFEADDKVLI
	GS_PIG	NAKGDQLEVAKRVVQ	GS_PIG	YFEADDKVLI
	GS_MOUSE	NAKGEQLEVAKRVVQ	GS_MOUSE	YFEADDKVLI
	GS_RAT	NAKGEQLEVAKRVVQ	GS_RAT	YFEADDKVLI
	GS_HAMSTER	NAKGEQLEVAKRVVQ	GS_HAMSTER	YFEADDKVLI
	GS_FROG	NTKAEQIEITKRIVH	GS_FROG	YFEADDKVLI
	GS_DROME	SRADSDILVTELTTT	GS_DROME	YFEANDKVLI
	GS_C.E.	GVQEATVQRILMVCT	GS_C.E.	YDEANDKVLI
	Ancestral sequence		Ancestral sequence	
	X	EIEDEKIQAIKQRVN		LSPADENALL
	Y	GAQEEQIQVIKRVVT		YDEANDKVLI
	Z	EREEEPIQAIRQRVN		QSEADKNVLL
	z	EREEEPIQAIRQRVN		QEEADKKVLL

Figure 11.3 Type I divergence-site candidates and ancestral sequence inference. (a) Cases where amino acids are conserved in the G_q cluster but not in the G_s cluster. (b) Cases where amino acids are conserved in the G_s cluster but not in the G_q cluster. X, the ancestral sequence for the G_q cluster; Y, the ancestral sequence for the G_s cluster; Z, the common ancestral sequence for the G_s and $G_i o$ cluster; z, the common ancestral sequence for the G_q and G12 cluster.

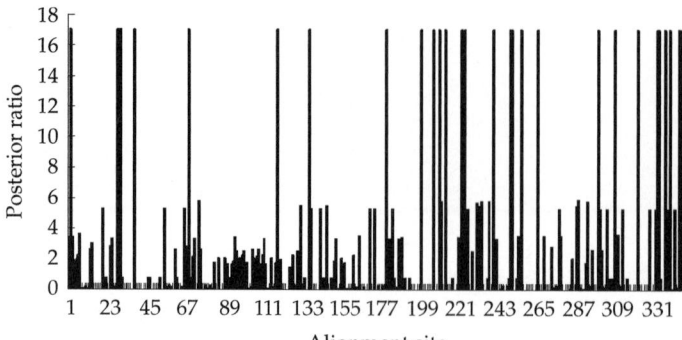

Figure 11.4 Type II divergence site-specific profiles for the G_q and G_s clusters.

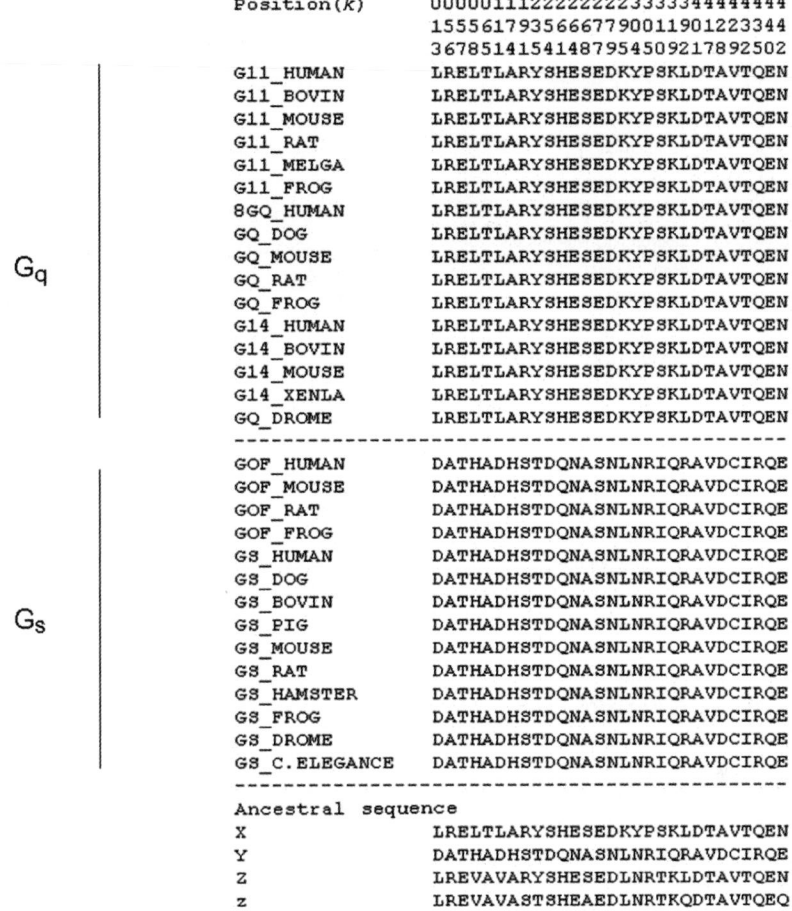

Figure 11.5 Type II divergence-site candidates and ancestral sequence inference. X, Y, Z and z are as described for Figure 11.3.

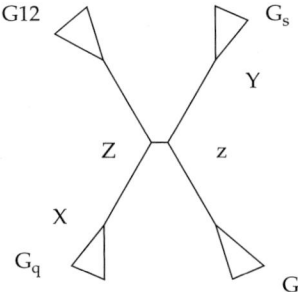

Figure 11.6 The ancestral inference points for the G-protein α subunit family. X, Y, Z and z are as described for Figure 11.3.

states of these 11 residues are all of the conserved G_q-type, whereas the variable G_s-type residues are the derived characters that are only specific to G_s proteins. The only difference is that type I functional divergence at the six five residues likely occurred after the common ancestor of G_s (Y), whereas those at the second five residues occurred before node Y. On the other hand, among 10 residues in group b—conserved in G_s but variable in G_q—two residues (positions 200 and 396) showed the conserved G_s-type amino acid in all four major internal nodes, whereas another two residues (positions 170 and 384) showed a similar pattern except for the common ancestor of G_q (X). Interestingly, at two residues (positions 214 and 398), the conserved G_s-type is recently derived, which is specific to the G_s cluster. For the rest of residues, one may not be able to determine the trend of functional divergence, due to the statistical

Table 11.1 Overview of the amino acid changes at the 29 sites in type II functional divergence.

Position	G_q	G_s	Property change
13	L	D	Hydrophobic/ −
56	R	A	+ /Hydrophobic
57	E	T	− /Hydrophilic
58	L	H	Hydrophobic/ +
65	T	A	Hydrophilic/hydrophobic
111	L	D	Hydrophobic/ −
174	A	H	Hydrophobic/ +
191	R	S	+ /Hydrophilic
235	Y	T	Hydrophobic/hydrophilic
254	S	D	Hydrophilic/ −
261	H	Q	+ /Hydrophilic
264	E	N	− /Hydrophilic
268	S	A	Hydrophilic/hydrophobic
277	E	S	− /Hydrophilic
279	D	N	− Hydrophilic
295	K	L	+ /Hydrophobic
304	Y	N	Hydrophobic/hydrophilic
305	P	R	Hydrophilic/ +
310	S	I	Hydrophilic/hydrophobic
319	K	Q	+ /Hydrophilic
392	L	R	Hydrophobic/ +
401	D	A	− /Hydrophobic
417	T	V	Hydrophilic/hydrophobic
428	A	D	Hydrophobic/ −
429	V	C	Hydrophobic/hydrophilic
432	T	I	Hydrophilic/hydrophobic
435	Q	R	Hydrophilic/ +
440	E	Q	− /Hydrophilic
442	N	E	Hydrophilic/ −

uncertainty of phylogenetic inference or ancestral sequence inference.

In the same manner, we have examined the ancestral amino acid residues for type II functional divergence (Figure 11.5). Among 29 predicted residues, 17 of them (e.g. position 013) show that these amino acid property shifts at conserved residues may occur in the evolutionary trend that can be simply represented as being from the internal node z (ancestral type) to Y (G_s-type). In contrast, four residues (e.g. position 065) show the evolutionary trend of type II functional divergence from the internal node Z (ancestral type) to X (G_q-type); see Figures 11.5 and 11.6.

11.8 Other related methods and new developments

Over the years, many measurements have been proposed to identify functional sites in protein families based on changes in site-specific evolutionary rates, which are well summarized in a recent report (Abhiman et al., 2006). Several groups have developed software or databases implementing some of the measurements; for example, SDPpred (Kalinina et al., 2004), FunShift (Abhiman and Sonnhammer, 2005a), BADASP (Edwards and Shields, 2005), and the Evolutionary Trace viewer (Morgan et al., 2006). Most of the software has been implemented with functionalities to identify type I and II sites (or variants with similar definitions).

The FunShift database (Abhiman et al., 2006; Abhiman and Sonnhammer, 2005a, 2005b) is of special interest as it is the first large-scale study on functional divergence (function shift) between subfamilies in protein families. This study was focused on the enzyme families from the Pfam database. The benefit of the enzyme families is the ability to use Enzyme Commission (EC) numbers as function tokens for subfamilies. The EC number is believed to be one of the most reliable functional annotations currently available. The BETE (Sjolander, 1998) method was used in the studies to define subfamilies in protein families, which has the benefit of having minimum human involvement. In the initial report, the BETE method divided 4225 Pfam families into 13 599 subfamilies that were larger than four sequences. The number increased to 7300 Pfam families and 151 934 subfamilies on release of FunShift 12. Only the subfamilies with more than four sequences, each with EC numbers, with all the EC number information available at all four levels, and with all sequences having the same domain composition, were considered for analysis. Pairs of subfamilies in a Pfam family were denoted as either Same_EC if their EC numbers were the same or Diff_EC if otherwise. This large-scale study of functional divergence in the Pfam families was largely made possible by this semi-automatic process of dividing and annotating the subfamilies.

Three types of measurement, the Conservation Shift Measure (CSM), Rate Shift Measure (RSM), and Alpha Shift Measure (ASM), were used in the FunShift studies. Conservation-Shifting Sites (CSS) are the sites where two different types of amino acid are each conserved in one of the subfamilies in a subfamily pair. By definition, CSS is similar to type II sites. CSS were identified using an entropy-based method previously described in Sjolander (1998). Briefly, for subfamilies p and q, the cumulative relative entropy of a site is the sum of the relative entropy of the site in both subfamilies (REp and REq). The cumulative relative entropy values are standardized over all the positions and transformed into Z scores. The CSS are the sites with Z scores greater than 0.5. CSM is CSS normalized to the alignment length. Rate-Shifting Sites (RSS) are the sites where the substitution rates in the two subfamilies are different. RSS can be viewed as more general cases of type I sites. RSS were identified using the likelihood ratio test (LRT) program (Knudsen and Miyamoto, 2001) with a threshold of under 4.0 for the U values. RSM is RSS normalized to the alignment length. The ASM is a heuristic measurement based on the α factor of the gamma distribution of substitution rates of the whole subfamily pair (α_3) and each of the two subfamilies (α_1, α_2); ASM = $\alpha_3-(\alpha_1+\alpha_2)/2$. It has been shown that slowly evolving proteins are more likely to have a high rate variation among the sites and a smaller α factor (Yang and Kumar, 1996; Zhang and Gu, 1998). Four methods were tested for calculating the α factor: GZ-Gamma (G; Gu and Zhang, 1997), method of moments (M; Tamura and Nei, 1993), Sullivan (S; Sullivan et al., 1995), and Yang (Y; Yang and Kumar, 1996). The ASM based on the GZ-Gamma method (ASM_G) outperforms other methods over the whole sensitivity/specificity range and is generally the most accurate.

RSS and CSS can be used to pinpoint the functional residuals responsible for functional divergence, similar to the application of type I and II sites. In addition, the RSM, CSM, and ASM measures can be used to predict functional divergence between subfamilies in a pair. These measures were put to the test for the Same_EC and Diff_EC subfamily pairs. A significant increase was observed in CSM and RSM in the Diff_EC subfamily pairs compared with the Same_EC subfamily pairs. The discrimination between Same_EC and Diff_EC pairs was more obvious from the RSM curve than the CSM curve (noted as pRSS and pCSS in the initial report). The discrimination was also found to be more pronounced for large families and single-domain families. In the most recent report (Abhiman et al., 2006), the ASM (α factor calculated by the GZ-Gamma method) was shown to outperform CSM and RSM in sensitivity (it was only slightly outperformed by RSM when the false-positive rate was below 0.2). Furthermore, an linear discriminant analysis combination of ASM_G, CSM, and RSM displayed even better sensitivity in discriminating Same_EC and Diff_EC subfamily pairs in a cross-validation test. On the other hand, even the best prediction measure in these studies still showed a substantial rate of false positives. For example, for a sensitivity of 80%, the false-positive rates were greater than 20% in most of the tests. EC number annotation, family size, and protein-domain composition were considered to be the possible reasons for the lack of a clear separation between Same_EC and Diff_EC in the tests.

In general, these and other studies showed that subfamilies with functional divergence did have detectable differences in the distribution of site-specific evolutionary rates in general. Meanwhile, the combination of large-scale functional annotations such as EC numbers with protein-family functional divergence studies has opened a new direction for functional divergence studies in the genomic era.

11.9 Concluding remarks

With the combination of functional divergence analysis and ancestral sequence inference we are able to trace the evolutionary trends of two types of functional divergence of amino acid residues after gene duplication. These pieces of evolutionary information are useful for making testable hypotheses about functional divergence between protein subfamilies, such as subtypes of G-protein α subunits, which can be verified by further experimentation. Note that the methods developed are usually based on protein sequence data. Bielawski and Yang (2004) developed a maximum-likelihood method for detecting functional divergence at the codon level, which is useful for young duplicate genes.

11.12 Acknowledgement

This work was partially supported by National Institutes of Health grants.

References

Abhiman, S. and Sonnhammer, E.L. (2005a) FunShift: a database of function shift analysis on protein subfamilies. *Nucleic Acids Res.* **33**: D197–D200.

Abhiman, S. and Sonnhammer, E.L. (2005b) Large-scale prediction of function shift in protein families with a focus on enzymatic function. *Proteins* **60**: 758–768.

Abhiman, S., Daub, C.O., and Sonnhammer, E.L. (2006) Prediction of function divergence in protein families using the substitution rate variation parameter alpha. *Mol. Biol. Evol.* **23**: 1406–1413.

Bielawski, J.P. and Yang, Z. (2004) A maximum likelihood method for detecting functional divergence at individual codon sites, with application to gene family evolution. *J. Mol. Evol.* **59**: 121–132.

Dayhoff, M., Schwartz, R., and Orcutt, B. (1978) A model of evolutionary change in proteins, In *Atlas of Protein Sequence and Structure* (Dayhoff, M., ed.), pp. 345–352. National Biomedical Research Foundation, Washington DC.

Edwards, R.J. and Shields, D.C. (2005) BADASP: predicting functional specificity in protein families using ancestral sequences. *Bioinformatics* **21**: 4190–4191.

Eisen, J.A. and Fraser, C.M. (2003) Phylogenomics: intersection of evolution and genomics. *Science* **300**: 1706–1707.

Force, A., Lynch, M., Pickett, F.B., Amores, A., Yan, Y.L., and Postlethwait, J. (1999) Preservation of duplicate genes by complementary, degenerative mutations. *Genetics* **151**: 1531–1545.

Gao, X., Vander Velden, K.A., Voytas, D.F., and Gu, X. (2005) SplitTester: software to identify domains responsible for functional divergence in protein family. *BMC Bioinformatics* **6**: 137.

Gaucher, E.A., Gu, X., Miyamoto, M.M., and Benner, S.A. (2002) Predicting functional divergence in protein evolution by site-specific rate shifts. *Trends Biochem. Sci.* **27**: 315–321.

Golding, G.B. and Dean, A.M. (1998) The structural basis of molecular adaptation. *Mol. Biol. Evol.* **15**: 355–369.

Gribaldo, S., Casane, D., Lopez, P., and Philippe, H. (2003) Functional divergence prediction from evolutionary analysis: a case study of vertebrate hemoglobin. *Mol. Biol. Evol.* **20**: 1754–1759.

Gu, J., Wang, Y., and Gu, X. (2002) Evolutionary analysis for functional divergence of Jak protein kinase domains and tissue-specific genes. *J. Mol. Evol.* **54**: 725–733.

Gu, X. (1999) Statistical methods for testing functional divergence after gene duplication. *Mol. Biol. Evol.* **16**: 1664–1674.

Gu, X. (2001) Maximum-likelihood approach for gene family evolution under functional divergence. *Mol. Biol. Evol.* **18**: 453–464.

Gu, X. (2003) Functional divergence in protein (family) sequence evolution. *Genetica* **118**: 133–141.

Gu, X. (2006) A simple statistical method for estimating type-II (cluster-specific) functional divergence of protein sequences. *Mol. Biol. Evol.*, in press.

Gu, X. and Zhang, J. (1997) A simple method for estimating the parameter of substitution rate variation among sites. *Mol. Biol. Evol.* **14**: 1106–1113.

Gu, X. and Vander Velden, K. (2002) DIVERGE: phylogeny-based analysis for functional-structural divergence of a protein family. *Bioinformatics* **18**: 500–501.

Jones, D.T., Taylor, W.R., and Thornton, J.M. (1992) The rapid generation of mutation data matrices from protein sequences. *Comput. Appl. Biosci.* **8**: 275–282.

Jordan, I.K., Bishop, G.R., and Gonzalez, D.S. (2001) Sequence and structural aspects of functional diversification in class I alpha-mannosidase evolution. *Bioinformatics* **17**: 965–976.

Kalinina, O.V., Novichkov, P.S., Mironov, A.A., Gelfand, M.S., and Rakhmaninova, A.B. (2004) SDPpred: a tool for prediction of amino acid residues that determine differences in functional specificity of homologous proteins. *Nucleic Acids Res.* **32**: W424–W428.

Kaziro, Y., Itoh, H., Kozasa, T., Nakafuku, M., and Satoh, T. (1991) Structure and function of signal-transducing GTP-binding proteins. *Annu. Rev. Biochem.* **60**: 349–400.

Kimura, M. (1983) *The Neutral Theory of Molecular Evolution*. Cambridge University Press, Cambridge.

Knudsen, B. and Miyamoto, M.M. (2001) A likelihood ratio test for evolutionary rate shifts and functional divergence among proteins. *Proc. Natl. Acad. Sci. USA* **98**: 14512–14517.

Landgraf, R., Fischer, D., and Eisenberg, D. (1999) Analysis of heregulin symmetry by weighted evolutionary tracing. *Protein Eng.* **12**: 943–951.

Landgraf, R., Xenarios, I., and Eisenberg, D. (2001) Three-dimensional cluster analysis identifies interfaces and functional residue clusters in proteins. *J. Mol. Biol.* **307**: 1487–1502.

Lichtarge, O., Bourne, H.R., and Cohen, F.E. (1996) An evolutionary trace method defines binding surfaces common to protein families. *J. Mol. Biol.* **257**: 342–358.

Lopez, P., Forterre, P., and Philippe, H. (1999) The root of the tree of life in the light of the covarion model. *J. Mol. Evol.* **49**: 496–508.

Madabushi, S., Gross, A.K., Philippi, A., Meng, E.C., Wensel, T.G., and Lichtarge, O. (2004) Evolutionary trace of G protein-coupled receptors reveals clusters of residues that determine global and class-specific functions. *J. Biol. Chem.* **279**: 8126–8132.

Morgan, D.H., Kristensen, D.M., Mittelman, D., and Lichtarge, O. (2006) ET Viewer: an application for predicting and visualizing functional sites in protein structures. *Bioinformatics*, in press.

Neer, E.J. (1995) Heterotrimeric G proteins: organizers of transmembrane signals. *Cell* **80**: 249–257.

Ohno, S. (1970) *Evolution by Gene Duplication*. Springer Verlag, Berlin.

Rastogi, S. and Liberles, D.A. (2005) Subfunctionalization of duplicated genes as a transition state to neofunctionalization. *BMC Evol. Biol.* **5**: 28.

Saitou, N. and Nei, M. (1987) The neighbor-joining method: a new method for reconstructing phylogenetic trees. *Mol. Biol. Evol.* **4**: 406–425.

Simon, M.I., Strathmann, M.P., and Gautam, N. (1991) Diversity of G proteins in signal transduction. *Science* **252**: 802–808.

Sjolander, K. (1998) Phylogenetic inference in protein superfamilies: analysis of SH2 domains. *Proc. Int. Conf. Intell. Syst. Mol. Biol.* **6**: 165–174.

Sullivan, J., Holsinger, K.E., and Simon, C. (1995) Among-site rate variation and phylogenetic analysis of 12S rRNA in sigmodontine rodents. *Mol. Biol. Evol.* **12**: 988–1001.

Tamura, K. and Nei, M. (1993) Estimation of the number of nucleotide substitutions in the control region of mitochondrial DNA in humans and chimpanzees. *Mol. Biol. Evol.* **10**: 512–526.

Wang, Y. and Gu, X. (2001) Functional divergence in the caspase gene family and altered functional constraints: statistical analysis and prediction. *Genetics* **158**: 1311–1320.

Yang, Z. and Kumar, S. (1996) Approximate methods for estimating the pattern of nucleotide substitution and the variation of substitution rates among sites. *Mol. Biol. Evol.* **13**: 650–659.

Yang, Z., Kumar, S., and Nei, M. (1995) A new method of inference of ancestral nucleotide and amino acid sequences. *Genetics* **141**: 1641–1650.

Zhang, J. and Nei, M. (1997) Accuracies of ancestral amino acid sequences inferred by the parsimony, likelihood, and distance methods. *J. Mol. Evol.* **44** (suppl. 1): S139–S146.

Zhang, J. and Gu, X. (1998) Correlation between the substitution rate and rate variation among sites in protein evolution. *Genetics* **149**: 1615–1625.

CHAPTER 12

Reconstruction of ancestral proteomes

Toni Gabaldón and Martijn A. Huynen

12.1 Introduction

The advent of molecular genetics has brought a new dimension to the field of evolutionary biology. Not only did the access to gene and protein sequences from various organisms facilitate the assessment of their evolutionary relationships but it also paved the way for the inference of ancestral properties of the compared molecules. Recent advances in the use of extant sequences to infer past states include the development of computational methods that specifically reconstruct ancestral DNA and protein sequences (Pupko et al., 2002). The use of such methods, reviewed extensively in other chapters in this volume, has been extremely useful in elucidating how certain molecular functions came about. In some cases, the deduced properties of ancestral molecules have been used to infer complex phenotypes of extinct organisms, such as the night vision of early dinosaurs (Chang et al., 2002) or the thermophilic lifestyle of ancestral bacteria (Gaucher et al., 2003). In most cases, however, a single molecule does not provide sufficient information to test evolutionary scenarios that involve complex properties.

A different view of the biology of ancestral organisms can be obtained from the reconstruction of the protein repertoire encoded in their genomes, a technique we call ancestral proteome reconstruction. Such an approach is now feasible thanks to the availability of a growing number of completely sequenced genomes stored in databases. As compared to ancestral sequence reconstruction, ancestral proteome reconstruction provides a broader, less-detailed, view about the functional properties of an extinct organism. Both views are complementary, providing information at different levels: that of the molecular properties of a given protein encoded in a gene and that of the functions and pathways encoded in a genome. Although ancestral proteome reconstruction is still in its infancy, its use has already provided important insights into very ancient events, such as the properties of the so-called last universal common ancestor (or LUCA) or the origin and evolution of certain eukaryotic organelles such as mitochondria and peroxisomes. Here we provide an overview of the current approaches for ancestral proteome reconstruction and illustrate their use with some examples.

12.2 What a proteome can tell

Proteins do not work as isolated entities, instead they perform their function through interactions with other proteins as part of a pathway or a complex and, ultimately, as part of a proteome. The term proteome was first used to describe the total set of proteins encoded by a genome (Wasinger et al., 1995). It can also refer to the subset of proteins that belongs to a specific subcellular compartment—for example, the mitochondrial proteome—or that is expressed in a particular cell type under a given set of environmental conditions. In principle, the description of a proteome can include information on splice variants and post-translational modifications of proteins. However, for the purpose of this chapter we will not consider such modifications and a simple list of protein-coding genes will be used to describe a given proteome. This operational definition allows us to directly infer whole proteomes from genome sequence data, something that is nowadays done automatically after completion of a whole-genome sequencing project (Pruess et al., 2005).

The proteome constitutes in itself a first-level approximation of an organism's biology. For instance, the cellular metabolism of a given species can be reconstructed by mapping the biochemical functions encoded in its proteome on to known metabolic pathways. This provides information on what metabolites will be needed or rendered by that organism, which in turn can be used to infer its lifestyle and even some characteristics of the surrounding environment. Similarly, the presence of certain signaling cascades, secreted proteins or specific transporters might give clues on how the cells interact with their environment. This type of functional inference is currently exploited to study the biology of organisms that are poorly characterized experimentally but for which the complete genome sequence is available. Considering the increasing speed and ease at which new genomes are sequenced, many from poorly understood species, it is expected that this way of characterizing an organism's biology will only gain importance over the years. This is especially true for metagenomics projects targeting organisms that cannot be grown in synthetic cultures (Tringe and Rubin, 2005). Recent examples of how functional inference through predicted proteome analyses has helped to characterize poorly understood organisms include the analysis of deficiencies in amino acid metabolism of the obligate pathogen *Tropheryma whipplei* (Renesto et al., 2003) and the inference of the metabolic properties of several bacteria from environmental acid-mine biofilms (Tyson et al., 2004).

In this context extinct organisms can be regarded as a similar case: we cannot characterize them experimentally but, if we could have information on the proteins coded by their genomes, the same logic could in principle be applied for functional inference, as described above. Ideally one would like to sequence completely the genome of the extinct organism of interest. Sequencing of the DNA of extinct species has now become feasible thanks to the combined use of current high-throughput sequencing techniques and metagenomic approaches. For instance, such procedures have been used in the analyses of DNA extracted from 40 000-year-old bones of *Ursus spelaeus*, an extinct species of cave bear that lived in Europe in the Late Pleistocene era (Noonan et al., 2005), and of DNA samples from a mammoth preserved for more than 27 000 years in the Siberian permafrost (Poinar et al., 2005). Although these recent developments in paleogenomics are promising, their use is limited to relatively young specimens (less than 50 000 years old) that have been preserved under exceptional circumstances which attenuate DNA deterioration. Most ancient genomes have likely disappeared forever and therefore other means are needed to gain insight on them. The next section describes current computational approaches for reconstructing ancestral proteomes in terms of their protein content.

12.3 Computational approaches for the reconstruction of ancestral proteomes

12.3.1 Comparing proteomes

Methodologically, ancestral proteome reconstruction resembles ancestral sequence reconstruction in that it constitutes an inference of ancestral character states. This inference is based on a sample of present states, the modern proteomes, and a given evolutionary framework, a correct phylogeny of the species considered and a model of evolution of the characters. In ancestral proteome reconstruction the characters of interest are the different proteins, which can display two different states: present or absent.

A crucial step in this process is the assignment of equivalences among the proteins in the different proteomes. Similar to a multiple protein alignment that aims to place homologous residues on top of each other, a multiple proteome comparison relies on establishing correct equivalences among the proteins present in the different genomes. Assessing which protein in a proteome is equivalent to one in another proteome is far from trivial. The starting point is the identification of homologous proteins through sequence search algorithms such as Smith–Waterman (Smith and Waterman, 1981) or psi-BLAST (Altschul *et al.*, 1997). However, this level is usually insufficient to define correct equivalences among proteomes because it cannot account for processes such as gene loss or gene duplication. The use of orthologous, rather than just homologous, relationships seems to be the most appropriate way of establishing correspondences among proteins in a multiple proteome comparison. Orthologs are defined as homologous proteins derived by speciation, in contrast to paralogs, which are homologous proteins derived by duplication (Fitch, 1970).

There are several methods to establish orthologous relationships among proteins of different proteomes (Figure 12.1). The simplest operational definition for orthology when comparing two species is the so-called best bidirectional hit (BBH; Huynen and Bork, 1998), which considers as orthologs every pair of proteins from different proteomes which are reciprocally the best hit of each other in a sequence search. This naïve definition of orthology works relatively well when comparing two organisms at close phylogenetic distance. At larger distances, however, the situation becomes more complicated. If gene duplications have occurred in each of the given two lineages subsequent to their divergence, only a many-to-many relationship will properly describe orthologous relationships and, accordingly, detection of the highest similarity will not identify the complete set of orthologs. To account for these issues and to cope with multiple genome comparisons, Tatusov and colleagues (1997) developed the notion of clusters of orthologous groups (COG$_s$). These clusters are delineated from the search of consistent patterns from within all possible pairwise comparisons of the proteomes considered.

Nevertheless, none of the above-mentioned methods use the original definition of orthology, which is a phylogenetic one. Indeed, to correctly assign orthologous and paralogous relationships we must be able to identify gene-duplication and speciation events, something that can only be determined through phylogenetic analysis. This classical way of inferring orthology is generally avoided in large-scale studies because the processes of multiple sequence alignment, tree reconstruction, and tree analysis are time-consuming and require manual analysis. There are promising steps, however, towards the automatic implementation of phylogeny-based orthology assignment on a large scale (Storm and Sonnhammer, 2002; Arvestad *et al.*, 2003; Gabaldón and Huynen, 2005).

12.3.2 Inferring protein content at ancestral nodes

Given a group of extant proteomes and a phylogenetic tree representing their evolutionary relationship we can use a maximum-parsimony approach to reconstruct the ancestral proteomes at

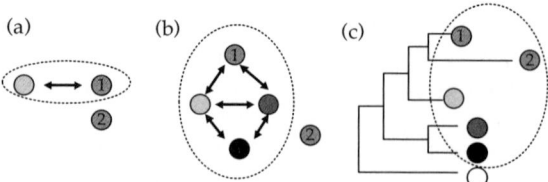

Figure 12.1 Methods for orthology assignment. Circles of different shades indicate proteins from different proteomes. Arrows represent reciprocal best hits. Proteins within dashed ovals belong to the same orthologous group according to the corresponding methodology. (a) Best bidirectional hit (BBH). All pairs of proteins with reciprocal best hits are considered orthologs. (b) Cluster of orthologous groups (COG)-like approach: proteins in the nodes of triangular networks of BBHs are considered as orthologs. New proteins are added to the orthologous group if they are present in BBH triangles that share an edge with a given cluster; for example, the black protein in the example will be added to the orthologous group because it forms a BBH triangle with two above it. Note that a BBH link with the first protein (1) is not required. A COG-like approach may add additional proteins from the same genome if they are more similar to each other than to proteins in other genomes or they form BBH triangles with members of the cluster. In the example this is not the case for the protein 2. (c) Phylogenetic approach: all proteins that derive from a common ancestor by speciation are considered members of the same orthologous group. Note that a one-to-many orthology relationship emerges because of a recent duplication in lineage leading to the yellow proteome. In the example, only phylogenetic reconstruction includes both yellow genes in the cluster. In the example, this orthology relationship was missed by the COG-like approach because of an extensive sequence divergence (long branch in the tree) that occurred after the duplication event.

every internal node of the tree. Such an approach consists of choosing, among all possible scenarios, the one that implies the minimal number of character changes on the given tree. In the simplest approximation, when horizontal gene transfer (HGT) is not considered, only two processes can alter the states of the characters: gene loss and gene genesis. In this scenario a given orthologous group is assumed to have originated (gene genesis) in the last common ancestor of all species represented in that group. Since only vertical descent is assumed, this protein family has necessarily been present in all internal nodes connecting that ancestor to the modern species representing that family. Finally, gene-loss events are assumed to have occurred at the branches leading to the clades that completely lack that orthologous group (Figure 12.2).

Although the evolution of most protein families can be well explained by a non-HGT model, this assumption may lead to reconstructions that involve unrealistic high numbers of gene losses in cases where patchy distribution patterns occur over large phylogenetic distances. A more likely scenario allows for the use of some HGT events to explain such distribution patterns. The central question here is where to set the limit where a single HGT event is preferred over multiple gene losses. To address this issue, Snel and co-workers (2002) introduced the variable HGT penalty, which is defined as the number of additional gene-loss events that are necessary to explain a given distribution if HGT is not included in the model. By varying this penalty they could differentiate between gene distributions that are to varying degrees best explained by HGT rather than parallel gene loss. In a later study, Kunin and Ozounis (2003a) investigated the optimal HGT penalty in prokaryotes by exploiting two constraints: first that the expected ratio between HGT and gene loss should correspond to the observed ratio of these events and second that the genome size in prokaryotes remains constant and thus processes of gain and loss should compensate each other. Interestingly both constraints produced similar optimal HGT penalties at about two to three gene losses.

Gene-Trace (Kunin and Ouzounis, 2003b), a more recent algorithm, implements an additional penalty value, the so-called loss threshold, for assuming an absence on a parental node if that protein is indeed present in a given number of descendants. If the difference between the number of potential gains and losses is smaller than this loss threshold, family absence is assigned to the node. Here, the use of two thresholds can lead to the situation in which some nodes are unassigned. These cases are filled in during a second round of the algorithm that, starting from the root of the tree, provides the parental assignment to the unassigned nodes.

12.3.3 Phylogenomic approaches for ancestral proteome reconstruction

An alternative approach to reconstruct ancestral proteomes exploits the large-scale analysis of sequence-based phylogenies. This approach is conceptually different from the maximum-parsimony approach in that it actually investigates the phylogeny, rather than just the phylogenetic distribution, of every single protein family to ascertain whether it was present or absent at a given ancestral node. The assumption behind this methodology is that the phylogeny of proteins that are vertically derived from that node would be similar to the species phylogeny. In such cases, when a given set of proteins from different species is monophyletic and the tree topology is consistent with a vertical descent from a common ancestral sequence, the presence of that protein family in the last common ancestor of the species considered can be assumed. This methodology was applied for the first time in the reconstruction of the ancestral protomitochondrial proteome, a study that involved the automatic reconstruction and analysis of more than 20 000 trees (see below and Gabaldón and Huynen (2003)). The advantage of the phylogenomic approach for the reconstruction of ancestral proteome is that orthologous relationships are readily obtained from the phylogenetic analysis, facilitating subsequent treatment of the data, for example for finding similar evolutionary patterns (Gabaldón and Huynen, 2005). The drawback of this methodology, besides requiring large computational resources, is that some evolutionary relationships might be missed due to weak phylogenetic signals. This is especially true for very ancient proteomes.

12.4 Practical examples of ancestral proteome reconstruction

12.4.1 The last universal common ancestor

One of the first ancestral organisms to be studied with the help of comparative genomics was the

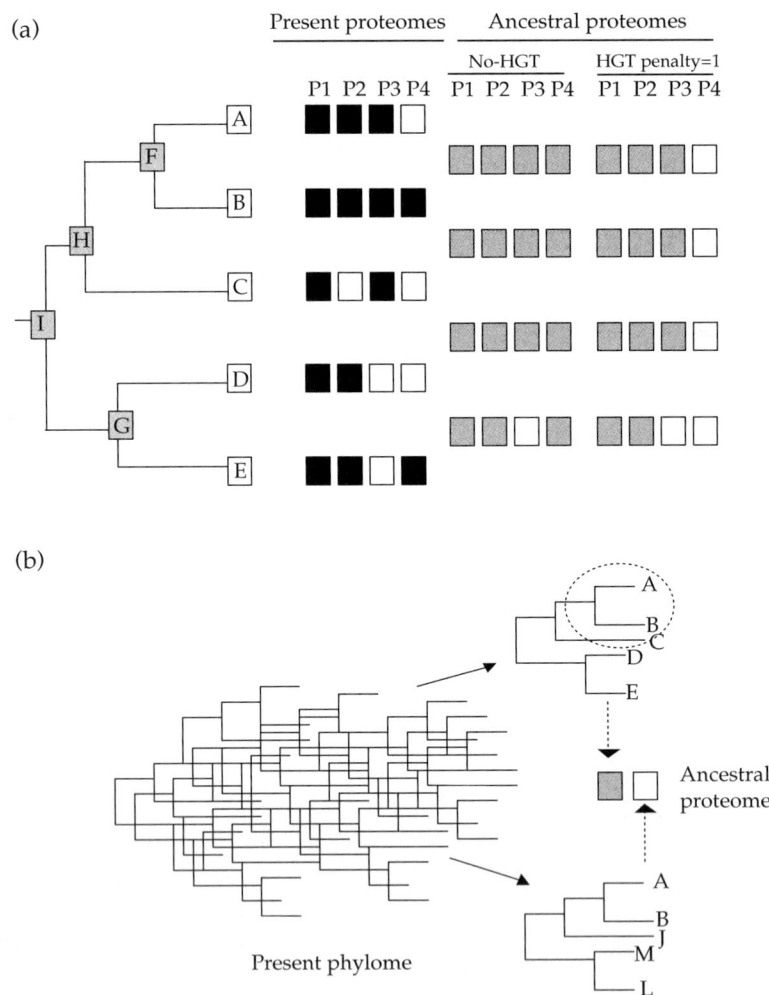

Figure 12.2 Schematic representation of main approaches for ancestral proteome reconstruction. (a) Maximum-parsimony approach: a phylogenetic tree (left) represents the evolutionary relationships between the extant and ancestral proteomes (white and gray squares on the tree, respectively). The presence (black squares) or absence (white squares) of four proteins (P1–P4) in the present proteomes is indicated in the first column next to the tree. Ancestral proteomes can be then reconstructed by assuming a non-HGT scenario (middle column), or a scenario with a given HGT penalty (right-hand column). Resulting inference of presence and absence of the proteins under consideration is indicated by gray and white squares, respectively. (b) Phylogenomic approach: the complete set of phylogenetic trees (phylome) of the protein families under investigation is scanned to detect a certain topology. In this case the proteome of the common ancestor of species A, B, and C is reconstructed. Protein families whose phylogeny is consistent with a vertical inheritance from that ancestor (tree at the top) are included in the reconstructed proteome (gray square), whereas those whose topology indicates a different origin (tree at the bottom) are regarded as absent in the reconstructed proteome (white square).

first cellular lifeform, the so-called last universal common ancestor (Mushegian and Koonin, 1996; Kyrpides et al., 1999). The reconstruction of this ancestral proteome involved the comparison of fully sequenced genomes in terms of their content of protein-coding genes and a parsimonious reconstruction of the ancestral protein repertoire. Recent estimates that correct for HGTs and non-orthologous gene displacements suggest a simple last universal common ancestor with only 500–600 proteins (Koonin, 2003). Although that number of proteins might seem very small, it represents substantial complexity if we consider that the minimal proteomic set to sustain cellular life in a rich medium could comprise as few as 206 proteins (Gil et al., 2004). According to this reconstruction, the last common universal ancestor possessed a large variety of protein domains, suggesting a very ancient origin for most individual domains. In contrast, most domain combinations are kingdom- or lineage-specific and have, therefore, appeared in later stages during evolution. An already complex proteome in the last common universal ancestor is also supported by another type of analysis in which the history of architectural diversification was reconstructed (Caetano-Anolles and Caetano-Anolles, 2005). According to this work all gobular proteins appeared very early in evolution and in a defined order: first the α/β class that contains interspersed helices and β-sheets. The α-helices and β-sheets were then segregated within the structure of proteins ($\alpha + \beta$ class) and subsequently confined to separate molecules (all-α and all-β classes).

12.4.2 Reconstruction of the mitochondrial ancestor's proteome

Mitochondria are eukaryotic organelles that result from the endosymbiosis of an α-proteobacterium (Gray et al., 1999). Several hypotheses have been proposed that explain the initial endosymbiosis in terms of different metabolic properties of the host and the endosymbiont (Figure 12.3). First, the serial endosymbiotic theory, as proposed by Margulis (1981), argued that the initial rationale for this endosymbiosis was based on the mutually beneficial exchange of ATP and glycolysis end products between the host and the endosymbiont. However, this idea was questioned when the analyses of the phylogenetic distribution of the mitochondrial ADP/ATP exchanger revealed a eukaryotic origin for this transporter. More recently, two alternative scenarios for the origin of mitochondria have been proposed, one in which the protomitochondrion would have been an oxygen scavenger (Kurland and Andersson, 2000) and another in which it would have been a hydrogen-producing, facultatively anaerobic organism (Martin and Müller, 1998).

To gain insight into the metabolic capacities of the mitochondrial ancestor we used a phylogenomic approach to reconstruct its proteome (Gabaldón and Huynen, 2003). First, we compared six α-proteobacterial genomes against a set of 74 fully sequenced genomes, including those of nine eukaryotes. Subsequently the complete set of phylogenetic trees, the so-called phylome (Sicheritz-Ponten and Andersson, 2001), for every α-proteobacerial genome was reconstructed. These phylomes were scanned by an algorithm that selected trees whose topology indicated a vertical descent of the eukaryotic protein from an α-proteobacterial ancestor. A total of 630 orthologous groups displayed such pattern. This set represents a minimal estimate of the protomitochondrial proteome, since many proteins were probably lost from the eukaryotes during the early evolution of mitochondria while others would have been missed because of weak phylogenetic signals.

By mapping the functions encoded in the reconstructed proteome on to metabolic maps, the protomitochondrial metabolism can be reconstructed. The emerging picture is that of an aerobic endosymbiont catabolizing compounds provided by the host. Although the presence of many pathways from the aerobic metabolism would seem to support the oxygen-scavenger hypothesis, some caution must be taken. First, although one anaerobic species was included (*Encephalotozoon cuniculi*) no hydrogenosomal eukaryote was included in the analyses and therefore the absence of pathways suggesting a hydrogen-producing endosymbiont does not rule out a hydrogenosomal origin for the mitochondria. Secondly, the conservation of a diverse set of protomitochondrial

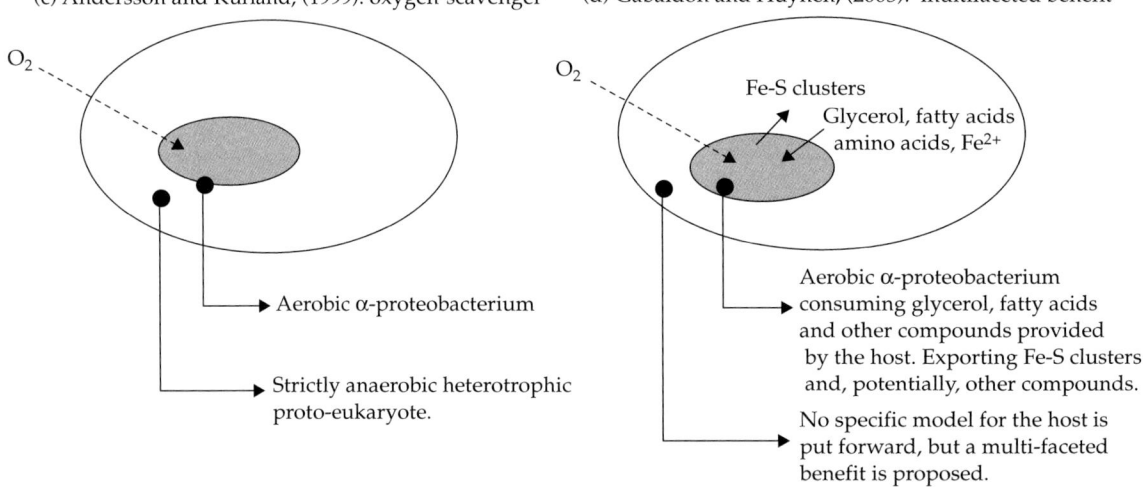

Figure 12.3 Alternative metabolic scenarios for the initial host–endosymbiont relationships during mitochondrial endosymbiosis as proposed by several authors. Big and small ovals represent the host and the protomitochondrion, respectively. Small arrows indicate exchange of metabolites between the host and the endosymbiont; the ability of the protomitochondrion to consume oxygen from the environment is indicated by a dashed arrow. Deduced properties of the host and the protomitochondrion are indicated below each figure. (a) In the initial model for the endosymbiotic origin of mitochondria the production of ATP by the endosymbiont was considered to be rationale for this association. (b) Martin and Müller (1998) proposed a symbiosis between a methanogenic archaebacterial host that would have consumed hydrogen produced by a facultatively aerobic endosymbiont. (c) Andersson and Kurland (1999) proposed a model in which the consumption of oxygen by the protomitochondrion would have benefited a strictly anaerobic host, for which the presence of this molecule in the environment was toxic. (d) The protomitochondrial proteome reconstruction shown in this chapter suggests an aerobic endosymbiont consuming a variety of compounds provided by the host, and in return the host could benefit from the production by the protomitochondrion of Fe-S clusters and other products.

pathways, among which the assembly of FeS clusters in modern eukaryotes, suggests a multi-faceted benefit for the host cell of the endosymbiosis, rather than a single crucial pathway. Although, in the absence of a reconstructed proteome for the host, it is difficult to define a specific symbiotic scenario, it seems to have been more complex than previously proposed.

Going further back in evolution, Boussau and colleagues (2004) reconstructed the proteome of the ancestor of all α-proteobacteria, containing 3100 orthologous groups. This early ancestor was probably a free-living aerobic and motile bacterium, with a metabolism that included glycolysis and broad biosynthetic capacity. Since the exact position of the protomitochondrion within the α-proteobacterial phylogeny has not been solved (Esser et al., 2004), it is unclear to what extent the protomitochondrion had already diverged from that ancestral α-proteobacterium. Nevertheless, these 3100 orthologous groups can be considered as an upper limit for the content of the protomitochondrial proteome. The reconstruction of such ancestral proteomes, combined with data from modern mitochondrial proteomes allows the identification of gains and losses of genes associated with specific stages of the evolution of mitochondria (Gabaldón and Huynen, 2004).

12.4.3 Origins of the peroxisomal proteome

Another ubiquituous eukaryotic organelle whose evolution has been investigated through ancestral proteome reconstruction is the peroxisome (de Duve, 1982). These organelles, which together with related organelles like glycosomes are also referred to as microbodies, contain enzymes involved in, among other processes, hydrogen peroxide detoxification and lipid metabolism (Titorenko and Rachubinski, 2001), including β-oxidation of fatty acids, methanol oxidation, and synthesis of cholesterol and bile acids. The microbody family includes plant-seed and leaf glyoxysomes (Nishimura et al., 1996) as well as glycosomes of kinetoplastida (Opperdoes and Michels, 1993) that possess specific metabolic pathways. Despite this diversity the presence of common sets of peroxisome biogenesis proteins (PEX) and enzymes of peroxide metabolism, fatty acid oxidation, and ether lipid biosynthesis (Parsons et al., 2001) suggest a common origin for all types of microbody. Most widespread hypotheses regarding peroxisomal origin are centered around the evolution from a symbiotic ancestral prokaryote (de Duve, 1982; Latruffe and Vamecq, 2000). One such hypothesis states that peroxisomes originated from an endosymbiotic ancestral prokaryote that protected the primitive eukaryotic cell from the rise of atmospheric oxygen levels (Latruffe and Vamecq, 2000). The endosymbiotic origin of peroxisomes has been proposed on the grounds of similarities with mitochondria and plastids such as the ability of peroxisomes to multiply by binary fission (Hoepfner et al., 2001) or the fact that peroxisomal proteins are post-translationally imported into the organelle (Lazarow and Fujiki, 1985). Moreover, the difficulty to explain how several enzymes of entire pathways would end up in the organelle if they were not the remains of an ancestral endosymbiont has also been interpreted as evidence for an endosymbiotic origin of peroxisomes.

Recent discoveries, however, have challenged this view. First, it has been observed that the phenotype of peroxisome-less mutants in the yeast *Saccharomyces cerevisiae* is reversible upon the introduction into the cell of the wild-type gene (Erdmann and Kunau, 1992). Second, accumulating evidence suggests that peroxisomes can be formed in the the endoplasmic reticulum membrane (Tabak et al., 2003). The lack of an organellar genome in peroxisomes precludes phylogenetic analyses of clearly vertically derived genes such as those performed in chloroplast and mitochondrial genomes. Therefore one has to resort to ancestral proteome reconstruction to address the origin and evolution of this organelle, assuming that its proteome has retained a significant evolutionary signal, as observed for yeast and human mitochondria (Karlberg et al., 2000; Gabaldón and Huynen, 2003). To this end Gabaldón et al. (2005) analyzed the phylogenies of a set of proteins with an experimentally determined persoxisomal location in *S. cerevisiae* and the rodent *Rattus norvegicus*.

This analysis revealed that the majority of peroxisomal proteins either have no prokaryotic homolog or that the eukaryotic protein family to which they belong does not originate from within a prokaryotic branch in the tree. Among the protein families with a clear bacterial outgroup, the most significant fraction pointed to an α-proteobacterial origin. At first sight this appears to support the endosymbiotic origin of peroxisomes from an α-proteobacterial ancestor. A closer look, however, suggested that these proteins are rather the result of a secondary re-targeting of proteins from the mitochondrion. Indeed, the reconstruction of several ancestral stages of the peroxisomal proteome revealed that this re-targeting of proteins from the mitochondrion and other compartments to the peroxisome might have been frequent throughout the evolution of the peroxisome. The reconstructed proteome of the last common ancestor of all microbodies already presents one such α-proteobacterium-derived protein, namely the multifunctional fatty acid-oxidation enzyme Fox2. Nevertheless, the ancestral core proteome is enriched in eukaryotic proteins and particularly in those that are homologous to proteins from the endoplasmic reticulum assisted decay system, suggesting the origin of the peroxisome from the endomembrane system, a model that perfectly fits within the last experimental results (Hoepfner et al., 2005).

12.5 Concluding remarks

Although it is a very recently developed methodology, ancestral proteome reconstruction has already proved to be a powerful tool to test hypotheses on extinct organisms and past evolutionary events. In this context, it complements perfectly ancestral sequence reconstruction, providing information about the context in which a given protein was functioning. Reversely, ancestral sequence reconstruction may provide useful information in the process of inferring the functions encoded in an ancestral proteome; note that, in this case, ancestral proteome reconstruction assumes that proteins in the same orthologous groups have the same function whereas ancestral sequence reconstruction can be used to specifically determine that ancestral function. Ideally, when testing an evolutionary scenario, one would combine both techniques to first have a broad view of an ancestral proteome and to then zoom in on the protein sequences that are key for the process under consideration.

The dynamics of proteome evolution may be analyzed by applying ancestral proteome reconstruction over successive evolutionary stages, something that, as we have seen in the case of peroxisomal protein import, can give hints about the function of pathways. The combination of these with other types of data may unveil unexpected processes. For instance, the combination of subcellular proteomics with ancestral proteome reconstruction in two of the above-mentioned studies suggests that relocation of proteins to other cellular compartments can be performed with relative ease on an evolutionary timescale. With more genomes to come and new developments in computational techniques it is expected that ancestral proteome reconstruction will become a central tool in evolutionary biology.

References

Altschul, S.F., Madden, T.L., Schaffer, A.A., Zhang, J., Zhang, Z., Miller, W., and Lipman, D.J. (1997) Gapped BLAST and PSI-BLAST: a new generation of protein database search programs. *Nucleic Acids Res.* **25**: 3389–3402.

Andersson, S.G.E. and Kurland, C.G. (1999) Origins of mitochondria and hydrogenosomes. *Curr. Opin. Microbiol.* **2**: 535–541.

Arvestad, L., Berglund, A.C., Lagergren, J., and Sennblad, B. (2003) Bayesian gene/species tree reconciliation and orthology analysis using MCMC. *Bioinformatics* **19** (suppl. 1): I7–I15.

Boussau, B., Karlberg, E.O., Frank, A.C., Legault, B.A., and Andersson, S.G. (2004) Computational inference of scenarios for alpha-proteobacterial genome evolution. *Proc. Natl. Acad. Sci. USA* **101**: 9722–9727.

Caetano-Anolles, G. and Caetano-Anolles, D. (2005) niversal sharing patterns in proteomes and evolution of protein fold architecture and life. *J. Mol. Evol.* **60**: 484–498.

Chang, B.S., Jonsson, K., Kazmi, M.A., Donoghue, M.J., and Sakmar, T.P. (2002) Recreating a functional ancestral archosaur visual pigment. *Mol. Biol. Evol.* **19**: 1483–1489.

de Duve, C. (1982) Peroxisomes and related particles in historical perspective. *Ann. NY Acad. Sci.* **386**: 1–4.

Erdmann, R. and Kunau, W.H. (1992) A genetic approach to the biogenesis of peroxisomes in the yeast Saccharomyces cerevisiae. *Cell Biochem. Funct.* **10**: 167–174.

Esser, C., Ahmadinejad, N., Wiegand, C., Rotte, C., Sebastiani, F., Gelius-Dietrich, G. et al. (2004) A genome phylogeny for mitochondria among alpha-proteobacteria and a predominantly eubacterial ancestry of yeast nuclear genes. *Mol. Biol. Evol.* **21**: 1643–1660.

Fitch, W.M. (1970) Distinguishing homologous from analogous proteins. *Syst. Zool.* **19**: 99–113.

Gabaldón, T. and Huynen, M.A. (2003) Reconstruction of the proto-mitochondrial metabolism. *Science* **301**: 609.

Gabaldón, T. and Huynen, M.A. (2004) Shaping the mitochondrial proteome. *Biochim. Biophys. Acta* **1659**: 212–220.

Gabaldón, T. and Huynen, M.A. (2005) Lineage-specific gene loss following mitochondrial endosymbiosis and its potential for function prediction in eukaryotes. *Bioinformatics* **21** (suppl. 2): ii144–ii150.

Gabaldón, T., Snel, B., Van Zimmeren, F., Hemrika, W., Henk, T., and Huynen, M.A. (2005) The evolutionary origin of the peroxisome. In *Origin and Evolution of the Mitochondrial Proteome* (Gabaldón, T., ed.). Printpartners Ipskamp, Nijmegen.

Gaucher, E.A., Thomson, J.M., Burgan, M.F., and Benner, S.A. (2003) Inferring the palaeoenvironment of ancient bacteria on the basis of resurrected proteins. *Nature* **425**: 285–288.

Gil, R., Silva, F.J., Pereto, J., and Moya, A. (2004) Determination of the core of a minimal bacterial gene set. *Microbiol. Mol. Biol. Rev.* **68**: 518–537.

Gray, M.W., Burger, G., and Lang, B.F. (1999) Mitochondrial evolution. *Science* **283**: 1476–1481.

Hoepfner, D., Van Den Berg, M., Philippsen, P., Tabak, H.F., and Hettema, E.H. (2001) A role for Vps1p, actin, and the Myo2p motor in peroxisome abundance and inheritance in Saccharomyces cerevisiae. *J. Cell. Biol.* **155**: 979–990.

Hoepfner, D., Schildknegt, D., Braakman, I., Philippsen, P., and Tabak, H.F. (2005) Contribution of the endoplasmic reticulum to peroxisome formation. *Cell* **122**: 85–95.

Huynen, M.A. and Bork, P. (1998) Measuring genome evolution. *Proc. Natl. Acad. Sci. USA* **95**: 5849–5856.

Karlberg, O., Canback, B., Kurland, C.G., and Andersson, S.G. (2000) The dual origin of the yeast mitochondrial proteome. *Yeast* **17**: 170–187.

Koonin, E.V. (2003) Comparative genomics, minimal gene-sets and the last universal common ancestor. *Nat. Rev. Microbiol.* **1**: 127–136.

Kunin, V. and Ouzounis, C.A. (2003a) The balance of driving forces during genome evolution in prokaryotes. *Genome Res.* **13**: 1589–1594.

Kunin, V. and Ouzounis, C.A. (2003b) GeneTRACE-reconstruction of gene content of ancestral species. *Bioinformatics* **19**: 1412–1416.

Kurland, C.G. and Andersson, S.G. (2000) Origin and evolution of the mitochondrial proteome. *Microbiol. Mol. Biol. Rev.* **64**: 786–820.

Kyrpides, N., Overbeek, R., and Ouzounis, C. (1999) Universal protein families and the functional content of the last universal common ancestor. *J. Mol. Evol.* **49**: 413–423.

Latruffe, N. and Vamecq, J. (2000) Evolutionary aspects of peroxisomes as cell organelles, and of genes encoding peroxisomal proteins. *Biol. Cell* **92**: 389–395.

Lazarow, P.B. and Fujiki, Y. (1985) Biogenesis of peroxisomes. *Annu. Rev. Cell Biol.* **1**: 489–530.

Margulis, L. (1981) *Symbioses in Cell Evolution.* W.H. Freeman, San Francisco.

Martin, W. and Müller, M. (1998) The hydrogen hypothesis for the first eukaryote. *Nature* **392**: 37–41.

Mushegian, A.R. and Koonin, E.V. (1996) A minimal gene set for cellular life derived by comparison of complete bacterial genomes. *Proc. Natl. Acad. Sci. USA* **93**: 10268–10273.

Nishimura, M., Hayashi, M., Kato, A., Yamaguchi, K., and Mano, S. (1996) Functional transformation of microbodies in higher plant cells. *Cell Struct. Funct.* **21**: 387–393.

Noonan, J.P., Hofreiter, M., Smith, D., Priest, J.R., Rohland, N., Rabeder, G. et al. (2005) Genomic sequencing of Pleistocene cave bears. *Science* **309**: 597–599.

Opperdoes, F.R. and Michels, P.A. (1993) The glycosomes of the Kinetoplastida. *Biochimie* **75**: 231–234.

Parsons, M., Furuya, T., Pal, S., and Kessler, P. (2001) Biogenesis and function of peroxisomes and glycosomes. *Mol. Biochem. Parasitol.* **115**: 19–28.

Poinar, H.N., Schwarz, C., Qi, J., Shapiro, B., Macphee, R. D., Buigues, B. et al. (2005) Metagenomics to paleogenomics: large-scale sequencing of mammoth DNA. *Science* **311**: 392–394.

Pruess, M., Kersey, P., and Apweiler, R. (2005) The Integr8 project–a resource for genomic and proteomic data. *In Silico Biol.* **5**: 179–185.

Pupko, T., Pe'er, I., Hasegawa, M., Graur, D., and Friedman, N. (2002) A branch-and-bound algorithm for the inference of ancestral amino-acid sequences when the replacement rate varies among sites: application to the evolution of five gene families. *Bioinformatics* **18**: 1116–1123.

Renesto, P., Crapoulet, N., Ogata, H., La Scola, B., Vestris, G., Claverie, J.M., and Raoult, D. (2003) Genome-based design of a cell-free culture medium for *Tropheryma whipplei. Lancet* **362**: 447–449.

Sicheritz-Ponten, T. and Andersson, S.G. (2001) A phylogenomic approach to microbial evolution. *Nucleic Acids Res.* **29**: 545–552.

Smith, T.F. and Waterman, M.S. (1981) Identification of common molecular subsequences. *J. Mol. Biol.* **147**: 195–197.

Snel, B., Bork, P., and Huynen, M.A. (2002) Genomes in flux: the evolution of archaeal and proteobacterial gene content. *Genome Res.* **12**: 17–25.

Storm, C.E. and Sonnhammer, E.L. (2002) Automated ortholog inference from phylogenetic trees and calculation of orthology reliability. *Bioinformatics* **18**: 92–99.

Tabak, H.F., Murk, J.L., Braakman, I., and Geuze, H.J. (2003) Peroxisomes start their life in the endoplasmic reticulum. *Traffic* **4**: 512–518.

Tatusov, R.L., Koonin, E.V., and Lipman, D.J. (1997) A genomic perspective on protein families. *Science* **278**: 631–637.

Titorenko, V.I. and Rachubinski, R.A. (2001) The life cycle of the peroxisome. *Nat. Rev. Mol. Cell. Biol.* **2**: 357–368.

Tringe, S.G. and Rubin, E.M. (2005) Metagenomics: DNA sequencing of environmental samples. *Nat. Rev. Genet.* **6**: 805–814.

Tyson, G.W., Chapman, J., Hugenholtz, P., Allen, E.E., Ram, R.J., Richardson, P.M. *et al.* (2004) Community structure and metabolism through reconstruction of microbial genomes from the environment. *Nature* **428**: 37–43.

Wasinger, V.C., Cordwell, S.J., Cerpapoljak, A. *et al.* (1995) Progress with gene-product mapping of the mollicutes–Mycoplasma genitalium. *Electrophoresis* **16**: 1090–1094.

CHAPTER 13

Computational reconstruction of ancestral genomic regions from evolutionarily conserved gene clusters

Etienne G.J. Danchin, Eric A. Gaucher, and Pierre Pontarotti

13.1 Introduction

Reconstruction of ancestral genomic features can be considered on multiple evolutionary scopes and at different levels of biological sequence information. For instance, one could anticipate the reconstruction of genomic features for the last common ancestor of all species on Earth, last universal common ancestor or LUCA, whereas others would focus on reconstructing these features in the last common ancestor of vertebrates and/or arthropods. In an analogous manner, biological sequences themselves can be divided into subcategories as a function of their nature or their scale. It is possible to consider reconstructing ancestral genes, ancestral proteins, ancestral retro-elements, ancestral chromosomes, or even an ancestral genome. We present here our conceptual and computational approach for reconstructing gene clusters, with a particular emphasis on the major histocompatibility complex (MHC) region. We anticipate that our approach will be extended, and coincide with technological advancements allowing reconstructionists to synthesize ancient genomes in the laboratory.

13.2 Small-scale reconstructions

On the smaller scale, representing individual sequences (i.e. gene, protein, mobile element, etc.), reconstruction of ancestral biological sequences can go beyond the conceptual level and lead to a physical reconstruction of the deduced ancestral sequence. Indeed, several research articles relate physical reconstruction of biological sequences based on phylogenetic reconstructions to ancient organismal behaviors, as reviewed in various chapters in this book.

13.3 Larger-scale reconstructions

Alternatively, larger-scale biological sequence reconstructions are concerned with ancient chromosomes, genomic regions, and genomes. Fewer studies, however, have been presented on this scale (Blanchette et al., 2004). Moreover, they do not go beyond the conceptual level *in silico* because (for the moment) technology does not allow extension towards physical reconstructions. A logical step towards realizing an ancestral genome consists first of inferring the gene content of the ancestral organism.

13.3.1 Ancestral gene content reconstruction

Several authors have recently evaluated the number of genes or proteins most likely present in the ancestors of different animal phyla. Koonin et al. (2004) performed an in-depth comparative analysis of whole proteomes from seven different eukaryotic species. Based on identified clusters, and on a study of the evolution of these species, they inferred the gene set that was probably present in the last common ancestor of the eukaryotes to consist of at least 3413 gene families. In a similar manner, they also evaluated the gene set for each internal node of the phylogeny of these seven species and, for example, they estimated that the last common ancestor of all bilaterian species had at least 5313 gene families. Using a similar approach, Hughes and Friedman (2004) compared complete proteomes of various bilaterian species (insects, vertebrates, and nematodes), and estimated that approximately 2100 protein families were present in the last common ancestor of these taxa (*Urbilateria*).

It is interesting to note here that these two analyses provide very different estimates (more than 2-fold) of the ancestral bilaterian proteome size. This difference can be explained by the fact that the set of species used to define the size of the ancestral proteome was not the same for the two analyses. Moreover, the definition of gene families between the two analyses was slightly different, and also the methods used to deduce ancestral gene content from clusters of conserved genes were not identical.

Both these approaches evaluated clusters of putative orthologous groups of protein families by all-against-all pairwise comparisons of proteins between the different species, but did not systematically test the orthology relationships between these genes by phylogenetic analysis. Sequence similarity-based approaches can misguide in some instances where evolutionary relationships between genes are particularly complex whereas phylogenetic analysis tends to resolve such complex cases (Danchin, 2004; Jordan et al., 2004; Gouret et al., 2005). Nevertheless, as explained by the authors, phylogenetic analysis for genome-wide comparisons can also be erroneous and remains labor-intensive. Even if these two analyses are likely to include false positive and negatives, they represent the most reliable estimations of ancestral gene and protein sets to date.

These studies evaluate the putative gene or protein content in the ancestor of various phyla, at the largest scale possible, through comparative analysis. Although similar analyses have been performed for Bacteria (Kunin and Ouzounis, 2003), we focus here on ancestral eukaryotic genome content.

13.3.2 Reconstruction of ancestral genomic organization

Several methods and analyses have been developed to reconstruct ancestral genome organization. For example, Bourque and Pevzner (2002) developed a method to decipher ancestral gene orders based on the comparison of gene order between modern species. These authors then presented a follow-up reconstruction of the genomic organization of the rodent ancestor from mouse and rat based on comparison of conserved genomic blocks and their relative order (Bourque et al., 2004). This genomic reconstruction included both coding and non-coding chromosomal regions but did not consider genomic regions that had been duplicated. Nor did it give information about the organization of genes inside the genomic blocks. More recently, Bourque et al. (2005) expanded their original method and proposed a reconstruction of the ancestral genome organization of the murid rodent ancestor, and of the mammalian ancestor. This latest analysis provides an opportunity to reconstruct gene content and organization inside the ancestral genomic blocks by considering comparisons at the coding regions level. In parallel, and using a similar approach, Jaillon et al. (2004) proposed a reconstruction of the ancestral karyotype of the vertebrates through comparison between the teleost fish *Tetraodon nigroviridis* and the human genome.

These analyses predicted a putative genomic organization in mammal, rodent, and vertebrate ancestors at the whole-genome scale. However, both of the analyses used reciprocal best-BLAST (Altschul et al., 1997) hit approaches to decipher orthology relationships (known to be problematic) and neither study considered duplicated regions and genes. Due to the limited number of whole genomes available for comparison, these analyses certainly missed genes or regions that were lost multiple times in different lineages, and thus ancestral reconstructions lacked these elements. We surmise that increasing the number of genome comparisons will lead to greater resolution.

13.3.3 Reconstruction of ancestral genomic regions through comparisons of evolutionarily conserved gene clusters

The reconstruction of ancestral biological features achieved in our research group to date is at an intermediate scale between individual sequences (genes, proteins, mobile elements, etc.) and large-scale reconstruction (whole ancestral karyotypes, genomes, or proteomes). We proposed the reconstruction of genomic regions at the level of their ancestral gene content (Danchin et al., 2003; Danchin, 2004; Danchin and Pontarotti, 2004a, 2004b) through the comparison of evolutionarily conserved gene clusters. Thus far, our conceptual reconstructions have not included predictions on the organization of genes (i.e. order and orientation) inside the ancestral regions, but are rather predictions of ancestrally grouped genes irrespective of their relative organization inside the clusters.

Our initial analyses focused on reconstructing regions in the last common ancestor of the euchordates (Danchin and Pontarotti, 2004b; named *Ureuchordata*) and in the last common ancestor of the bilaterians (Danchin et al., 2003; Danchin, 2004; Danchin and Pontarotti, 2004a, 2004b; named *Urbilateria*). The most obvious way to expand these initial analyses of ancestral genomic information content is to compare the genomic *organization* of conserved regions that are suspected to have originated from a common ancestral region.

Reconstruction of ancestral genomic clusters as far back as the last common ancestor of all bilaterian species (*Urbilateria*) has been possible through the comparison of genomic regions whose gene composition was evolutionarily conserved between Protostomes (like *Drosophila melanogaster*) and Deuterostomes (like *Homo sapiens*). Evolutionarily conserved genomic regions were identified between Protostomes and Deuterostomes prior to reconstructing putative ancestral clusters. We first started from selected regions in the human genome for which we had evidence of evolutionary conservation in vertebrates. These selected regions of the human genome consisted of relatively well-conserved paralogous gene clusters that had been shown previously to originate from a common ancestral region after duplication (Abi-Rached et al., 2002; Vienne et al., 2003a). From these clusters, we next retrieved genes that appeared to constitute signatures of evolutionary conservation. These so-called signature genes had to fulfill several criteria, in that they must be present in at least one copy in one of the paralogous regions and the estimation of their duplication date should be in a consistent time window. Orthologs to these anchor genes were then searched for in the genomes of protostomian species (i.e. *Anopheles gambiae*, *Drosophila melanogaster*, and *Caenorhabditis elegans*) by a systematic phylogenetic analysis. We retrieved genomic locations of each protostomian gene having a human ortholog. For each protostomian genomic segment containing at least two orthologs and spanning less than 2 Mb, a statistical test was applied. The appropriate statistical test allows us to distinguish significant conservation from conservation by chance.

13.4 Choice of candidate regions

Our previous analyses of bilaterian ancestral genomic reconstructions relied on ancient duplicated clusters that today have remained structurally conserved. These clusters resulted from two rounds of duplication from a unique ancestral region after the divergence between cephalochordates (amphioxus, *Branchiostoma floridae*) and craniates (hagfishes plus vertebrates), and before the emergence of gnathostomata (jawed vertebrates). These paralogous regions retained significant conservation of gene content despite hundreds of millions of years of divergence from their common ancestral state.

The two sets of quadruplicated regions studied were the MHC and its paralogous regions, and the 8–10–4–5 regions. For both sets, data suggested the existence of an ancestral region (at least early in chordate history) from which they originate, and derived after *en bloc* duplications. Indeed, conservation of gene clustering can still be observed between the paralogous regions inside a given quadruplicated set (Abi-Rached et al., 2002; Vienne et al., 2003a). As a consequence, the two sets of four paralogous regions we observe today in vertebrate genomes may represent echoes of a conserved common ancestral cluster. In our objective towards reconstructing ancestral regions, our preliminary observations placed these quadruplicated regions as obvious candidates to look for further conservation in other species within the tree of life.

We hypothesized that these two sets of quadruplicated regions in vertebrates (Deuterostomes) may have diverged from a more ancient genomic cluster, possibly as distant as Protostomes. The remainder of this chapter will focus on the MHC and its three paralogous regions, since the strategy and approach used for the 8-10-4-5 regions are analogous.

13.4.1 The MHC and its paralogous regions

The MHC region is located in the human genome on chromosome 6p21.3. This genomic region of approximately 2 Mb contains genes that are involved in the immune response. For instance, PSMB8 and PSMB9 encode two subunits of the immunoproteasome (a multimeric complex which cleaves peptides to a specific size for presentation at the cell surface), and C4 encodes a subunit of the complement system (a 30-protein system involved in immunological response, anaphylaxis, and cell destruction). Other genes with no clear reported role in immunity are also present in this region. For example, retenoid X receptor (RXR) B is a co-activator that increases the DNA-binding activity of retinoic acid receptors (RARs) whereas PBX2 encodes a protein with a homeobox domain but whose function is not well documented.

Three other regions of the human genome (chromosomes 1p22–p11, 9q33–q34, and 19p13) contain clustered copies (paralogs) of some of the genes present in the MHC region. This observation was initially made by Kasahara et al. (1996, 1997), who defined three MHC-like regions in the human genome in addition to the original MHC region on chromosome 6p21.3. These three MHC-like regions have been predicted by Abi-Rached et al. (2002) to have been the result of two rounds of *en bloc* duplication from an ancestral region. A schematic representation of the MHC region as well as its three paralogous conserved regions is presented in Figure 13.1. These four paralogous clusters arose through duplication from their common ancestral region around 700 million years ago (Abi-Rached et al., 2002). During millions of years of evolution these regions may have undergone fixation of several rearrangements. Among these rearrangements, gene loss and translocations could be invoked to explain why not all members of quadruplicated genes are still present as four copies in the quadruplicated regions. For example, in the RXR family, one paralogous copy is found on each of chromosomes 6, 1, and 9 (respectively RXRB, RXRG, and RXRA) but no paralogous copy is present within the fourth region (on chromosome 19). The same type of loss pattern is also found for other genes not listed here. In some cases, losses can be more extended and leave only two remaining copies (as for AGPAT family; 1-acyl-glycerol-3-phosphate O-acyltransferases 1 and 2). Note that at this stage it is difficult to state whether singleton genes are the remains of quadruplicated genes that experienced multiple losses, or whether they represent a single-copy gene translocated into these regions after the *en bloc* duplications and subsequent divergence from the common ancestral region. An important point that must be specified is that the relative order of genes along the four regions of paralogy is not conserved between the MHC and any of its three paralogous regions. Thus, the only feature that characterizes these regions is a common clustering of paralogous genes regardless of their relative order.

13.5 Conservation in other species

Anchor genes representing signatures from the two sets of vertebrate quadruplicated regions (as defined above) were used to identify potentially conserved clusters in other species. The species that have been tested for conservation were chosen according to the following criteria: their genomes are completely sequenced, assembled, and annotated to allow retrieval of gene locations along the genome. The selected species were *Drosophila melanogaster*, *Anopheles gambiae* (two dipteran insects), and *Caenorhabditis elegans* (a nematode). These three species are all bilaterian species belonging to the protostomian group. Moreover, while still debated today for nematodes (Blair et al., 2002; Copley et al., 2004; Telford, 2004a, 2004b;

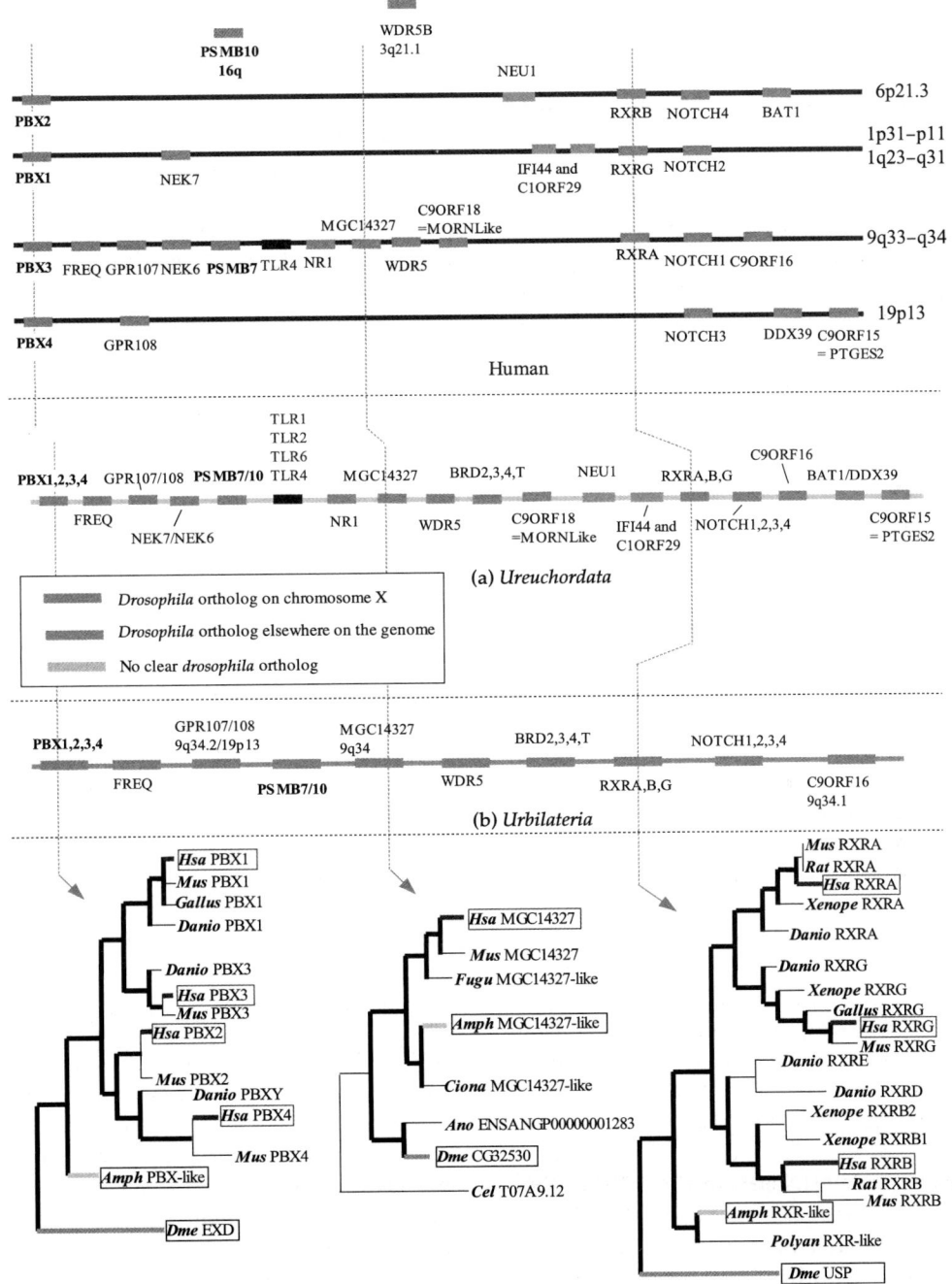

Figure 13.1 *Ureuchordata* and *Urbilateria* proto-MHC reconstructions. Top panel: distribution of the 18 conserved gene families between human and amphioxus on the human MHC and paralogous regions. (a) Minimal reconstruction of the putative ancestral region in *Ureuchordata*. (b) Reconstruction of a minimal region in *Urbilateria* based on conserved clustering in *Drosophila*. Bottom panel: three examples of phylogenetic trees for three gene families presenting different patterns of gene presence or absence. Note: the actual organization (i.e. order and orientation) of genes on the reconstructed ancestral regions is not known and probably rearranged; we chose to represent homologous genes in the same order on the various different regions so that their homology relationships are easier to read. *Hsa, H. sapiens; Amph*, amphioxus (*Branchiostoma floridae*); *Dme, D. melanogaster, Ano, Anopheles gambiae; Cel, C. elegans; Xenope, Xenopus laevis; Polyan, Polyandrocarpa misakiensis; Mus, Mus musculus; Danio, Danio rerio; Fugu, Takifugu rubripes; Gallus, Gallus gallus; Rat, Rattus norvegicus; Ciona, Ciona intestinalis.*

Wolf et al., 2004) they may all be in the same group of protostomians, the ecdysozoans. Additionally, we also used partial genomic information available for other species such as one cephalochordate (*Branchiostoma floridae*) and two urochordates (*Ciona intestinalis* and *Ciona savignyi*) in the case of the MHC and its paralogous regions.

The MHC region and its three paralogous regions in human are widely conserved in other primates (chimpanzee), mammals (mouse, cattle, and rat), and vertebrates (fugu and zebrafish; Flajnik and Kasahara, 2001). Outside the vertebrate lineage, conservation has been shown with parts of the cephalochordate amphioxus genome (Abi-Rached et al., 2002).

In addition, conservation has been reported (Trachtulec et al., 1997; Trachtulec and Forejt, 2001) with *D. melanogaster* chromosome X, *C. elegans* chromosome III, and parts of snake and *Schizosaccharomyces pombe* genomes. These latest instances of conservation were not confirmed by a phylogenetic analysis assessing orthologous relationships between genes of the various species considered. Moreover, the statistical test performed for these analyses did not consider heterogeneity of gene distribution along the genomes. Therefore, these examples needed to be confirmed and completed with new genomic data available. We present here a summary of our analyses (Danchin et al., 2003; Danchin and Pontarotti, 2004b). We confirm some of the previously reported examples of conservation, reject others, and expand the knowledge of conservation of gene organization inside these regions.

13.5.1 Conservation in euchordates

Previous work on MHC and its three paralogous clusters have suggested conservation of the cluster between jawed vertebrates and amphioxus (*B. floridae*; Flajnik and Kasahara, 2001; Abi-Rached et al., 2002; Vienne et al., 2003b). In combination with follow-up studies (Castro et al., 2004), we were able to identify conserved clusters of 18 families of orthologous genes between human and amphioxus. Moreover, the conservation of this clustering is statistically significant (Abi-Rached et al., 2002; Vienne et al., 2003b; Danchin and Pontarotti, 2004b). Altogether, these results demonstrate the existence of an MHC-like region in the amphioxus.

Furthermore, as the MHC and its three regions of paralogy are widely conserved among vertebrates, we can hypothesize that the conservation between vertebrates and cephalochordates reflects the existence of a proto-MHC region before the divergence between these lineages. This hypothesis is consistent with the fact that the four regions of paralogy found in vertebrates duplicated *en bloc* from a common ancestral region after separation between the cephalochordate (amphioxus) and craniate (hagfishes plus vertebrates) lineages, but before the emergence of the gnathostomata (jawed vertebrates; Abi-Rached et al., 2002).

Conserved gene clustering with the MHC and its three paralogous regions can thus be deciphered throughout the euchordate lineage (cephalochordates and craniates) and such conservation suggests that a proto-MHC cluster probably existed in the last common ancestor of these two lineages. We name the last common ancestor of all euchordates *Ureuchordata* (Danchin and Pontarotti, 2004b) in reference to *Urbilateria*, which is the last common ancestor of all bilaterian species. A reconstruction of this *Ureuchordata* proto-MHC region is considered further in the next sections, and in Danchin and Pontarotti (2004b), by comparing the genomic organization of cephalochordate and vertebrate MHC-like regions.

13.5.2 Conservation in chordates

Conservation in the euchordate lineage is clear, as shown in the previous section. Based on this observation, we investigated whether conservation could be revealed more widely in the chordate lineage. We thus compared genomic information from euchordate species with partial genomic information available for two urochordate species (*Ciona intestinalis* and *Ciona savignyi*), detailed in Danchin and Pontarotti (2004b). As the last common ancestor of species from the euchordate and urochordate lineages is the last common ancestor of all chordates, information on genomic organization from *Ciona* species would provide additional data for deciphering the ancestral genomic organization in *Ureuchordata*. Moreover, this could provide evidence for a more ancestral pre-existing region in the last common ancestor of all chordates.

The genomes of the two *Ciona* species are both fully sequenced and assembled into scaffolds of various lengths whose relative positions, however, are unknown to date. If significant conservation can be found in the genome of these urochordates, we do not expect it to extend further than scattered pieces of small conserved clusters. We nevertheless identified traces of conservation for the MHC-like regions between the amphioxus, human, and either one or both the two *Ciona* species. Indeed, several sets of *Ciona* orthologs are grouped on the same *Ciona* scaffolds in a manner similar to genes clustered in euchordate MHC-like regions (in human or amphioxus; Danchin and Pontarotti, 2004b). These sets consist of two scaffolds of three co-localized genes in *Ciona savignyi* in conjunction with one scaffold of four co-localized genes, one scaffold of three co-localized genes, and two scaffolds of two co-localized genes in *Ciona intestinalis* (Danchin and Pontarotti, 2004b).

Unfortunately, as none of the two Ciona genomes are sufficiently assembled, we could not statistically test the significance of this conservation between euchordates and urochordates. However, as *Ciona* genomic information becomes more advanced, we should be able to test whether this conservation is significant and may reveal the existence of a proto-MHC region at the base of chordate evolutionary history.

13.5.3 Conservation in bilateria

Conservation of an MHC-like region is clear and statistically significant inside the euchordate lineage between cephalochordates and vertebrates (Danchin and Pontarotti, 2004b). This conservation suggests the pre-existence of a proto-MHC region in *Ureuchordata*, the last common ancestor of all euchordates. In parallel, we revealed traces of conservation between urochordates, vertebrates, and cephalochordates (Danchin and Pontarotti, 2004b). The significance of this conservation could not be evaluated but may also indicate conservation of an MHC-like region in urochordates. We needed to identify conservation in Protostomes, however, to attain our goal of reconstructing ancestral genomic organization back to the origin of the Bilateria. As far as this is concerned, we showed statistically significant conservation of MHC and paralogous regions clustering in *Drosophila melanogaster* with both vertebrates (human) and cephalochordates (amphioxus). Moreover, we also showed statistically significant conservation of the clustering between *Drosophila melanogaster* and *Anopheles gambiae* (Danchin et al. 2003; Danchin and Pontarotti, 2004b). Altogether these results show conservation of an MHC-like genomic region organization between Deuterostomes and Protostomes. The last common ancestor of all these species is *Urbilateria*. Thus, the observed conserved gene clusters may represent orthologous regions that originated by speciation from a common ancestral region in *Urbilateria*.

13.6 Significance and hypotheses concerning conservations

We identified conservation of genes clustering in several species including Deuterostomes and Protostomes for the MHC and its three paralogous regions (and for the 8–10-4–5 quadruplicated regions (Danchin and Pontarotti, 2004a)). Deuterostomes and Protostomes separated more than 700 million years ago from the last common ancestor of bilaterian species (Douzery et al., 2004). We can address the question of significance and the expectation of observing conservation despite such evolutionary divergence between the species considered here.

Several hypotheses can be considered to explain conservation between such phylogenetically distant species. The first hypothesis, even if unlikely, is that the conserved genomic organization is due to chance and is not biologically significant. As described previously, the significances of the conserved clusters we deciphered were all evaluated by a statistical test. In all the cases we tested the following null hypothesis (H_0): the distribution of species B orthologs to species A genes (present in a given region X) is random along the genome of species B and does not reflect significant conservation. As further detailed in Danchin et al. (2003) and Danchin and Pontarotti (2004a, 2004b) the statistical test allowed the rejection of the null hypothesis in all cases except for comparisons with

C. elegans. These tests thus suggest that all the conservations (except for the nematode) were biologically significant (Danchin *et al.*, 2003; Danchin and Pontarotti, 2004a, 2004b). The null hypothesis of similarity in gene content by chance could be rejected with high significance, and this is particularly unexpected for clusters between chordates and dipteran since they diverged more than 700 million years ago. Genomes can have accumulated numerous rearrangements in the different species since their divergence from their last common ancestor. The more ancient the divergence is, the more likely these genomes are to be differentially organized.

Two alternative hypotheses can explain conservation when the null hypothesis has been rejected. This can either be the result of evolutionary conservation from an ancestral cluster, or be due to convergent evolution with positive selection driving similar genome organization/content.

13.6.1 Convergence with positive selection

An apparent role of shared ancestry between Deuterostomes and Protostomes may be the result of convergence with positive selection. Here, the genes considered in the two sets of conserved regions may not be ancestrally clustered, but rather, the genes grouped together within chordate (for Deuterostomes) and dipteran insect (for Protostomes) lineages separately.

Under this hypothesis, reconstruction of ancestral genomic regions should not be considered, as the conserved clusters we observe do not represent traces of the existence of clusters in the ancestor of the considered species. It is interesting, however, to investigate the evolutionary forces that could have favored these genes to independently cluster in two different lineages. We can imagine that particular positive selection acting on these genes favored their clustering into limited regions. Such a hypothesis could be tested if sufficient expression and functional data for these genes are available. Few functional or expression data for these genes in the different species considered here are currently available, and it is difficult to test this hypothesis at the moment. The logic of this argument is that convergence of location would likely be driven by co-expression if there were a positive selection pressure driving it.

13.6.2 Likelihood of the hypotheses

In our goal towards reconstructing ancestral genomic clusters in *Urbilateria*, it is necessary to consider the likelihood of the different hypotheses. Indeed, the reconstruction analysis is only possible under the hypothesis that the conserved clusters derived from a common ancestral region. As shown above, the hypothesis of similarity by chance can be rejected and thus the two alternative hypotheses that remain are the hypotheses of ancestral conservation and convergence with positive selection.

13.7 Reconstruction of ancestral regions

Based upon the hypothesis that the conservations between Protostomes and Deuterostomes that we observe constitute traces of inheritance from a common ancestral region, we propose conceptual reconstruction of the putative ancestral region from which they are derived. The general strategy for these reconstructions consists of inferring the presence of a given gene in the ancestral region based on its presence in both the corresponding conserved regions of the compared species. Using this approach, we necessarily provide a minimal reconstruction which only includes genes that are both still present in the two compared regions. As a consequence, genes ancestrally present in the region but that were lost in one or both of the compared conserved regions will not be included in these reconstructions. Similarly, genes initially present in the ancestral region but that were translocated to new locations after speciation are also absent from the reconstruction. Future comparisons to other phyla will help determine the ancestral presence or absence of uncertain genes discussed above. In all, the reconstructions we propose should be viewed as a minimal set of genes whose clustering is conserved throughout evolution and which are thus probably a remnant of ancestral gene clusters.

Reconstructions at different evolutionary scales can be considered, and we propose reconstruction of an ancestral MHC region, both in the ancestor of all euchordates and in the ancestor of all bilaterian species. For the MHC and its three paralogous regions we benefit from genomic-organization data in a wide variety of vertebrate species and from additional information concerning the amphioxus, as well as partial data on urochordates. Based on these data we can propose a reconstruction of the proto-MHC in the ancestor of all euchordates. In addition, comparisons with conserved cluster data from insects will generate inferences for *Urbilateria*.

13.7.1 Euchordates

We identified 18 families of orthologous genes whose clustering is conserved between vertebrates and cephalochordates (Danchin and Pontarotti, 2004b). We propose that these conserved clusters echo an ancestral cluster in which 18 ancestral genes were already grouped in the last common ancestor of all euchordates, namely *Ureuchordata*. As the duplications that gave rise to the gene families in vertebrates occurred after the separation between cephalochordates and gnathostomes (Abi-Rached *et al.*, 2002; Vienne *et al.*, 2003a; Danchin and Pontarotti, 2004b), we can deduce that genes were single-copy in the ancestor of these two species. A minimal reconstruction of the ancestral cluster in *Ureuchordata* is presented in Figure 13.1a.

13.7.2 *Urbilateria*

Comparisons of conserved gene clusters between vertebrates and insects allowed us to propose a putative proto-MHC cluster of 19 genes in the last common ancestor of bilaterians (Danchin *et al.*, 2003). Follow-up analyses based on additional data for conservation of an MHC-like region throughout euchordates proceeded with a comparison of the putative ancestral *Ureuchordata* reconstructed proto-MHC region to the *Drosophila* genome. Based on this analysis we identified 10 families of orthologous genes whose clustering is conserved between vertebrates, cephalochordates, and insects. From these 10 gene families we deduced a core ancestral cluster of 10 genes that was probably present in *Urbilateria*. These 10 genes remain clustered in modern species despite approximately 700 million years of evolution

for each derived cluster (Danchin and Pontarotti, 2004b). A representation of this ancestral cluster is illustrated in Figure 13.1b.

For genes present in the ancestral region reconstructed by Danchin *et al.* (2003) but not on the reconstructed region of Danchin and Pontarotti (2004b), these constitute good candidates to check for their presence in the cephalochordate MHC-like chromosomal region. Unfortunately no supplemental genomic data are available today for the amphioxus, but when such data are released we can consider testing this hypothesis. If this is demonstrated, such genes can be reintroduced in both the reconstructed ancestral *Ureuchordata* and *Urbilateria* proto-MHC regions, leading to a more accurate and complete reconstruction.

13.8 Discussion and perspectives

For the two sets of quadruplicated regions analyzed (MHC and 8-10-4-5), we developed a method combining phylogenetic analysis and statistical testing that allowed identification of a set of statistically significant conserved gene clusters between phylogenetically distant species. Based on these conservations, we then proposed reconstruction of the minimal gene content of the corresponding region in the last common ancestor of the compared species, *Urbilateria*. In order to make additional progress in the reconstruction of the genome of our distant Bilaterian ancestor, several points can be considered. The first one concerns improvement of the reconstruction methods and the development of an algorithm to evaluate the likelihood of presence/absence of a given gene in an ancestral region. An additional point undoubtedly consists of enriching the analysis by including new genomic information from phylogenetically informative species to improve the reliability and sensitivity of reconstructions. Lastly, we can also consider automation of the process with inclusion of the improved reconstruction method, the likelihood algorithm, and new species data. Such automation would allow high-throughput treatment of potential regions of evolutionary conservation, and thus provide advanced tools for the reconstruction of the genome of our distant bilaterian ancestor.

13.8.1 Improvements to the reconstruction method

To date, reconstructions of ancestral regions that we have proposed have been based on the manual examination of genes present in all the evolutionarily conserved clusters. The basic concept is that the minimal common set of genes present in all the regions of conserved syntenies we compare were ancestrally present in the regions from which it originated. Whereas the co-presence of genes in multiple regions of conserved synteny provides strong support for their ancestral presence, there are several drawbacks. The first and most important one is that this method does not provide any statistical value (or score) for the likelihood of ancestral presence or absence of a given gene in the deduced corresponding ancestral region. The second is that by using such an approach we necessarily miss genes that were translocated to a new position or lost. A statement about the ancestral presence of a gene in an ancestral region may require comparisons to a third species. Such cases will also require a weighting scheme for the presence of homologs in additional taxa. A likelihood value for the ancestral presence of the gene in the ancestral corresponding region should then be assigned. This likelihood value should vary as a function of the phylogenetic pattern of presence/absence of the gene in an evolutionarily conserved cluster in different species. Lastly, some cases require that we consider the potential of convergent translocation of genes into two regions of conserved synteny between two species while they were actually not present in the corresponding ancestral region. Once again a likelihood value should be assigned with the use of an appropriate statistical test in conjunction with comparisons of additional species. Several methods, using parsimony or likelihood have been developed to evaluate the probability of presence or absence of a given gene in an ancestral genome with regards to their phylogenetic patterns (Kunin and Ouzounis, 2003; Koonin et al., 2004). Such methods could be adopted to reconstruct ancestral gene content by adding a term for the conservation of genomic location.

13.8.2 Automation of the pipeline

We previously developed an automated computational platform, termed FIGENIX (Gouret et al., 2005), dedicated to biological sequence analysis. In a collaborative effort with Philippe Gouret and Virginie Lopez Rascol, we are developing a multi-agent system, named CASSIOPE, to find all statistically significant conserved clusters in other species and inside the query species (for possible regions of paralogy), test the convergence and inheritance hypotheses, and then propose a putative reconstruction of the corresponding ancestral region at various nodes of the bilaterian tree of life. The whole process will be based upon the following main steps:

- All the genes present in a queried genomic region are extracted, and homologs are automatically searched for in all the other species whose genomes are fully sequenced (orthologs) and reside within the genome of the search species (paralogs). At this step, the FIGENIX platform will be used to automatically detect homologs based upon robust phylogenetic reconstruction.
- Genomic locations of all homologs found in other species and in the search species are extracted. Selection of all genomic segments, defined by at least two homologs on the same DNA molecule (i.e. same chromosome) is performed, including a test of statistical significance with two maximizations (max significance or max cluster length).
- Ancestral clusters are reconstructed using parsimony according to the phylogenetic pattern of presence or absence of genes in the conserved clusters among different phyla.

With such automation of the process, it will be possible to progress faster toward reconstructing the genome of *Urbilateria*. Ultimately, ancestral reconstructions at higher levels of organization will be considered until the goal of whole-genome reconstruction is realized.

13.8.3 Beyond genomes

Additional ancestral features can be considered for reconstruction beyond the reconstruction of ancestral genomic clusters. By comparing the proteomes and gene sets of different bilaterian species in the context of the ancestral states in *Urbilateria*, it will be possible to decipher differential gene losses, gains, and duplications between these different phyla. Gene losses, gains, and duplications could then be correlated to gains, losses, and changes of biological capabilities in these lineages. These biological changes will in turn be related to environmental or geological changes at the planetary scale, as proposed by Benner *et al.* (2002). Once the ortholome (set of orthologous sequences between two or more species) has been deciphered with high reliability, reconstructions at other levels can be considered. Indeed, for example, ancestral interactomes and biological pathways could be deduced through comparisons between pathways and interaction networks from modern bilaterian species. In a similar manner, reconstruction of ancestral regulatory elements, of ancestral regulation networks, and of ancestral gene-expression patterns, could be expected as new information on expression is available for large-scale comparative studies. Thus, based upon sequence-level reconstruction, the biology of our distant ancestors can be reconstructed, and correlated with ecological and geological data. In addition to providing crucial information to understand how the genomes of bilaterian species evolved from a common ancestral genome, such information will shed light on biological mechanisms of modern species as well.

References

Abi-Rached, L., Gilles, A., Shiina, T., Pontarotti, P., and Inoko, H. (2002) Evidence of en bloc duplication in vertebrate genomes. *Nat. Genet.* **31**: 100–105.

Altschul, S.F., Madden, T.L., Schaffer, A.A., Zhang, J., Zhang, Z., Miller, W., and Lipman, D.J. (1997) Gapped BLAST and PSI-BLAST: a new generation of protein database search programs. *Nucleic Acids Res.* **25**: 3389–3402.

Benner, S.A., Caraco, M.D., Thomson, J.M., and Gaucher, E.A. (2002) Planetary biology–paleontological, geological, and molecular histories of life. *Science* **296**: 864–868.

Blair, J.E., Ikeo, K., Gojobori, T., and Hedges, S.B. (2002) The evolutionary position of nematodes. *BMC Evol. Biol.* **2**: 7.

Blanchette, M., Green, E.D., Miller, W., and Haussler, D. (2004) Reconstructing large regions of an ancestral mammalian genome in silico. *Genome Res.* **14**: 2412–2423.

Bourque, G. and Pevzner, P.A. (2002) Genome-scale evolution: reconstructing gene orders in the ancestral species. *Genome Res.* **12**: 26–36.

Bourque, G., Pevzner, P.A., and Tesler, G. (2004) Reconstructing the genomic architecture of ancestral mammals: lessons from human, mouse, and rat genomes. *Genome Res.* **14**: 507–516.

Bourque, G., Zdobnov, E.M., Bork, P., Pevzner, P.A., and Tesler, G. (2005) Comparative architectures of mammalian and chicken genomes reveal highly variable rates of genomic rearrangements across different lineages. *Genome Res.* **15**: 98–110.

Castro, L.F., Furlong, R.F., and Holland, P.W. (2004) An antecedent of the MHC-linked genomic region in amphioxus. *Immunogenetics* **55**: 782–784.

Copley, R.R., Aloy, P., Russell, R.B., and Telford, M.J. (2004) Systematic searches for molecular synapomorphies in model metazoan genomes give some support for Ecdysozoa after accounting for the idiosyncrasies of Caenorhabditis elegans. *Evol. Dev.* **6**: 164–169.

Danchin, E.G.J. (2004) *Reconstruction of ancestral genomic regions by comparative analysis of evolutionary conserved syntenies. Towards reconstructing the genome of the ancestor of all Bilaterian species (Urbilateria)*. PhD thesis, Aix-Marseille II, Marseilles.

Danchin, E.G.J. and Pontarotti, P. (2004a) Statistical evidence for a more than 800-million-year-old evolutionarily conserved genomic region in our genome. *J. Mol. Evol.* **59**: 587–597.

Danchin, E.G.J. and Pontarotti, P. (2004b) Towards the reconstruction of the bilaterian ancestral pre-MHC region. *Trends Genet.* **20**: 587–591.

Danchin, E.G., Abi-Rached, L., Gilles, A., and Pontarotti, P. (2003) Conservation of the MHC-like region throughout evolution. *Immunogenetics* **55**: 141–148.

Douzery, E.J., Snell, E.A., Bapteste, E., Delsuc, F., and Philippe, H. (2004) The timing of eukaryotic evolution: does a relaxed molecular clock reconcile proteins hand fossils? *Proc. Natl. Acad. Sci. USA* **101**: 15386–15391.

Flajnik, M.F. and Kasahara, M. (2001) Comparative genomics of the MHC: glimpses into the evolution of the adaptive immune system. *Immunity* **15**: 351–362.

Gouret, P., Vitiello, V., Balandraud, N., Gilles, A., Pontarotti, P., and Danchin, E.G. J. (2005) FIGENIX: intelligent automation of genomic annotation: expertise integration in a new software platform. *BMC Bioinformatics* **6**: 198.

Hughes, A.L. and Friedman, R. (2004) Differential loss of ancestral gene families as a source of genomic divergence in animals. *Proc. R. Soc. Lond. Ser. B Biol. Sci.* **271**: S107–S109.

Jaillon, O., Aury, J.M., Brunet, F., Petit, J.L., Stange-Thomann, N., Mauceli, E. *et al.* (2004) Genome duplication in the teleost fish *Tetraodon nigroviridis* reveals the early vertebrate proto-karyotype. *Nature* **431**: 946–957.

Jordan, I.K., Wolf, Y.I., and Koonin, E.V. (2004) Duplicated genes evolve slower than singletons despite the initial rate increase. *BMC Evol. Biol.* **4**: 22.

Kasahara, M., Hayashi, M., Tanaka, K., Inoko, H., Sugaya, K., Ikemura, T., and Ishibashi, T. (1996) Chromosomal localization of the proteasome Z subunit gene reveals an ancient chromosomal duplication involving the major histocompatibility complex. *Proc. Natl. Acad. Sci. USA* **93**: 9096–9101.

Kasahara, M., Nakaya, J., Satta, Y., and Takahata, N. (1997) Chromosomal duplication and the emergence of the adaptive immune system. *Trends Genet.* **13**: 90–92.

Koonin, E.V., Fedorova, N.D., Jackson, J.D., Jacobs, A.R., Krylov, D.M., Makarova, K.S., *et al.* (2004) A comprehensive evolutionary classification of proteins encoded in complete eukaryotic genomes. *Genome Biol.* **5**: R7.

Kunin, V. and Ouzounis, C.A. (2003) GeneTRACE-reconstruction of gene content of ancestral species. *Bioinformatics* **19**: 1412–1416.

Telford, M.J. (2004a) Animal phylogeny: back to the coelomata? *Curr. Biol.* **14**: R274–R276.

Telford, M.J. (2004b) The multimeric beta-thymosin found in nematodes and arthropods is not a synapomorphy of the Ecdysozoa. *Evol. Dev.* **6**: 90–94.

Trachtulec, Z. and Forejt, J. (2001) Synteny of orthologous genes conserved in mammals, snake, fly, nematode, and fission yeast. *Mamm. Genome* **12**: 227–231.

Trachtulec, Z., Hamvas, R.M., Forejt, J., Lehrach, H.R., Vincek, V., and Klein, J. (1997) Linkage of TATA-binding protein and proteasome subunit C5 genes in mice and humans reveals synteny conserved between mammals and invertebrates. *Genomics* **44**: 1–7.

Vienne, A., Rasmussen, J., Abi-Rached, L., Pontarotti, P., and Gilles, A. (2003a) Systematic phylogenomic evidence of en bloc duplication of the ancestral 8p11.21–8p21.3-like region. *Mol. Biol. Evol.* **20**: 1290–1298.

Vienne, A., Shiina, T., Abi-Rached, L., Danchin, E., Vitiello, V., Cartault, F., *et al.* (2003b) Evolution of the proto-MHC ancestral region: more evidence for the plesiomorphic organisation of human chromosome 9q34 region. *Immunogenetics* **55**: 429–436.

Wolf, Y.I., Rogozin, I.B., and Koonin, E.V. (2004) Coelomata and not Ecdysozoa: evidence from genome-wide phylogenetic analysis. *Genome Res.* **14**: 29–36.

IV

Experimental methodology and concerns

CHAPTER 14

Experimental resurrection of ancient biomolecules: gene synthesis, heterologous protein expression, and functional assays

Eric A. Gaucher

14.1 Introduction

The recent accumulation of DNA sequence data, combined with advances in evolutionary theory and computational power, have paved the way for innovative approaches to understand the origins, evolution, and distribution of life and its constituent biomolecules. One approach to understanding ancestral states follows a present-day-backwards strategy, whereby genomic sequences from extant (modern) organisms are incorporated into evolutionary models that estimate the extinct (ancient) character states of genes no longer present on Earth. These inferred ancestral gene sequences act as hypotheses that can be tested in the laboratory through the resurrection of the ancestral proteins themselves (Pauling and Zuckerkandl, 1963; Thornton, 2004). Results from functional assays of the protein products from these ancient genes permit us to accept/reject hypotheses about the sequences themselves, or about their interactions, binding specificities, environments, etc.

To date, approximately 20 narratives have emerged where specific molecular systems from extinct organisms have been resurrected for study in the laboratory (Sassi et al., 2007). These include digestive proteins (ribonucleases, proteases, and lysozymes) in ruminants and primates to illustrate how digestive function arose from non-digestive function in response to a changing global ecosystem (Jermann et al., 1995; Zhang, 2006), fermentive enzymes from fungi to illustrate how molecular adaptation supported mammals as they displaced dinosaurs as the dominant large land animals (Thomson et al., 2005), pigments in the visual system adapting to different environments (Chang et al., 2002; Shi and Yokoyama, 2003; Chinen et al., 2005), steroid hormone receptors adapting to changing function in steroid-based regulation of metazoans (Thornton et al., 2003), fluorescent proteins from ocean-dwelling invertebrates (Ugalde et al., 2004), enzyme cofactor evolution (Zhu et al., 2005), proteins from very ancient bacteria helping to define environments where the earliest forms of bacterial life lived (Miyazaki et al., 2001; Gaucher et al., 2003; Iwabata et al., 2005), among others.

This chapter will summarize the different approaches exploited by these studies. This includes the different strategies exploited when building ancient genes in the laboratory, the various systems used to express the encoded proteins of the ancient genes, and the different types of functional assay used to characterize the behaviors of the ancient biomolecules.

14.2 Constructing ancient sequences in the laboratory

14.2.1 Site-directed mutagenesis

A widely used approach to synthesize ancestral genes is site-directed mutagenesis. This is the optimal approach when an ancestral gene can be obtained by introducing a small number of mutations into a modern (extant) form of the gene. This approach has been used, for example, in the synthesis of ancestral alcohol dehydrogenase (ADH) and seminal ribonuclease family members discussed in other chapters of this book, as well as other families such as isocitrate dehydrogenase and eosinophil-derived neurotoxins (EDNs; Zhang and Rosenberg, 2002; Zhu et al., 2005).

In my experience, the key to a successful mutagenesis experiment lies in the design of primers. For instance, the QuikChange® site-directed mutagenesis protocol (Stratagene) recommends multiple conditions be met when designing primers. The primers should be between 25 and 45 bases in length to avoid unwanted secondary-structural elements and have melting temperatures above 75°C. Furthermore, the primers should have a minimal G + C content of 40%, terminate in one or more Cs and/or Gs, and the site-specific mutations should lie in the middle of the primer.

After proper primer design, mutations are achieved through thermal cycling using template DNA (backbone gene), a polymerase with high fidelity and the primer pairs. The parental (non-mutated) DNA can then be degraded with *Dpn*I due to the methylated/hemimethylated states of the template DNA if isolated from an organismal host. Amplification of the mutated DNA is achieved using standard cloning and transformation protocols.

14.2.2 Gene synthesis

Site-directed mutagenesis is a rapid and cost-effective way to generate mutants so long as the number of required mutations is manageable. When the number of mutations is large, however, this approach holds less value. Under these circumstances, an alternative approach is to synthesize the complete ancestral gene using oligonucleotides and primer-extension reactions. Again, primer design is critical to achieve successful gene synthesis. Two approaches for primer design have been used to synthesize ancestral genes and are discussed below. Both approaches follow the same general protocol but differ in the size of oligonucleotides used to synthesize the ancestral genes (Figure 14.1).

Incorporation of short oligonucleotides: elongation factors
The first example of total synthesis for an enzyme-encoded gene was presented more than 20 years ago. These authors used 66 different oligonucleotides ranging in size from 10 to 22 bases to synthesize a ribonuclease S gene (Nambiar et al., 1984). Although advances in molecular biology techniques and reagents have greatly enhanced this approach, the basic principles guiding it are still practical. In short, overlapping primers are used to generate a backbone of the gene, a DNA polymerase uses deoxynucleotide triphosphates (dNTPs) as a substrate to insert bases, thereby

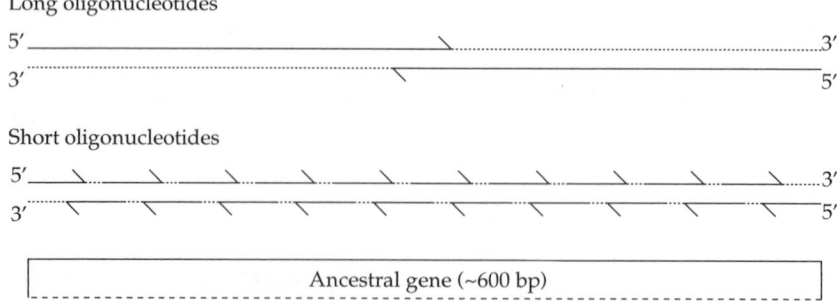

Figure 14.1 Schematic of gene synthesis using long and short oligonucleotides for ancestral genes. Long oligonucleotides are approximately 300 bases in length and short nucleotides are typically 50 bases in length. The primers overlap across 15–20 bases regardless of whether using long or short oligonucleotides. Arrows indicate 5′ → 3′ directionality. To produce full-length genes, or large fragments of the gene, primers are mixed in a PCR reaction and subjected to one extension cycle, thereby allowing the polymerase to synthesize the second strand of the non-base-paired regions of the gene (shown as dashed lines). Amplification of the gene product is achieved using flanking primers and standard PCR conditions and cycles.

filling in the gaps of the backbone, and then PCR is used to amplify the whole gene.

Total synthesis of ancient bacterial elongation factors followed this procedure (see Chapter 2 in this volume). The ancestral genes were synthesized in two steps by thermal reactions using complementary 50-mer oligonucleotides with 15–20-bp overlap. Nineteen primers were used to synthesize the 3' end of the gene while an additional 18 primers were used to synthesize the 5' end of the gene. Each reaction consisted of 10 × PCR buffer, 10 mM dNTPs, 5 μM each primer, and 2.5 units of *Pfu* polymerase in a 25-μl reaction (*Pfu* has greater proofreading ability and higher processivity than *Taq* polymerase). The reactions were heated to 94°C for 3 min, cooled to 60°C (1°C/10 s) for 10 min, and heated to 72°C for 5 min. The products of the two reactions then served as template DNA for a PCR reaction. This reaction consisted of 10 × PCR buffer, 10 mM dNTPs, 10 μM each of two flanking primers, 2.5 μl of a 1:10 dilution of each of the two template DNA reactions, 2.5 units of IDProof polymerase (ID Labs Biotechnology). This reaction mixture was amplified by PCR using the following settings: 94°C for a 1-min hot-start, 94°C for 30 s, 60°C for 45 s, and 72°C for 90 s, with 10 cycles to minimize errors during the amplification process. Genes were cloned in the Topo-TA vector system (Invitrogen). Errors resulting from primer synthesis and/or PCR were fixed using the standard site-directed mutagenesis protocol discussed above. In my experience, it seems inevitable that mutations are present in the completely synthesized gene and that most of these mutations arise from errors during synthesis of oligonucleotides.

Incorporation of long oligonucleotides: rhodopsins
Longer oligonucleotides offer some advantages over short oligonucleotides when designing ancestral genes. For instance, short oligonucleotides require on the order of 30–100 overlapping fragments (depending on gene size), and it may be difficult to optimize the GC content, melting temperature, etc., for each primer. Since regions of overlap are required to be about 20 bp in length when building genes, regardless of whether employing short or long oligonucleotides, fewer long oligonucleotides are required to cover an entire synthetic gene. Incorporating long oligonucleotides therefore minimizes the total amount of overlap required and reduces the total number of bases to be synthesized. This strategy is potentially more economical, but this remains to be determined. In this regard, the accuracy of oligonucleotide synthesis is substantially diminished when the DNA fragment is longer than approximately 50–60 bases. If a 300-mer oligonucleotide contains numerous incorrect bases, than the overall cost of synthesizing an ancestral gene will be higher than anticipated due to the increased cost of repairing the gene via site-directed mutagenesis.

Regardless of this concern, long oligonucleotides have been successfully employed to synthesize an ancestral rhodopsin gene inferred from vertebrate species (Chang *et al.*, 2002). The entire ancestral gene (1114 bp long) was synthesized from five long oligonucleotides and amplified using PCR. A detailed discussion of this approach is presented elsewhere in the book (see Chapter 15).

14.2.3 Codon optimization

It is well known that codon preference differs between species and that there is a correlation between codon usage of highly expressed genes and tRNA molecules within individual species. Although the exact nature of this correlation is unclear, it is important to take it into account when constructing ancient sequences. Expression of an ancient metazoan gene in microorganisms such as *Escherichia coli* can result in misincorporation of certain amino acids during protein translation if the metazoan codons are not optimized for bacterial expression. For instance, the AGG codon for arginine is frequently used in metazoan species such as *Drosophila* and humans. In *E. coli*, however, this codon is used less frequently than the most common stop codon. Incorporating this codon in an ancient gene inferred from metazoan sequences but expressed in *E. coli* may result in misincorporation of amino acids, translational frameshifts, and an overall reduction in translational efficiency.

Interestingly, observing differences in codon usage does not require an inter-domain comparison of species, such as bacteria and eukaryotes. Even the expression of a bacterial gene in a heterologous bacterial species can be affected by codon usage. Figure 14.2a shows the differences in expression profiles for a *Taq* polymerase (from *Thermus aquaticus*) translated in *E. coli* using wild-type *Thermus* codons compared with a synthesized *Thermus* polymerase codon optimized for expression in *E. coli*. Such results highlight the importance of considering codon

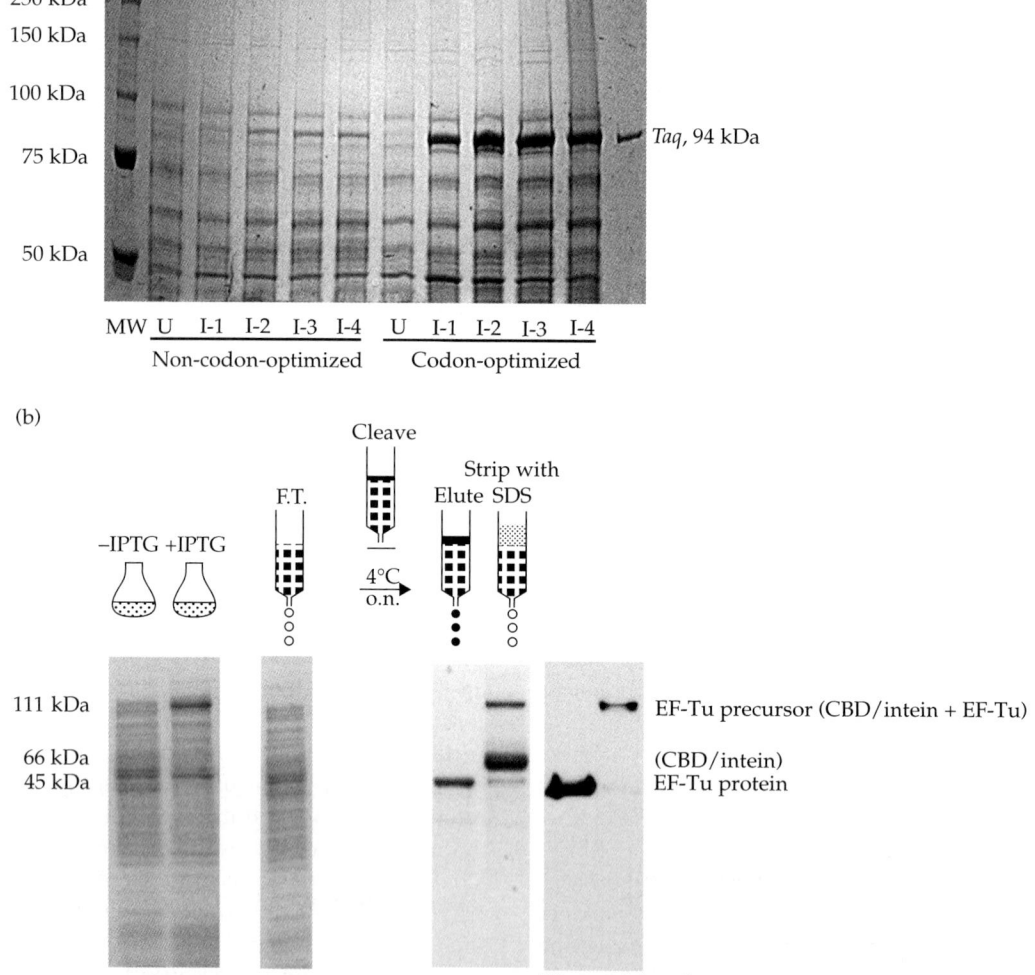

Figure 14.2 Heterologous protein expression and purification. (a) The effects of codon-optimized and non-codon-optimized heterologous gene expression of *Thermus aquaticus* polymerase (*Taq*) in E. coli cells. Molecular weight markers are shown (MW). Lanes include non-induced cells (U) and induced cells after 1, 2, 3, and 4 h of expression (I-1–I-4). (b) Procedure for protein purification of modern and ancestral elongation factor (EF) proteins. EF protein is expressed as a construct with a chitin-binding domain (CBD) and an intein domain (a self-cleaving peptide that releases the EF protein). Shown is an example of non-induced cells (lane 1), induced cells (lane 2), flow-through (F.T.) of the column indicating strong binding of the construct to the column (lane 3), release of EF protein upon activation of the intein domain (lane 4), protein present on column after elution (lane 5), and Western-blot hybridization using an antibody that recognizes a conserved epitope in modern and ancestral EF proteins (right-hand column). IPTG, isopropyl β-D-thiogalactoside; o.n., overnight; SDS, sodium dodecyl sulfate.

optimization when expressing ancient genes in heterologous species, as this optimization may be required to obtain sufficient quantities of accurately translated protein for functional assays.

14.3 Heterologous expression

Regardless of whether ancestral genes are synthesized using site-directed mutagenesis, overlapping PCR, or the currently popular outsourcing to a gene-synthesis company (often charging US$0.60–1.30 per synthesized base pair), the majority of resurrection studies are more concerned with the ancient protein than the ancient DNA itself. Upon completion of gene synthesis, ancestral genes are typically translated in an expression host *in vivo*. The use of specific expression hosts is often determined by known protein-folding/-activation requirements (redox potential, molecular chaperones, post-translational modifications, propensity to precipitate, etc.) and whether ancestral protein behaviors are determined through *in vitro* or *in vivo* cellular assays.

14.3.1 Ancestral bacterial gene expressed in bacterial cells

The ancestral elongation factor genes discussed earlier in this chapter were cloned into and expressed from the TYB11 vector in *E. coli* (IMPACT System; intein-mediated purification with an affinity chitin-binding tag), and purified according to the manufacturer's instructions (New England Biolabs). The proteins were eluted from a chitin-affinity column in a buffer consisting of 20 mM Tris/HCl (pH 8.5), 500 mM NaCl, 10 mM $MgCl_2$, 5 µM GDP, and 1 µM phenylmethylsulfonyl fluoride (PMSF), and stored at $-20°C$. The samples were filtered and concentrated using Centricon YM-30 (Amicon). SDS/PAGE verified the isolation of a single band of appropriate size, approximately 44 kDa (Figure 14.2b). Western blots were performed using a monoclonal antibody (mAb 900) that recognizes a conserved epitope from extant elongation factor (EF) proteins and present in the ancestral proteins.

14.3.2 Ancestral yeast gene expressed in yeast cells

Isogenic strains of *Saccharomyces cerevisiae*, BY4741 (MATa his3Δ1 leu2Δ0 met15Δ0 ura3Δ0) and BY4742 (MATα his3Δ1 leu2Δ0 lys2Δ0 ura3Δ0; A.T. C.C.), were used the create deletions of the two primary alcohol dehydrogenase alleles ADH2 and ADH1, respectively (Thomson et al., 2005). Primers were designed to amplify the URA3 gene with 50 bp of sequence identity to either the ADH1 or ADH2 5′ and 3′ untranslated regions. The URA3-containing vector pRS316 was used as a template and cycled 30 times at 94°C for 30 s, 55°C for 45 s, and 72°C for 1 min. Alleles were disrupted according to a published protocol (Baudin et al., 1993). The ADH1/ADH2 double-deletion strain (YMT-1D) was made by mating the single-deletion strains, followed by tetrad dissection.

The double-deletion strain YMT-1D was transformed with the pRS411-ADH2, pRS415-ADH1, or pRS415-ADH$_A$s vectors carrying wild-type ADH2, wild-type ADH1, and ancestral ADHs, respectively. Cells were grown in 2% glucose yeast minimal medium to mid-logarithmic phase and an OD_{600} value of 0.6. Extracts were applied to a Cibracon blue-coupled agarose column and protein was eluted with a gradient of NADH (0–200 µM). Fractions were pooled and applied to a Superdex 200 gel-filtration column to remove excess $NAD^+(H)$ and to collect tetrameric ADH protein (Figure 14.3).

14.3.3 Ancestral vertebrate gene expressed in mammalian cells

Synthetic G-protein-coupled receptor genes are generally expressed in mammalian cells in tissue culture where pharmacological and cellular physiological effects can be correlated with structural changes introduced by mutation. In the case of ancestral rhodopsins, large quantities of the opsin apoprotein are produced in monkey kidney cells by transfection, where transcription is under the control of the human adenovirus major-late promoter, or in stable cell lines (Chang et al., 2002). The apoprotein in the plasma membrane is regenerated with the chromophore 11-*cis*-retinal to

158 ANCESTRAL SEQUENCE RECONSTRUCTION

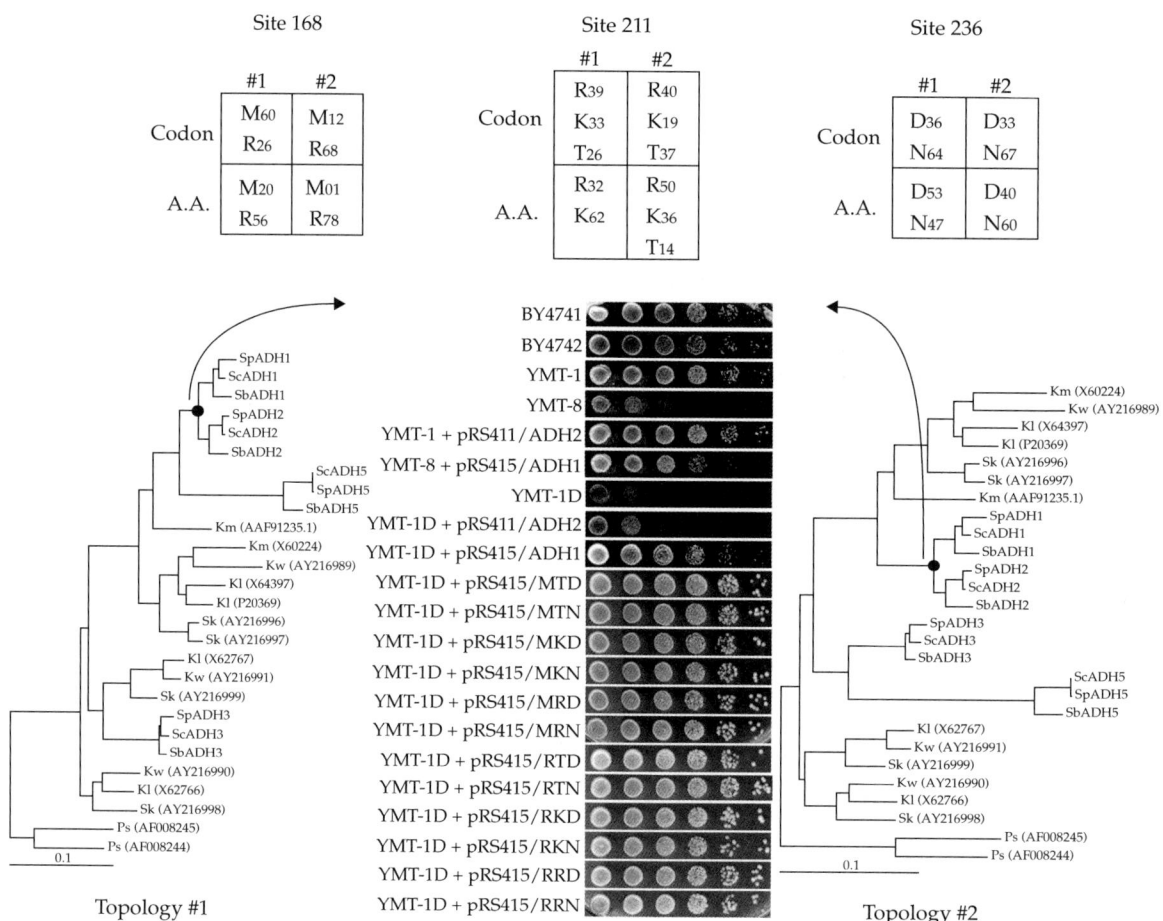

Figure 14.3 Accounting for ambiguity in reconstructed alcohol dehydrogenases (ADHs). The top panel presents differences in inferred residues for ancient ADH when considering analogous nodes from two competing phylogenies (topologies 1 and 2) and different evolutionary analyses (codon- and amino acid- (A.A.) based analyses and their posterior probabilities for individual residues). Three sites were considered ambiguous and 12 variants were synthesized to account for this ambiguity. The ability of ancient alcohol dehydrogenase variants to rescue knockouts of endogenous ADH1 and/or ADH2 in yeast S. cerevisiae as measured by a pin-stamp assay is presented in the middle of the bottom panel. BY4741 and BY4742 are isogenic strains discussed in the text. YMT-1 and YMT-8 are knockouts of ADH2 and ADH1 respectively. YMT-1D is a double knockout of ADH1 and ADH2. Vector pRS411 carries a wild-type ADH2 gene, while vector pRS415 carries either a wild-type ADH1 gene or the ancestral variants of the ADH1/ADH2 ancestral node highlighted on the phylogenies. As seen, the double knockout does not grow well and wild-type ADH2 does not rescue the wild-type phenotype. The double knockout is, however, rescued by wild-type ADH1 and all ancestral ADHs. Reprinted from Gaucher et al. (2003) Inferring the paleoenvironment of ancient bacteria on the basis of resurrected proteins. *Nature* **425**: 285–288.

form rhodopsin. The recombinant rhodopsin is then solubilized with detergent treatment and purified using an affinity adsorption method (see Chapter 15).

14.3.4 Ancestral monkey gene expressed in bacterial cells

EDN (or RNase 2) and eosinophil cationic protein (ECP; or RNase 3) are ribonucleases belonging to the RNase A family and are hypothesized to have arisen from a duplication event about 31 million years ago in the evolutionary lineage of hominoids and Old World monkeys (Zhang and Rosenberg, 2002). Since the inferred ancestral protein representing the common ancestor of EDNs and ECPs can be assayed *in vitro*, and ribonucleases can be heterologously translated in bacteria, the ancient primate genes can be expressed in *E. coli*.

Ancestral EDN/ECP sequences were constructed by site-directed mutagenesis using the owl monkey EDN sequence as the backbone (Zhang and Rosenberg, 2002). The ancient sequences were cloned into the pFCTS bacterial expression vector (IBI). The pFCTS vector adds the octapeptide DYKDDDDK (the FLAG epitope) to the recombinant protein, which permits its isolation and detection using the M2 anti-FLAG monoclonal antibody. These authors have demonstrated that the FLAG octapeptide does not interfere with the folding or catalytic activity of recombinant ribonucleases. Recombinant proteins were isolated from bacterial cultures after induction with isopropyl β-D-thiogalactoside (IPTG; 1 mM for EDNs, and 1 µM for human ECP). After harvest and sucrose lysis (EDNs) or harvest and cell lysis by freeze–thaw and sonication (human ECP), recombinant proteins were concentrated and isolated by M2 monoclonal antibody–agarose affinity chromatography (IBI).

14.4 Functional assays

Biological diversity is generally achieved through modifications to chromosomal architecture, gene expression, and protein structure. Transposition of mobile genetic elements is one mechanism that leads to chromosomal rearrangements which can subsequently generate novel proteins via domain-swapping or novel expression patterns of endogenous genes by modifying regulatory elements. Gene-expression patterns and protein structures can also be modified by point mutations to genomic DNA. For protein structure, mutations in the primary structure (amino acid sequence) are propagated through the secondary structure (helices, strands, etc.) and up to the tertiary structure (protein folds) where biological function is carried out. Here, function is based on the biomolecule's ability to contribute to the organismal success of its host through physical properties such as thermodynamics and protein stability, substrate recognition and catalysis, cofactor binding, and protein–protein interactions, among others. Ancestral sequence reconstructions and resurrections provide an opportunity to determine the specific physical behaviors of biomolecules that are responsible for functional divergence and thus biological diversity (Table 14.1).

The activation of mobile genetic elements such as transposons and retroposons throughout a genome can have a major impact on chromosomal architecture. Although mostly inactive during a species' lifetime, these elements can be activated during prolonged cellular stress and lead to mutant phenotypes that contribute to biological innovations such as speciation events. To date, three resurrection studies have verified that inactive, or dead, transposons and retroposons found in extant organisms (fish, frog, and mouse) were indeed active in their ancestral forms (Adey et al., 1994; Ivics et al., 1997; Miskey et al., 2003). These studies monitored transcription/expression and transposition of the ancient elements using a chloramphenicol acetyltransferase assay and ability of a neomycin-resistance gene to integrate into a host genome.

The contributions of protein structure and behavior to organismal fitness are more discernable than the contributions of mobile genetic elements, and this is reflected by ancestral resurrection studies. These studies often attempt to determine the behavior of an ancestral biomolecule whose descendents exhibit diverse biological behaviors, whereby diversity is generated from the divergence of paralogous and/or orthologous sequences.

Functional divergence of substrate-binding and catalysis of *paralogous* proteins has stimulated many resurrection studies. For instance, the kinetic interconversion of ethanol and acetaldehyde by alcohol dehydrogenase paralogs in yeast has been the focus of one study (Thomson *et al.*, 2005), whereas the binding of different substrates to an ancient steroid receptor in invertebrates (as measured by the ability of steroids such as estradiol, progesterone, testosterone, and others to active an ancient receptor) and mammalian proteases (as measured by the ability of ancient chymases to degrade various angiotensins) have been the foci of others (Chandrasekharan *et al.*, 1996; Thornton *et al.*, 2003). Further, DNA-binding properties of ancient *Pax* paralogs in metazoans (as measured against various DNA substrates), optimal activation of ancient zebrafish visual pigment paralogs

Table 14.1 Examples of heterologous expression and functional assays of resurrected genes.

Gene family	Expression system	Reference	Functional assay
Digestive ribonucleases	Bacterial expression, assay in vitro	Stackhouse et al. (1990)	Single-stranded RNA hydrolysis, thermostability
Digestive ribonucleases	Bacterial expression, assay in vitro	Jermann et al. (1995)	Single- and double-stranded RNA hydrolysis, thermostability
Lysozyme	Bacterial expression, assay in vitro	Malcolm et al. (1990)	Bacterial cell-wall degradation, thermostability
L1 retroposons in mouse	Mouse cells	Adey et al. (1994)	Chloramphenicol acetyltransferase assay
Chymase proteases	Insect cell expression, assay in vitro	Chandrasekaran et al. (1996)	Angiotensin degradation
Tc1/mariner transposons	Human HeLa cells	Ivics et al. (1997)	Transposition of neomycin-resistance gene
Immune RNases	Bacterial expression, assay in vitro	Zhang and Rosenberg (2002)	RNA hydrolysis and ability to reduce viral infectivity
Pax transcription factors	Drosophila, in vivo	Sun et al. (2002)	DNA-binding assay and in vivo influence on development
SWS1 visual pigment	Monkey cell expression, assay in vitro	Shi and Yokoyama (2003)	Absorption spectra
Vertebrate rhodopsins	Monkey cell expression and assay	Chang et al. (2002)	Absorption spectra
Fish opsins (blue, green)	Monkey cell expression, assay in vitro	Chinen et al. (2005)	Absorption spectra
Steroid hormone receptors	Hamster cells	Thornton et al. (2003)	Transcriptional activation of estrogen-response element
Yeast alcohol dehydrogenase	Yeast	Thomson et al. (2005)	Binding kinetics to ethanol, acetaldehyde and cofactors
Green fluorescent proteins	Bacterial expression, assay in vitro and in vivo	Ugalde et al. (2004)	Color emission spectra
Isocitrate dehydrogenase	Bacteria	Zhu et al. (2005)	Cofactor-binding kinetics and chemostat competition
Isopropylmalate dehydrogenase	Bacterial expression, assay in vitro	Miyazaki et al. (2001)	Thermostability
Isocitrate dehydrogenase	Bacterial expression, assay in vitro	Iwabata et al. (2005)	Thermostability
Elongation factors	Bacterial expression, assay in vitro	Gaucher et al. (2003)	Thermostability

(as measured by spectral absorption), and ribonuclease paralogs in monkeys (as measured by RNA hydrolysis and ability to reduce viral infectivity) have been studied (Sun et al., 2002; Chinen et al., 2005; Zhang, 2006). In each case, resurrection of the last common ancestor preceding the duplication event produced an active ancestral biomolecule whose behavior was consistent within the context of the descendent sequences.

Functional divergence of substrate binding and catalysis among *orthologs* has also inspired ancestral resurrection studies. For instance, the ability of ancient artiodactyl ribonucleases to act on different substrates has been analyzed (measured by the ability to hydrolyze single- and double-stranded RNA), as has the behavior of ancient fluorescent proteins (measured by color emission spectra; Jermann et al., 1995; Ugalde et al., 2004). In addition to the study of paralogous visual pigments, ancestral forms of the orthologous vertebrate visual pigments rhodopsin and SWS1 have been separately resurrected and their absorption spectra determined (Chang et al., 2002; Shi and Yokoyama, 2003).

Several other resurrection studies of ancient sequences inferred from orthologous genes have exploited the notion that protein function is dictated by protein folding and that tertiary-structure-based assays provide an opportunity to monitor changes in protein function. For instance, thermostability profiles of ancient isopropylmalate dehydrogenase, isocitrate dehydrogenase, and elongation factor proteins have been determined to infer the environmental temperature of ancient bacteria (Miyazaki et al., 2001; Gaucher et al., 2003; Iwabata et al., 2005). Further, the thermodynamic properties of ancient ribonucleases and lysozymes were correlated to the enzymatic properties of these ancient proteins (Malcolm et al., 1990; Jermann et al., 1995).

The above examples highlight the broad biomolecular properties that can be measured for resurrected DNA and protein. Substrate recognition, enzymatic catalysis, protein–protein/protein–DNA interactions, stabilization of transition states through cofactor binding, thermodynamics and protein stability, gene expression, genetic transposition, fluorescence via intramolecular interactions, and protein activation in response to visible-light stimulation are some of the properties *in vitro* and *in vivo* that can be exploited to determine ancient biomolecular behaviors.

14.5 Discussion

The utility of ancestral resurrection studies relies on an ability to connect ancient sequences to ancient molecular behaviors. This connection often consists of determining which of two extant molecular properties was present in an ancestral biomolecule. For instance, one can ask whether an ancient alcohol dehydrogenase converted ethanol to acetaldehyde (similar to the modern ADH1 enzyme) or whether it converted acetaldehyde to ethanol (similar to the modern ADH2 enzyme). Under this condition, an ancient duplication event would have resulted in neofunctionalization whereby one paralog retained the general molecular behavior of the ancestral form, while the other paralog acquired a novel biochemical behavior. Such a scenario is also consistent with other resurrection studies including, but not limited to, isocitrate/isopropylmalate dehydrogenase, steroid hormone receptors, and chymases.

Alternatively, a condition that involves subfunctionalization may be required to explain the evolution of ribonuclease paralogs. Here, modern paralogs display lineage-specific properties against a variety of RNA substrates. Conversely, the ancestral forms of ribonucleases display broad specificities in terms of hydrolysis of single- and double-stranded RNA, as well as immunosuppression. It appears that ancient ribonucleases were generalists in regards to RNA hydrolysis and that formation of paralogs though ancient duplication events allowed the generalists to become more specialized over time (honing some molecular behaviors, while shedding others). Since the precise selective pressures that shaped ancient ribonucleases are unknown, we cannot definitively know whether the broad molecular behaviors of ancient ribonucleases were simply promiscuous behaviors or were rather guided by natural selection. The inferences drawn from these ancient behaviors can, however, be supported by other lines of evidence such as those drawn from chemical theory, geology, animal physiology, and paleontology.

A third condition consists of resurrecting so-called dead genes that were once active but whose transcription/translation was subsequently abolished, through either mutations in regulatory elements and/or nonsense mutations in the coding regions of the gene. Examples of this condition include studies on ancient promoters and inactivated transposons. For instance, the genomes of various vertebrate species contain inactivated transposons with stop codons in unique positions but whose sequences are highly similar—suggesting that these elements were active in recent history. Reconstruction and resurrection of these elements demonstrated that the ancestral forms of the biomolecules were indeed active, as measured by their ability to transpose throughout a host genome. In a similar study, the inactivated promoter of an extant murine retroposon was determined to be active in its ancestral form. In all, these studies distinctively fulfill the resurrection moniker.

A fourth condition consists of inferring an ancient sequence whose orthologous descendents encode a large range of molecular phenotypes. For instance, visual pigments such a rhodopsins exhibit absorption maxima between 480 and 510 nm in modern vertebrates. Although this quantitative difference may be negligible for an enzymatic reaction, the difference in absorption maxima for rhodopsins can determine whether a visually acute terrestrial animal is active during the day, dusk/dawn, or night. For aquatic animals, this difference can determine whether an organism lives in shallow-, mid-, or deep-range waters. As such, the absorption maximum of an ancestral rhodopsin can suggest a specific environmental niche for the host of the ancestral protein. Fluorescent proteins provide another example of this fourth condition since modern orthologs display a broad range of emission wavelengths (from red to green to non-fluorescent blue). Similarly, the thermostabilities of modern proteins exhibit a large range of temperature optima (5°C for proteins from psychrophiles to 100°C for proteins from hyperthermophiles) and these are linearly correlated to the environmental temperature of their respective hosts. As such, thermostability of ancient proteins is a good proxy to infer niche-specific temperatures of ancient organisms.

Functional divergence among homologous sequences is the *raison d'être* that supports ancestral reconstruction and resurrection studies. There is clearly little utility in resurrecting ancient biomolecules in the absence of functional divergence and biological diversity at the molecule level. The discussion above presents just a few of evolutionary scenarios that generate this functional divergence. These include neo- and subfunctionalization after gene duplication, gene inactivation and pseudogenization, and species-specific behaviors among orthologous sequences. For each scenario, an experiment in paleogenetics has provided insight into the evolutionary path that produced functional divergence.

14.6 Conclusions

The overview presented in this chapter reflects the diversity of approaches used for gene synthesis, heterologous expression, and functional assays of ancestral genes. We anticipate that these approaches will be further developed as the field itself progresses. For instance, a greater understanding of the statistical methods (and their associated biases) employed to infer ancestral sequences will require more sophisticated approaches to account for site-specific ambiguity during synthesis of the ancestral genes themselves (see Chapter 15). We also anticipate that functional assays will become more sophisticated as molecular reconstructionists try to answer higher-order questions regarding evolutionary paths and trajectories (Lunzer et al., 2005; Weinreich et al., 2006). These include experiments attempting to replay the molecular tape of life by integrating ancestral genes in modern organisms and allowing them to evolve in the laboratory. Further, this sophistication will be achieved as the field enters into an evolutionary synthetic biology whereby the evolutionary models and approaches of ancestral sequence reconstruction guide the engineering requirements of synthetic biology (see Chapter 2).

References

Adey, N.B., Tollefsbol, T.O., Sparks, A.B., Edgell, M.H., and Hutchison, C.A. (1994) Molecular resurrection of an extinct ancestral promoter for mouse L1. *Proc. Natl. Acad. Sci. USA* **91**: 1569–1573.

Baudin, A., Ozier-Kalogeropoulos, O., Denouel, A., Lacroute, F., and Cullin, C. (1993) A simple and efficient method for direct gene deletion in *Saccharomyces cerevisiae*. *Nucleic Acids Res.* **21**: 3329–3330.

Chandrasekharan, U.M., Sanker, S., Glynias, M.J., Karnik, S.S., and Husain, A. (1996) Angiotensin II-forming activity in a reconstructed ancestral chymase. *Science* **271**: 502–505.

Chang, B.S., Jonsson, K., Kazmi, M.A., Donoghue, M.J., and Sakmar, T.P. (2002) Recreating a functional ancestral archosaur visual pigment. *Mol. Biol. Evol.* **19**: 1483–1489.

Chinen, A., Matsumoto, Y., and Kawamura, S. (2005) Reconstitution of ancestral green visual pigments of zebrafish and molecular mechanism of their spectral differentiation. *Mol. Biol. Evol.* **22**: 1001–1010.

Gaucher, E.A., Thomson, J.M., Burgan, M.F., and Benner, S.A. (2003) Inferring the palaeoenvironment of ancient bacteria on the basis of resurrected proteins. *Nature* **425**: 285–288.

Ivics, Z., Hackett, P.B., Plasterk, R.H., and Izsvak, Z. (1997) Molecular reconstruction of Sleeping beauty, a Tc1-like transposon from fish, and its transposition in human cells. *Cell* **91**: 501–510.

Iwabata, H., Watanabe, K., Ohkuri, T., Yokobori, S., and Yamagishi, A. (2005) Thermostability of ancestral mutants of *Caldococcus noboribetus* isocitrate dehydrogenase. *FEMS Microbiol. Lett.* **243**: 393–398.

Jermann, T.M., Opitz, J.G., Stackhouse, J., and Benner, S.A. (1995) Reconstructing the evolutionary history of the artiodactyl ribonuclease superfamily. *Nature* **374**: 57–59.

Lunzer, M., Milter, S.P., Felsheim, R., and Dean, A.M. (2005) The biochemical architecture of an ancient adaptive landscape. *Science* **310**: 499–501.

Malcolm, B.A., Wilson, K.P., Matthews, B.W., Kirsch, J.F., and Wilson, A.C. (1990) Ancestral lysozymes reconstructed, neutrality tested, and thermostability linked to hydrocarbon packing. *Nature* **345**: 86–89.

Miskey, C., Izsvak, Z., Plasterk, R.H., and Ivics, Z. (2003) The Frog Prince: a reconstructed transposon from Rana pipiens with high transpositional activity in vertebrate cells. *Nucleic Acids Res.* **31**: 6873–6881.

Miyazaki, J., Nakaya, S., Suzuki, T., Tamakoshi, M., Oshima, T., and Yamagishi, A. (2001) Ancestral residues stabilizing 3-isopropylmalate dehydrogenase of an extreme thermophile: experimental evidence supporting the thermophilic common ancestor hypothesis. *J. Biochem. (Tokyo)* **129**: 777–782.

Nambiar, K.P., Stackhouse, J., Stauffer, D.M., Kennedy, W.P., Eldredge, J.K., and Benner, S.A. (1984) Total synthesis and cloning of a gene coding for the ribonuclease S protein. *Science* **223**: 1299–1301.

Pauling, L. and Zuckerkandl, E. (1963) Chemical paleogenetics molecular restoration studies of extinct forms of life. *Acta Chem. Scand.* **17**: S9–S16.

Sassi, S.O., Benner, S.A., and Gaucher, E.A. (2007) Molecular paleosciences. Systems biology from the past. In *Advances in Enzymology and Related Areas of Molecular Biology: Protein Evolution*, vol. 75, pp. 1–132 (Toone, E., ed.). Wiley, Chichester.

Shi, Y. and Yokoyama, S. (2003) Molecular analysis of the evolutionary significance of ultraviolet vision in vertebrates. *Proc. Natl. Acad. Sci. USA* **100**: 8308–8313.

Stackhouse, J., Presnell, S.R., Mcgeehan, G.M., Nambiar, K.P., and Benner, S.A. (1990) The ribonuclease from an extinct bovid ruminant. *FEBS Lett.* **262**: 104–106.

Sun, H., Merugu, S., Gu, X., Kang, Y.Y., Dickinson, D.P., Callaerts, P., and Li, W.H. (2002) Identification of essential amino acid changes in paired domain evolution using a novel combination of evolutionary analysis and *in vitro* and in vivo studies. *Mol. Biol. Evol.* **19**: 1490–1500.

Thomson, J.M., Gaucher, E.A., Burgan, M.F., De Kee, D.W., Li, T., Aris, J.P., and Benner, S.A. (2005) Resurrecting ancestral alcohol dehydrogenases from yeast. *Nat. Genet.* **37**: 630–635.

Thornton, J.W. (2004) Resurrecting ancient genes: experimental analysis of extinct molecules. *Nat. Rev. Genet.* **5**: 366–375.

Thornton, J.W., Need, E., and Crews, D. (2003) Resurrecting the ancestral steroid receptor: ancient origin of estrogen signaling. *Science* **301**: 1714–1717.

Ugalde, J.A., Chang, B.S., and Matz, M.V. (2004) Evolution of coral pigments recreated. *Science* **305**: 1433.

Weinreich, D.M., Delaney, N.F., Depristo, M.A., and Hartl, D.L. (2006) Darwinian evolution can follow only very few mutational paths to fitter proteins. *Science* **312**: 111–114.

Zhang, J. (2006) Parallel adaptive origins of digestive RNases in Asian and African leaf monkeys. *Nat. Genet.* **38**: 819–823.

Zhang, J.Z. and Rosenberg, H.F. (2002) Complementary advantageous substitutions in the evolution of an antiviral RNase of higher primates. *Proc. Natl. Acad. Sci. USA* **99**: 5486–5491.

Zhu, G.P., Golding, G.B., and Dean, A.M. (2005) The selective cause of an ancient adaptation. *Science* **307**: 1279–1282.

CHAPTER 15

Dealing with model uncertainty in reconstructing ancestral proteins in the laboratory: examples from archosaur visual pigments and coral fluorescent proteins

Belinda S.W. Chang, Mikhail V. Matz, Steven F. Field, Johannes Müller, and Ilke van Hazel

15.1 Introduction

Resurrecting ancestral proteins in the laboratory can be a powerful tool in studies of protein structure and function as they can offer a rare glimpse into the evolutionary history of molecular function (Malcolm et al., 1990; Adey et al., 1994; Chandrasekharan et al., 1996; Dean and Golding, 1997; Bishop et al., 2000; Chang and Donoghue, 2000; Sun et al., 2002; Zhang and Rosenberg, 2002; Thornton, 2004). Another, perhaps even more intriguing reason for reconstructing ancestral proteins lies in the hope of achieving a better understanding of the biology of ancient animals that may have possessed these proteins (Jermann et al., 1995; Messier and Stewart, 1997; Galtier et al., 1999; Benner, 2002; Gaucher et al., 2003). Proteins that are involved in sensory systems might be particularly revealing with respect to the physiology and behavior of ancient animals that can no longer be studied directly in the laboratory (Nei et al., 1997; Boissinot et al., 1998; Chang et al., 2002). Moreover, experimental tests of laboratory-recreated ancestral proteins would provide information different from that obtained through studies of fossils. Although interpretations based on recreations of single molecules are of course limited, under the best of circumstances one may hope to test some of the theories of ancient animal biology derived from other methods such as paleontological studies (Chang et al., 2002).

Reconstructions of the past depend entirely on the accuracies and limitations of the statistical methods and models employed. However, even in cases of deep divergences, when the accuracy of the reconstruction may be low, the experimental outcome remains valuable, as the effects of altering specific amino acids on a protein's structure and function can be interesting, independently of whether or not the sequence represents the true ancestor. The general approach of using site-directed mutagenesis methods to alter amino acids in order to assess shifts in function is one of the most widely employed in studying protein function.

Advances in phylogenetic methods of ancestral reconstruction, particularly the development of likelihood/Bayesian models that incorporate many different aspects of sequence evolution, have led to a plethora of models and methods available for use in phylogenetic reconstruction in recent years (Thorne, 2000; Huelsenbeck and Bollback, 2001; Whelan et al., 2001; Nielsen, 2005). This chapter briefly discusses some of the models available for use in ancestral reconstruction, then describes ways to address variation in reconstructed ancestral sequences when the intent is to recreate proteins experimentally in the laboratory. The primary purpose of this chapter is to focus on efficient experimental strategies to explore variation in ancestral sequence reconstructions. The statistical basis of this variation is addressed in detail in Chapter 8 in this volume. The experimental strategies described here are illustrated with two examples, ancestral rhodopsins in archosaurs and green fluorescent protein (GFP)-like proteins in corals.

15.2 Likelihood/Bayesian methods of ancestral reconstruction

Ancestral reconstruction methods based on a likelihood/Bayesian framework, such as those implemented in PAML (Yang, 1997), use as an optimality criterion a likelihood score, calculated according to a specified model of evolution (Felsenstein, 2004). Optimization of the likelihood score can be used to specify topology and parameters such as branch lengths, character-state frequencies, and ancestral states. Bayesian methods can also be used to estimate posterior probabilities of ancestral states. This can be done using the maximum-likelihood topology, branch lengths, and model parameters as priors (Yang et al., 1995), or alternatively the posterior probabilities can be calculated by taking into account the uncertainty in the maximum-likelihood topology and parameters using a Markov chain Monte Carlo approach (Huelsenbeck and Bollback, 2001). These likelihood/Bayesian approaches can have considerable advantages over parsimony (Koshi and Goldstein, 1996; Lewis, 1998). In using an explicit model of molecular evolution, stochastic methods allow for the incorporation of knowledge of the mechanisms and constraints acting on coding sequences, as well as the possibility of comparing the performance of different models, ultimately resulting in the development of more realistic models (Goldman, 1993).

With stochastic methods, it is important to explore different models of molecular evolution to determine how robust the reconstruction results are. Oversimplified or unrealistic models can lead to incorrect or otherwise misleading phylogenetic reconstructions (Cao et al., 1994; Huelsenbeck, 1997; Buckley, 2002), emphasizing the importance of model selection. Likelihood models can be generally divided into three different types: nucleotide-, amino acid-, and codon-based models. Nucleotide models range from the simplest, such as Jukes–Cantor (Jukes and Cantor, 1969), which assumes equal base frequencies and rates of transitions and transversions, to much more complex models allowing unequal base frequencies (Felsenstein, 1981), transition/transversion bias (Kimura, 1980), among-site rate heterogeneity (Yang, 1994), and/or non-stationary base composition (Galtier and Gouy, 1998).

The simplest amino acid model is the Poisson, which assumes equal amino acid frequencies and rates of substitution among amino acids. Models have also been developed that allow unequal amino acid frequencies (Hasegawa and Fujiwara, 1993), and among-site rate heterogeneity (Yang, 1994), in addition to a general time reversible (GTR) model for amino acids, which allows for unequal rates of substitutions in the rate matrix for all the different classes of amino acids (Yang, 1997). Rate matrices have been calculated for a number of data-sets, including those of Dayhoff (Dayhoff et al., 1978; Kishino et al., 1990) and Jones (Jones et al., 1992; Cao et al., 1994) for globular proteins, and mitochondrial transmembrane proteins (Adachi and Hasegawa, 1996). This allows a substantial reduction in the number of parameters in the model of evolution. Models have also been developed that allow replacement rates to be proportional to the frequencies of both the replaced and resulting residues ($+gwF$ model; Goldman and Whelan, 2002).

Codon-based models have been the subject of much recent development, particularly in the context of detecting positive selection, or changes in selective constraint in a phylogenetic context (Bielawski and Yang, 2003; Nielsen, 2005). These models can be among the most complex models, and have the potential to incorporate both nucleotide and amino acid information. The original codon-based models assumed equal non-synonymous to synonymous rate ratios among sites and lineages (Goldman and Yang, 1994; Muse and Gaut, 1994). Subsequent models have allowed that ratio to vary across lineages (Yang, 1998), or sites in the protein (Nielsen and Yang, 1998; Yang et al., 2000; Wong et al., 2004), or both (Yang et al., 2005; Zhang et al., 2005), as well as across amino acids with different physiochemical characters such as charge, polarity, or volume (Sainudiin et al., 2005; Wong et al., 2006).

Finally, the use of genome-based approaches has enabled more extensive investigations of sources of systematic bias, or inconsistency in currently implemented methods of phylogenetic analyses (Phillips et al., 2004; Philippe et al., 2005a, 2005b; Jeffroy et al., 2006) and identified new effects difficult to detect in smaller data-sets, such as site-specific changes in evolutionary rates among lineages, or heterotachy (Lopez et al., 2002; Misof et al., 2002; Baele et al., 2006). However, these issues are only just being investigated and addressed (Kolaczkowski and Thornton, 2004; Philippe et al., 2005b; Steel, 2005; Thornton and Kolaczkowski, 2005; Lockhart et al., 2006).

Given the diversity of models now available, the exploration of different models for use in ancestral-state inference is critical. In certain cases likelihood ratio tests can be used to statistically compare two models of evolution that are nested with respect to one another, in order to determine whether the more complex model fits the sequence data significantly better than the simpler model (Navidi et al., 1991; Felsenstein, 2004). However, under many circumstances the models being compared are not nested, and an important alternative is to directly compare the results of ancestral reconstruction variants synthesized in the laboratory.

15.3 Laboratory synthesis of ancestral proteins

How can the variability in results from ancestral reconstruction by different models be most efficiently addressed in attempting to reconstruct ancestral proteins in the laboratory? We can distinguish two types of variability in ancestral-reconstruction results. The first is due to sites for which there are alternative reconstructions with significant posterior probabilities under a single model of evolution. However, recreation of only the most probable variant for each site does not significantly improve the chance of the whole sequence being correct, and moreover, some have recently argued that such a strategy may lead to biases in estimating the protein's properties (Williams et al., 2006). A more desirable approach would be to allow the ambiguous positions to vary and experimentally determine whether this would affect the protein phenotype. If the phenotype is robust to such variations, it may be assumed that the errors of ancestral sequence prediction would not matter. Ideally, to address this issue, the best approach would be to experimentally recreate the predicted posterior distribution of ancestral sequences. This would mean synthesizing a library of genes in which the proportion of each variant at each site would be equal to the posterior probability of this variant according to the prediction. Unfortunately, the construction of a degenerate gene in which the variants are represented at unequal pre-defined proportions represents a significant technical challenge. In the current version of the gene-synthesis protocol (see below), site variation is introduced through the use of commercially synthesized degenerate oligonucleotides, which, in turn, are achieved by including more than one type of nucleotide precursor at a particular step of the oligonucleotide synthesis. It is usually assumed that by controlling the proportions of the precursors in the mixture one may manipulate the proportions of the corresponding incorporation products, but this represents a significant technical challenge not routinely offered for commercially synthesized oligonucleotides. In addition to these problems, there is also the question of how many sequences would be considered adequate sampling from the posterior. These issues are addressed in detail in another chapter (see Chapter 8), and will not be discussed further here.

The second type of variability in ancestral reconstructions results from sites for which the reconstructions vary when different models of evolution are employed. Investigating the effects of this type of model variation is more easily addressed experimentally, and will be the focus of the experimental strategies discussed further here. In cases where the number of variable sites is not too high, the results of different models can be synthesized in their entirety from an initial variant using site-directed mutagenesis methods, and the various proteins assayed for any changes in function. The major advantage of this strategy is that proteins resulting from different models can be compared directly. This strategy/method was employed for reconstructing the ancestral archosaur visual pigment protein, described in detail below. The disadvantage of this approach is that it is only feasible if the number of variable sites in the protein is small. If the number of variable sites exceeds an amount that can be easily incorporated using mutagenesis techniques, as is the case with the ancestral coral GFP-like proteins, then a different strategy needs to be employed.

In situations where the number of variable sites is high, an efficient strategy is to incorporate the variable sites directly during the gene-synthesis, instead of mutating the sites after the first protein has already been synthesized. Degenerate oligonucleotides are incorporated into the gene-synthesis methods that allow for variation found in the ancestral reconstructions. This means that sites found to be variable in the reconstructions are incorporated in a random combinatorial fashion into the synthesis of the gene. This method utilizes an array of overlapping oligonucleotides 30–35 bases long to assemble both strands of the synthesized gene by means of ligation, followed by PCR amplification of the target product using flanking oligonucleotides as primers. The degenerate sites should be positioned as far as possible from the ligation points (see example in Figure 15.1). Although very simple, the method has the important advantage that the relatively short oligonucleotides can be ordered commercially, in contrast to other techniques that rely on longer oligonucleotides (Ferretti *et al.*, 1986). Moreover, the oligonucleotides do not need to be modified (for example, they do not require 5′ phosphates), which further decreases the cost of the project.

Finally, a gene-synthesis strategy that incorporates so many ligation points per gene is particularly useful because of the fact that the ligation efficiency is significantly diminished by the presence of mismatches in the vicinity of the ligation site. In our protocol the separation between the ligation sites is only 16–17 bases, which means that almost three-quarters of the gene length is actually proofread at the ligation step since DNA ligase is sensitive to mismatches up to at least 6 bases from the ligation site (Roth *et al.*, 2004). As a result, the mutated clones in our experiments comprised less than 50% of the total number, and even in those clones the mutations were likely to be PCR errors rather than gene-assembly artifacts. The advantage of this approach is that a large number of variable sites can be incorporated into the synthesis fairly easily, and many ancestral variants efficiently obtained in only one synthesis step. As long as enough variants are assayed so that a sufficient number of variable sites are represented, this approach can be extremely efficient in determining the average phenotype of an ancestral

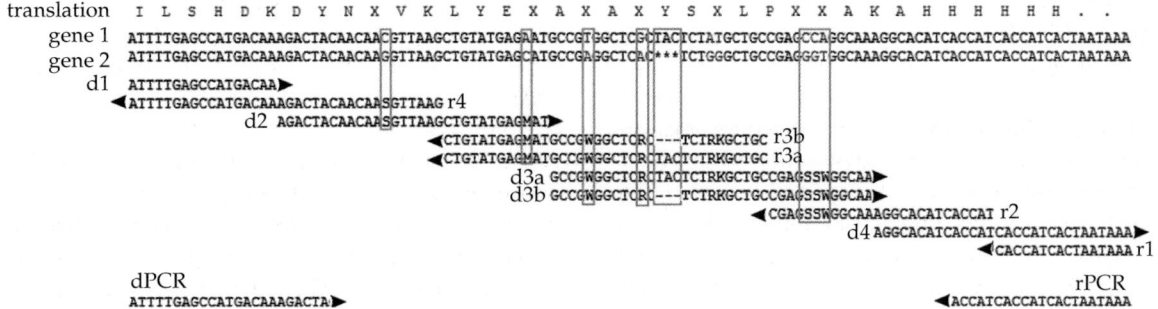

Figure 15.1 A practical example of oligonucleotide design for degenerate gene synthesis. The fragment that is being synthesized is the actual 3′-terminal portion of the highly degenerate gene incorporating transitional mutations between the ancestral identified gene 1 and the extant gene 2 (see text, Example 2). dPCR and rPCR are the PCR primers that would be used to amplify the final product. Arrowheads indicate 3′-termini of the oligonucleotides; the sequences show the corresponding portion of the sense DNA strand. Antisense oligonucleotides (on the right-hand end of sequences, starting with r) actually have the sequence complementary to the one shown. Note that to model a three-nucleotide deletion (asterisks in gene 2) two sets of sense and antisense oligonucleotides are prepared—with and without the deletion, which are then mixed in the synthesis reaction in equal proportions.

protein in the face of a large number of reconstruction variants. Unlike site-directed mutagenesis methods which can be used to synthesize the results of any one model in its entirety, the use of degenerate oligonucleotides in gene synthesis means that no one clone is likely to have the results of any one model because the variability is incorporated randomly among the sites. However, this is not necessarily a disadvantage, as the best model may vary across sites, and it is not clear that any one model should give the most accurate reconstruction for all sites.

15.4 Ancestral archosaur visual pigment

Archosaurs are a major branch of diapsid reptiles that include modern-day birds, crocodiles, and alligators (Figure 15.2). In addition to these extant taxa, the archosaur ancestors also gave rise to impressive reptiles now long vanished including several lineages of dinosaurs and pterosaurs. The closest relatives of archosaurs can be traced back into the Upper Permian, and together with the former constitute the clade Archosauriformes (Reisz and Müller, 2004; Müller and Reisz, 2005). Archosaurs in particular are thought to have originated in the Early Triassic; more specifically, the fossil record, our current understanding of archosauriform phylogeny, and the consideration of errors in stratigraphic dating techniques make it possible to constrain the date of archosaur origin to a time frame of only a few million years, i.e. between 251–243 million years ago (Müller and Reisz, 2005). Shortly after their origin, archosauriform reptiles rapidly became one of the dominant components of terrestrial ecosystems; however, what little is known of the paleobiology and paleoecology of these archosaur ancestors is inferred from the fossil record, and by analogy to their living descendants. It remains largely a mystery why they diverged so spectacularly from other diapsid reptiles such as lepidosaurs, which gave rise to modern snakes and lizards. Clearly, more knowledge about the physiology and behaviour of early archosaurs is needed, but this is difficult to gather from fossilized hard parts alone. For example, additional information about the visual adaptations of basal archosaurs would be highly desirable; in addition to its general relevance for terrestrial animals, the visual system may be of special importance for archosaurs due to their role as large predators in their respective fossil ecosystems.

The laboratory recreation of ancestral proteins involved in sensory pathways such as vision offers the opportunity to experimentally study a protein from an ancient animal to understand better the aspects of its physiology and behavior that are difficult to achieve by more traditional methods. Sensory proteins are particularly well suited for this type of study, as small changes in biochemical function can have profound consequences for the sensory capabilities of an animal (Wilkie et al., 1998; Hunt et al., 2001).

Visual pigments form the first critical step in the primary visual transduction cascade (Menon et al., 2001). They are composed of an opsin protein moiety to which a retinal chromophore is covalently attached. It is this chromophore, 11-cis-retinal or its derivatives, that isomerizes in response to light, inducing a conformational change in its associated opsin protein, activating the second-messenger G-protein transducin, and triggering the biochemical cascade of events in retinal photoreceptors which constitute the signal that light has been perceived (Baylor, 1996).

In order to achieve the greater photosensitivity required for vision at low light levels, the visual system has many specializations, including a visual pigment expressly adapted for this purpose, rhodopsin. This visual pigment is found in rod photoreceptors, which are active only under dim light conditions. The ability of an animal to see well at night is thus determined, at least in part, by the functional properties of the rhodopsin contained within its rod photoreceptors. An ancestral archosaur rhodopsin sequence was inferred using phylogenetic methods and synthesized in the laboratory in order to investigate the nocturnal visual capabilities of these ancient animals (Chang et al., 2002).

In this study, the ancestral archosaur rhodopsin amino acid sequence was inferred on a phylogeny reflecting systematic relationships among vertebrates for which rhodopsin sequences were available (Figure 15.3). This phylogeny has been

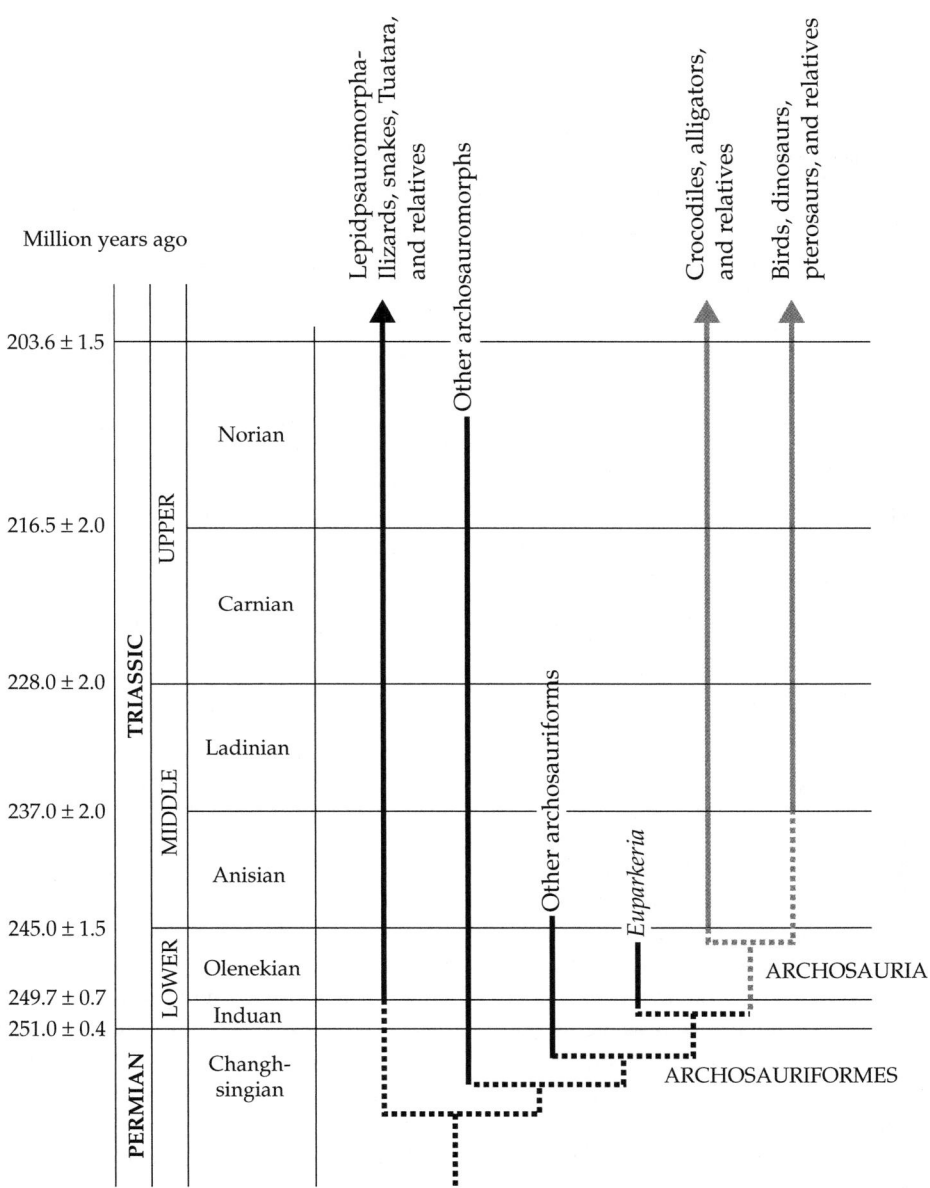

Figure 15.2 Phylogeny depicting the relationships among archosaurs (indicated in grey) and their origins from diapsid outgroups, mapped on a stratigraphy of the Late Permian and Triassic. Phylogeny, stratigraphy, and the fossil record suggest that the divergence between the two major archosaur lineages occurred in the Early Triassic, between 243 and 251 million years ago, while archosauriforms as a whole can be traced back into the Upper Permian. Current evidence also suggests that the split between archosauromorphs and lepidosauromorphs took place in the Late Permian (Reisz and Muller, 2004; Muller and Reisz, 2005).

slightly updated to reflect the current understanding of vertebrate systematics (Garcia-Moreno et al., 2002). Ancestral sequences were estimated for the archosaur node on this updated phylogeny using empirical Bayesian methods as implemented in PAML (Yang, 1997), with likelihood ratio tests performed where possible to determine the relative fit of the models (Navidi et al., 1991; Felsenstein, 2004). For the ancestral archosaur node, the amino acid reconstructions of the three best-fitting

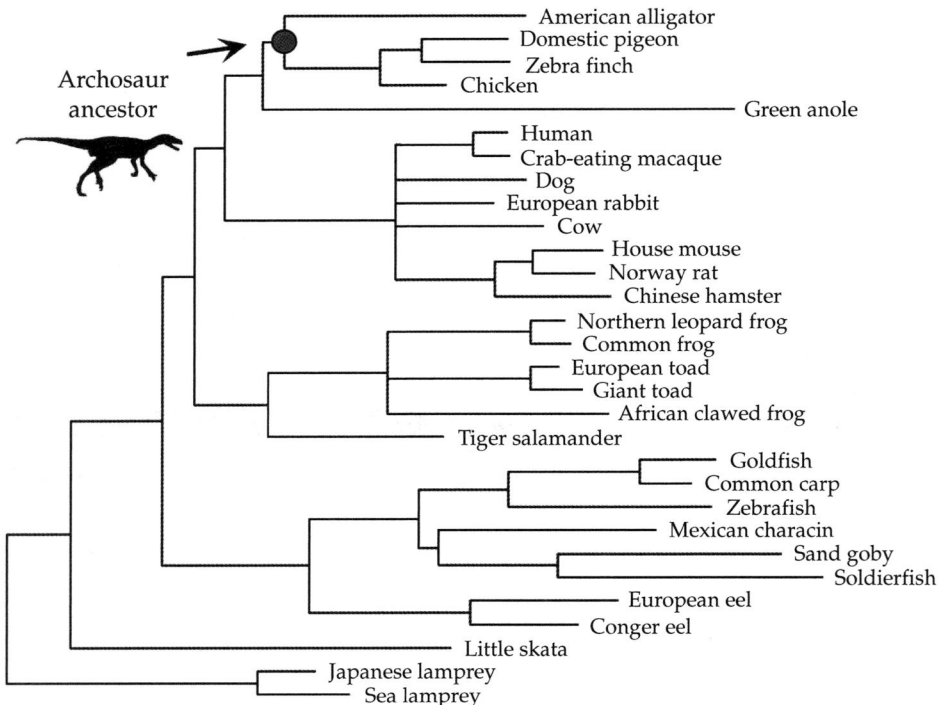

Figure 15.3 Vertebrate rhodopsin phylogeny used in inferring the sequence of the indicated ancestral archosaur node. Evolutionary relationships among vertebrates reflect current understanding of divergences among major lineages (Chang et al., 2002 Recreating a functional ancestral archosaur visual pigment. *Mol. Biol. Evol.* **19**: 1483–1489, by permission of Molecular Biology Evolution; Garcia-Moreno *et al.*, 2002).

models agreed at all but two sites, where one reconstruction differed from the other two. A third site for which reconstructions were found to differ on the original phylogeny now all agree for the three best-fitting models, although the alternative reconstruction remains among the possibilities with a much lower posterior probability (Figure 15.4).

The artificial gene corresponding to the inferred ancestral archosaur rhodopsin protein sequence was chemically synthesized in long fragments of up to 230 bases, and placed into a mammalian expression vector (pMT). Alternative reconstructions (T213I, T217A, and V218I; Figure 15.4) were introduced into the synthetic ancestral archosaur gene using site-directed mutagenesis methods, as implemented in the QuikChange® protocol (Stratagene). These ancestral genes were transiently transfected into monkey kidney (COS-1) cells, harvested, regenerated with 11-*cis*-retinal in the COS-cell membranes, solubilized, and immunoaffinity-purified using the 1D4 monoclonal antibody (Ferretti *et al.*, 1986; Chang *et al.*, 2002).

The purified ancestral proteins were tested for their ability to activate transducin in a fluorescence assay, and their absorption spectra measured. Among the variants of the ancestral archosaur rhodopsin, there were no significant differences in either of these assays. All variants were able to activate transducin (original archosaur rhodopsin, 86% activation rate with respect to a bovine rhodopsin control; variants, 79–83%), and all showed a red-shifted absorption maxima relative to bovine rhodopsin of approximately 508 nm (Figure 15.5).

All variants of the ancestral archosaur rhodopsin expressed in the laboratory not only activated the second messenger in the visual transduction cascade, transducin, at least as well as a mammalian rhodopsin, but they also showed a slightly red-shifted absorption maxima characteristic of a few modern birds (Chang *et al.*, 2002). These functional characteristics of the recreated ancestral archosaur protein imply that, at least for these aspects of rhodopsin function, the ancestral archosaur would have been able to see as well at night as a modern-day mammal,

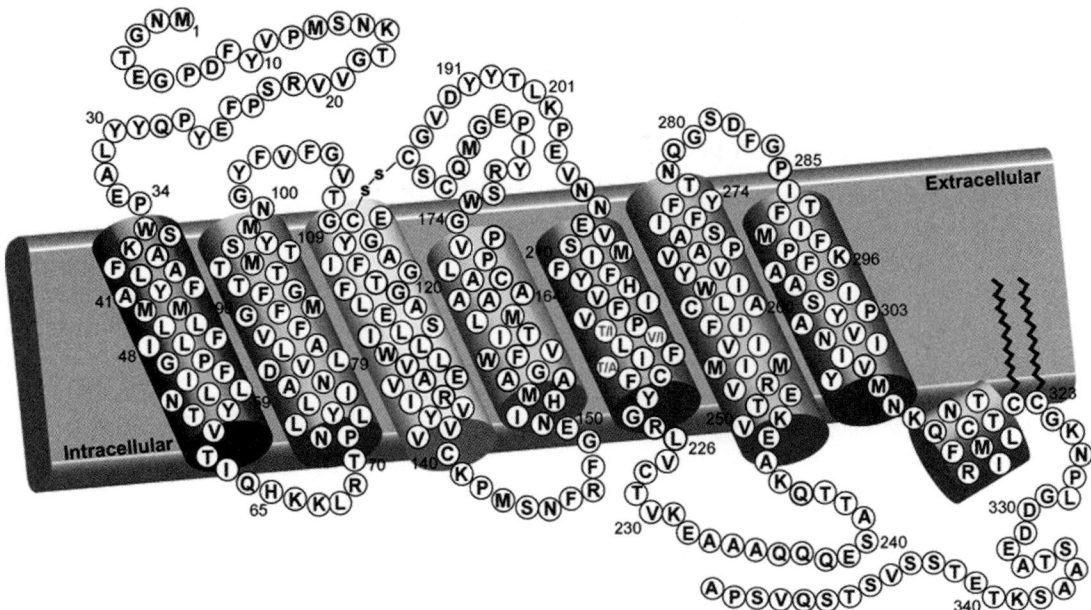

Figure 15.4 Inferred ancestral archosaur rhodopsin protein sequence, drawn in schematic form indicating the putative transmembrane domains, disulfide linkage, and palmitoylation sites. The three sites for which alternative reconstructions were experimentally introduced are indicated with slashes (in the fifth transmembrance domain); models tested agree at all other sites. For these sites at which alternative reconstructions were investigated, posterior probabilities of models employed are as follows: site 213 [F61 + G] T (0.83), I (0.078), M (0.035), V (0.028), A (0.03); [HKY + G] I (0.584), M (0.016), T (0.340), V (0.036), A (0.021); [Jones + F + G] V (0.951), I (0.048). Site 217 [F61 + G] A (0.833), T (0.102), I (0.012), M (0.004), S (0.034), V (0.015); [HKY + G]; [Jones + F + G] T (0.834), I (0.059), M (0.008), A (0.091), V (0.007). Site 218 [F61 + G] V218 (0.981), I (0.019); [HKY + G] V (0.916), I (0.083); [Jones + F + G] V (0.951), I (0.048).

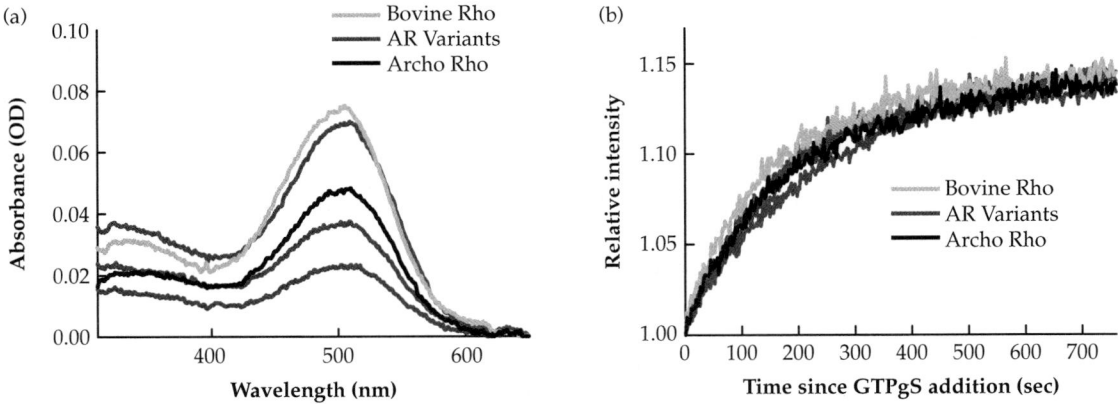

Figure 15.5 Functional assays of all variants of ancestral archosaur rhodopsin (Rho). (a) Dark absorption spectra, recorded at 25 C in a UV-visible spectrophotometer. (b) Rate of transducin activation as measured by increases in fluorescence intensity, recorded at 25 C in a spectrofluorimeter. For all assays, similarly expressed and purified bovine rhodopsin was used as a control (Chang et al., 2002). Recreating a functional ancestral archosaur visual pigment. *Mol. Biol. Evol.* **19**: 1483–1489, by permission of Molecular Biology Evolution. Original ancestral archosaur rhodopsin is indicated by black lines, variants in dark gray, and bovine rhodopsin in light gray.

which is surprising given theories of an extended nocturnal phase thought to have occurred in early mammals (Crompton et al., 1978).

15.5 Ancestral GFP-like coral proteins

Coral fluorescent proteins are homologous to the GFP (Matz et al., 1999; Field et al., 2006). These proteins share a remarkable ability to produce a chromophore moiety autocatalytically within their own globule, using their own side chains as substrates (Heim et al., 1994; Matz et al., 2002). GFP-like proteins are very convenient for experimental evolutionary studies. They are small (about 230 amino acid residues long) and can be expressed easily in a functional form in a variety of heterologous systems including bacteria. The phenotype, which is simply the color of fluorescence, can be precisely quantified in the bacterial colonies growing on a solid medium. This provides an excellent opportunity for high-throughput screening of expression libraries, or phenotypic characterization of mutants or products of degenerate gene synthesis. Here we discuss the details of the degenerate gene-synthesis method, as well as two illustrative examples from our evolutionary studies of GFP-like protein function in corals.

15.6 Degenerate gene synthesis

15.6.1 Oligonucleotides

The artificially designed ancestral gene should be divided into overlapping oligonucleotide fragments of about 30–35 bases in length (Figures 15.1 and 15.6). These oligonucleotides can be ordered from any reliable commercial service. No additional purification or modification is required; the smallest offered synthesis scale (usually 25 nmol) is sufficient.

15.6.2 Phosphorylation

In a 0.5-ml tube, combine 5 μl of 2× buffer for T4 ligase, 4 μl of the oligonucleotide mixture (all oligonucleotides that comprise the gene in a concentration of 0.1 μM each, except the two 5'-terminal ones that will not be ligated by their 5'-ends; see Figure 15.6), and 1 μl of T4 polynucleotide kinase (New England Biolabs). We used the buffer provided within the pGEM-T PCR

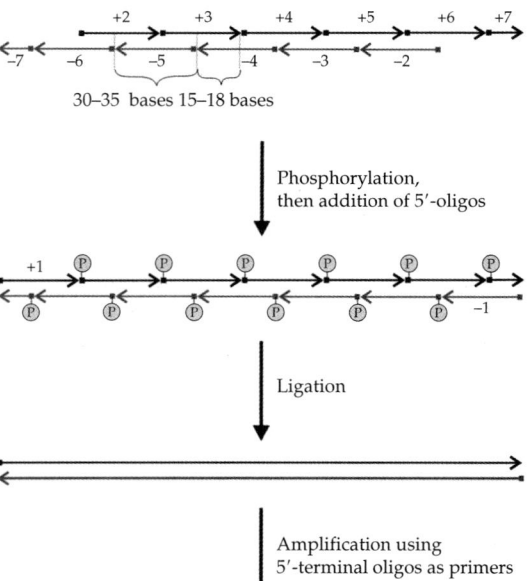

Figure 15.6 Schematic outline of the described gene-synthesis strategy. Oligonucleotides corresponding to plus and minus DNA strands are shown as black and gray arrows, respectively. Arrowheads correspond to free 3' termini, squares to free 5' termini. For simplicity of representation, the scheme shows the synthesis of a short fragment about 210–250 bp in length; however, the strategy will work for the longer genes as well.

cloning kit (Promega): it is similar in composition to the standard T4 polynucleotide kinase buffer, but already contains ATP in appropriate concentration. Incubate the reaction at 37°C for 30 min, then at 65°C for 20 min to deactivate the enzyme.

15.6.3 Ligation

To the completed phosphorylation reaction, add 5 μl of 2 × ligation buffer (Promega), 4 μl of the terminal oligonucleotide mixture (Figure 15.6; 0.1 μM each) and 1 μl of the T4 DNA ligase (New England Biolabs). Incubate the reaction for 2 h at 37°C.

15.6.4 PCR amplification of the ligated products

It is important to use a polymerase or polymerase mixture exhibiting proofreading activity, to minimize PCR errors. In our experiments, we used Advantage 2 polymerase mixture (BD Biosciences Clontech) with provided buffer. To perform the amplification, combine the following in an 0.5-ml thin-walled PCR tube: 2 μl of the ligation reaction, 2 μl of each of the 5′-terminal oligonucleotides (or specifically designed diluted to 1 μM, 2 μl of the 10 × reaction buffer, 2 μl of 5 mM dNTP mixture, 12 μl deionized water, and 0.5 μl of the Advantage 2 polymerase mix. Perform cycling according to the following program: 45 s at 94°C, 1 min at the annealing temperature (depends on the sequence of the primers), 1 min at 72°C (add 1 min per each 1000 bp of the synthesized gene over 1500 bp); run for 15–20 cycles. The accumulation of the PCR product should be monitored to keep the number of PCR cycles to the necessary minimum. The product of amplification should become visible on a standard agarose gel after 15–20 cycles, when one-tenth of the reaction volume is loaded into the well. The PCR product is then cloned into pGEM-T (Promega) to obtain bacterial expression libraries.

15.7 Example 1: resurrecting ancestral proteins using alternative evolutionary models

The corals of the Faviina suborder exhibit four basic colors of fluorescence, each determined by a specific type of GFP-like protein: cyan (blue-shifted version of green), two slightly different shades of green, and red. Cyan proteins possess the same chromophore as greens, the blue shift being due to the modified molecular environment of the chromophore (Gurskaya et al., 2001; Henderson and Remington, 2005). Red proteins, however, possess the chromophore which is the 'extended version' of the green structure (Mizuno et al., 2003), requiring one additional autocatalytic reaction for its synthesis (Figure 15.7). To see how such a complex feature might have evolved, we first applied degenerate gene-synthesis methods to recreate the ancestral proteins at the nodes descending from the common ancestor of all colors (ALL ancestor) to the common ancestor of all the red proteins (Red ancestor; Figure 15.8).

The prediction of the ancestral sequences was done using three alternative maximum-likelihood models: amino acid-based JTT (Jones et al., 1992), codon-based M5 (Yang et al., 2000), and nucleotide-based GTR + G3 (Tavare, 1986). A small number of sites were predicted differently under different models. These differences were not due to model biases but rather to the fact that these sites were poorly predictable in general: no disagreement was observed between models when all three of them generated the site prediction with posterior probability exceeding 0.80. When planning ancestral gene synthesis, the codons corresponding to these ambiguous sites were designed to be degenerate, to incorporate the alternative predictions. As a result, the designed genes for ALL, Red/green, Pre-red and Red ancestors contained nine, six, four, and six degenerate codons, respectively.

Some 500–1000 fluorescent clones from each of the four combinatorial libraries were visually surveyed using a Leica MZ FLIII fluorescence stereomicroscope with the optical filters providing excitation in the 400–450-nm range and emission from 475 nm and up (long-pass filter). Twenty-four clones from each library were sequenced and plated for spectroscopy. The fraction of clones containing no additional mutations was 0.54–0.75 (for different ancestral genes). Among these clones there were variations at all the degenerate sites. The common ancestor of all colors (ALL ancestor) turned out to be short-wave green. Most interestingly, all clones

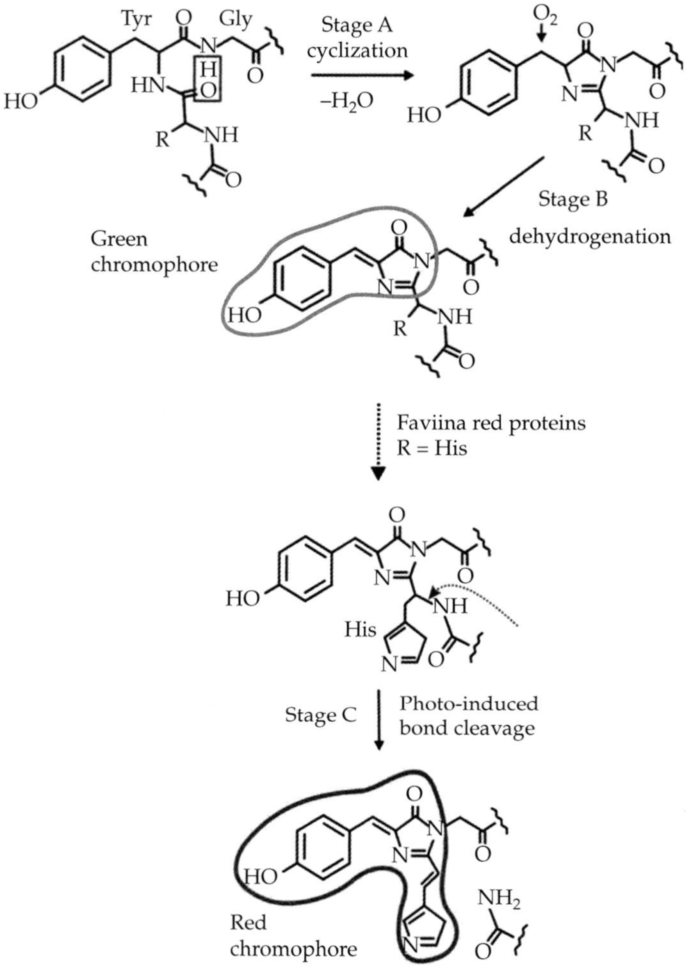

Figure 15.7 Chromophore formation pathways in green (stages A and B) and red (stages A–C) GFP-like fluorescent proteins from the corals of Faviina suborder (Heim et al., 1994; Mizuno et al., 2003). To synthesize the chromophore, a GFP-like protein uses its own backbone and side chains as substrates and also catalyzes all the biosynthesis reactions.

corresponding to the two possible common ancestors of red and green proteins (Red/green and Pre-red) showed an intermediate green/red phenotype: although the majority of the expressed protein remained green, a small fraction was able to complete the third chromophore maturation stage resulting in a minor peak of red emission.

These results (Ugalde et al., 2004) indicate that the evolution of red emission color, which corresponds to an increase in functional and structural complexity (Shagin et al., 2004), progressed through a series of intermediate stages. The follow-up analysis of color evolution in fluorescent proteins, which includes studies of selection pressure across individual sites and mutagenesis experiments, has recently been published (Field et al., 2006).

15.8 Example 2: evolutionary structure–function study

Our second example of the use of degenerate gene synthesis is slightly different. To acquire further insight into the evolutionary pathway of red fluorescence from green, the minimal set of mutations, both necessary and sufficient for the green-to-red phenotype transformation among the 37 mutations separating the green common ancestor of all Faviina colors (ALL ancestor; see Figure 15.8) and the least divergent of the extant red proteins was determined. Since the evolution of red from green included intermediate stages, more than one mutation must be responsible. However, the identification of the correct combination of

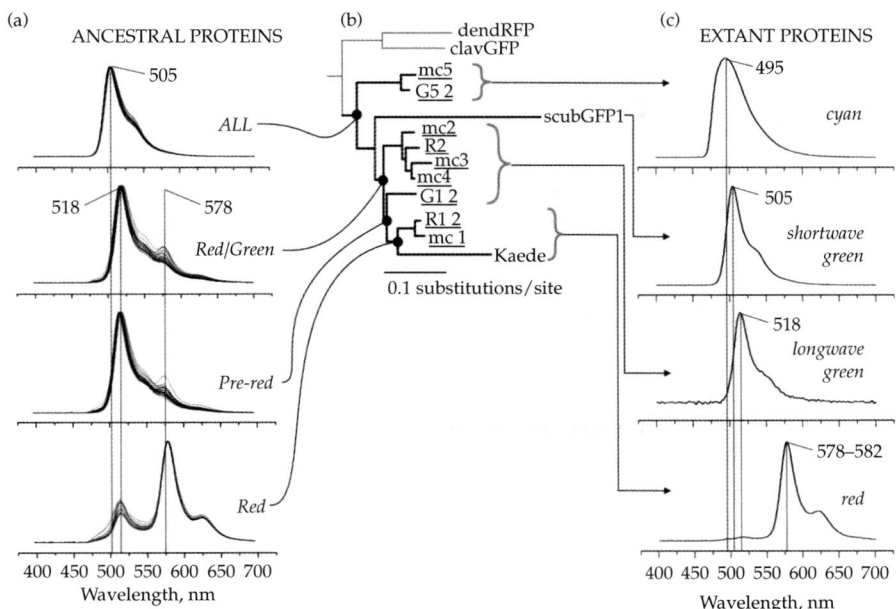

Figure 15.8 (a) Fluorescence spectra of the reconstructed ancestral proteins. Multiple curves correspond to clones bearing variations at degenerate sites. (b) Phylogeny of GFP-like proteins from the great star coral *Montastraea cavernosa* (underlined sequence names) and closely related coral species. The red and green proteins from soft corals (dendRFP and clavGFP) represent an outgroup. (c) Fluorescence spectra of extant proteins.
(d) Phylogenetic tree of GFP-like proteins from the great star coral drawn on a Petri dish using bacteria expressing extant and ancestral genes, under UV-A light.

mutations among the huge number of all possible combinations presents a problem (for 37 mutations it exceeds 10^{10}). Degenerate gene synthesis was used to generate a library of all possible combinations, from which red clones were to be identified by fluorescence screening of expressed proteins. With 37 mutations, about 16% of all codons would have to contain variations, representing a dramatic jump in degeneracy in comparison to the previous example, where the most degenerate gene contained only nine (3.8%) variable codons (see Example 1, above). Screening a sufficient number of clones by fluorescence to cover all of the possible 10^{10} combinations was simply not possible. This would have been ideal, as it would have allowed for the identification of perfectly red clones, which could then be sequenced to determine the minimal number of mutations that produces the red phenotype. However, since our screening capabilities were much more modest (10^5–10^6 clones at most), a number of clones that exhibit incomplete reddening were screened (similar to Red/green and Pre-red ancestors on Figure 15.8). Their mutational composition was compared to the green clones from the same library to identify sites that tend to be found preferentially in one of the states in the redder clones, but not in green clones. The comparison to green clones in this case provides a necessary control for the mutation bias.

The degenerate gene was successfully synthesized, and sequencing confirmed that the resulting library was composed of genes with the transitional mutations in all possible combinations (Figure 15.9a). As expected, redder phenotypes were much less frequent than green ones (note that the frequencies of phenotypes in the sample presented on Figure 15.9a do not reflect the true situation due to the clone selection bias). A total of 28 reddish and 67 green clones were selected for sequencing (note that only part of this data-set is represented on Figure 15.9). It immediately appears that some variable sites indeed are much more conserved in the redder clones and thus are likely to be responsible for converting the phenotype. One prominent example is a histidine residue at GFP position 65 (position 11 in the alignment on Figure 15.9a), which is strictly conserved in all proteins demonstrating even a slight red fluorescence. This result is

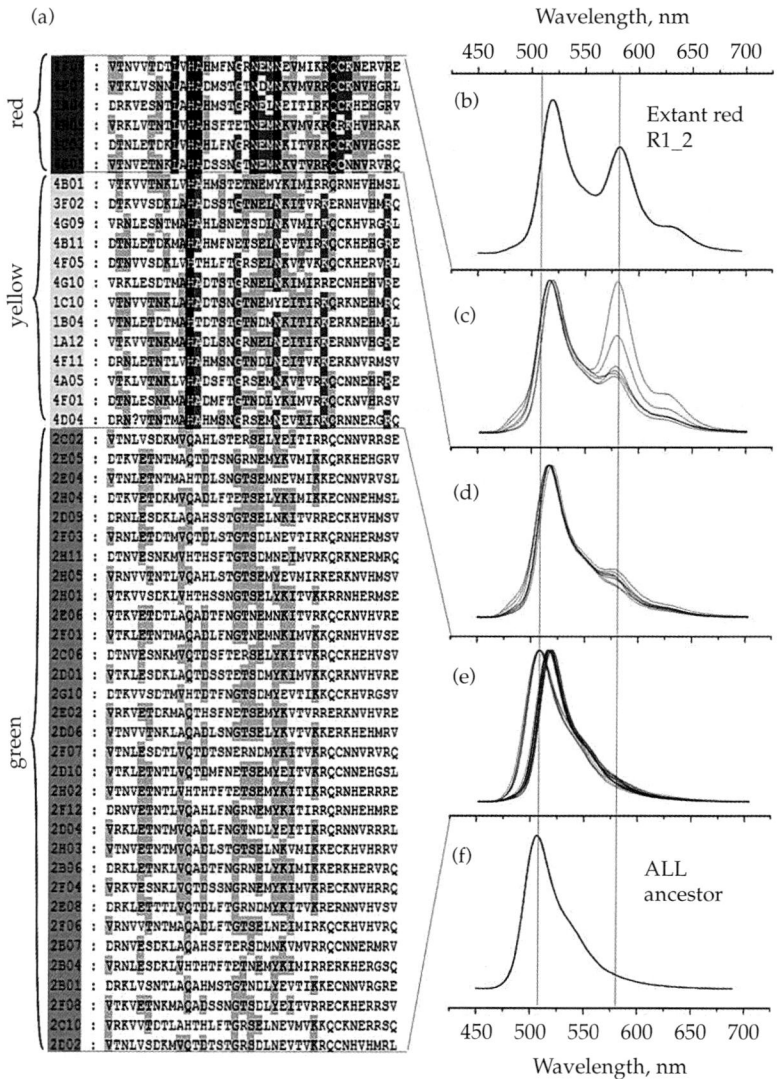

Figure 15.9 Degenerate gene synthesis to recreate a library of possible transitional variants between the green fluorescent ALL ancestor (see Figure 15.4) and the extant red fluorescent protein R1_2. (a) Alignment of only the 37 variable positions in 51 sequences from the degenerate gene library grouped according to the fluorescence color of the resulting proteins, classified as red, yellow, and green. Shading reflects the degree of within-color-group conservation. (b) Fluorescence of the extant red R1_2, measured at the intermediate stage of maturation to evaluate the efficiency of the chromophore synthesis. (c–e) Fluorescence spectra of red, yellow, and green synthetic proteins at the equivalent maturation stage. (f) Fluorescence of the ancestral green protein. On panels b–f, two vertical lines mark positions of the ancestral green and the final red fluorescence peaks.

not suprising since the side chain of His-65 forms an integral part of the red chromophore (see Figure 15.7). The Fisher exact test was used for a more rigorous comparison of site state distribution in green and red data-sets, which made it possible to rank all the variable sites in order of increasing P value that presumably corresponds to their importance for the red phenotype. The mutations were then introduced one by one into the green ALL ancestor in the ranking order until the protein became fully red, which required 15 mutations (interestingly, the last one had the P value of 0.06). Then each of the mutations was individually reversed back to the ancestral state to confirm that it was indeed essential for the red color. This procedure eliminated three mutations, which may have appeared as high ranking in our list due to artifacts of limited sampling and multiple comparisons. As a result, we arrived at the subset of as many as 12 mutations that are needed to evolve red fluorescence from the ancestral green. Such a substantial number of mutations required for the evolution of a novel feature is remarkable: in previously investigated cases of evolution of novelties in proteins (Bridgham et al., 2006; Weinreich et al., 2006) there were no more than five mutations responsible. This result confirmed our belief that the red GFP-like proteins represent a truly unique model for studying evolution of complex protein phenotypes.

Acknowledgements

This work was funded by the National Science and Engineering Council of Canada (B.S.W.C.), the National Institutes of Health (M.V.M.), the Deutsche Forschungsgemeinschaft (J.M.), and the Canadian Institutes of Health Research (I.v.H.).

References

Adachi, J. and Hasegawa, M. (1996) Model of amino acid substitution in proteins encoded by mitochrondrial DNA. *J. Mol. Evol.* **42**: 459–468.

Adey, N.B., Tollefsbol, T.O., Sparks, A.B., Edgell, M.H., and Hutchison, III, C.A. (1994) Molecular resurrection of an extinct ancestral promoter for mouse L1. *Proc. Natl. Acad. Sci. USA* **91**: 1569–1573.

Baele, G., Raes, J., Van de Peer, Y., and Vansteelandt, S. (2006) An improved statistical method for detecting heterotachy in nucleotide sequences. *Mol. Biol. Evol.* **23**: 1397–1405.

Baylor, D. (1996) How photons start vision. *Proc. Natl. Acad. Sci. USA* **93**: 560–565.

Benner, S.A. (2002) The past as the key to the present: resurrection of ancient proteins from eosinophils. *Proc. Natl. Acad. Sci. USA* **99**: 4760–4761.

Bielawski, J.P. and Yang, Z. (2003) Maximum likelihood methods for detecting adaptive evolution after gene duplication. *J. Struct. Funct. Genomics* **3**: 201–212.

Bishop, J.G., Dean, A.M., and Mitchell-Olds, T. (2000) Rapid evolution in plant chitinases: molecular targets of selection in plant-pathogen coevolution. *Proc. Natl. Acad. Sci. USA* **97**: 5322–5327.

Boissinot, S., Tan, Y., Shyue, S.K., Schneider, H., Sampaio, I., Neiswanger, K. et al. (1998) Origins and antiquity of X-linked triallelic color vision systems in New World monkeys. *Proc. Natl. Acad. Sci. USA* **95**: 13749–13754.

Bridgham, J.T., Carroll, S.M., and Thornton, J.W. (2006) Evolution of hormone-receptor complexity by molecular exploitation. *Science* **312**: 97–101.

Buckley, T.R. (2002) Model misspecification and probabilistic tests of topology: evidence from empirical data sets. *Syst. Biol.* **51**: 509–523.

Cao, Y., Adachi, J., Yano, T.-a., and Hasegawa, M. (1994) Phylogenetic place of guinea pigs: no support of the rodent-polyphyly hypothesis from maximum-likelihood analyses of multiple protein sequences. *Mol. Biol. Evol.* **11**: 593–604.

Chandrasekharan, U.M., Sanker, S., Glynias, M.J., Karnik, S.S., and Husain, A. (1996) Angiotensin II—forming activity in a reconstructed ancestral chymase. *Science* **271**: 502–505.

Chang, B.S.W. and Donoghue, M.J. (2000) Recreating ancestral proteins. *Trends Ecol. Evol.* **15**: 109–114.

Chang, B.S.W., Jonsson, K., Kazmi, M., Donoghue, M.J., and Sakmar, T.P. (2002) Recreating a functional ancestral archosaur visual pigment. *Mol. Biol. Evol.* **19**: 1483–1489.

Crompton, A.W., Taylor, C.R., and Jagger, J.A. (1978) Evolution of homeothermy in mammals. *Nature* **272**: 333–336.

Dayhoff, M.O., Schwartz, R.M., and Orcutt, B.C. (1978) A model of evolutionary change in proteins. In *Atlas of Protein Sequence and Structure* (Dayhoff, M.O., ed.), pp. 345–352. National Biomedical Research Foundation, Washington DC.

Dean, A.M. and Golding, G.B. (1997) Protein engineering reveals ancient adaptive replacements in isocitrate dehydrogenase. *Proc. Natl. Acad. Sci. USA* **94**: 3104–3109.

Felsenstein, J. (1981) Evolutionary trees from DNA sequences: a maximum likelihood approach. *J. Mol. Evol.* **17**: 368–376.

Felsenstein, J. (2004) *Inferring Phylogenies.* Sinauer Associates, Sunderland, MA.

Ferretti, L., Karnik, S.S., Khorana, H.G., Nassal, M., and Oprian, D.D. (1986) Total synthesis of a gene for bovine rhodopsin. *Proc. Natl. Acad. Sci. USA* **83**: 599–603.

Field, S.F., Bulina, M.Y., Kelmanson, I.V., Bielawski, J.P., and Matz, M.V. (2006) Adaptive evolution of multicolored fluorescent proteins in reef-building corals. *J. Mol. Evol.* **62**: 332–339.

Galtier, N. and Gouy, M. (1998) Inferring pattern and process: maximum-likelihood implementation of a nonhomogeneous model of DNA sequence evolution for phylogenetic analysis. *Mol. Biol. Evol.* **15**: 871–879.

Galtier, N., Tourasse, N., and Gouy, M. (1999) A non-hyperthermophilic common ancestor to extant life forms. *Science* **283**: 220–221.

Garcia-Moreno, J., Sorenson, M.D., and Mindell, D.P. (2002) Congruent avian phylogenies inferred from mitochondrial and nuclear DNA Sequences. *J. Mol. Evol.* **57**: 27–37.

Gaucher, E.A., Thomson, J.M., Burgan, M.F., and Benner, S.A. (2003) Inferring the paleoenvironment of ancient bacteria on the basis of resurrected proteins. *Nature* **425**: 285–288.

Goldman, N. (1993) Statistical tests of models of DNA substitution. *J. Mol. Evol.* **36**: 345–361.

Goldman, N. and Yang, Z. (1994) A codon-based model of nucleotide substitution for protein-coding DNA sequences. *Mol. Biol. Evol.* **11**: 725–736.

Goldman, N. and Whelan, S. (2002) A novel use of equilibrium frequencies in models of sequence evolution. *Mol. Biol. Evol.* **19**: 1821–1831.

Gurskaya, N.G., Savitsky, A.P., Yanushevich, Y.G., Lukyanov, S.A., and Lukyanov, K.A. (2001) Color transitions in coral's fluorescent proteins by site-directed mutagenesis. *BMC Biochemistry* **2**: 6.

Hasegawa, M. and Fujiwara, M. (1993) Relative efficiencies of the maximum likelihood, maximum parsimony, and neighbor-joining methods for estimating protein phylogeny. *Mol. Phylogenet. Evol.* **2**: 1–5.

Heim, R., Prasher, D.C., and Tsien, R.Y. (1994) Wavelength mutations and posttranslational autoxidation of green fluorescent protein. *Proc. Natl. Acad. Sci. USA* **91**: 12501–12504.

Henderson, J.N. and Remington, S.J. (2005) Crystal structures and mutational analysis of amFP486, a cyan fluorescent protein from *Anemonia majano*. *Proc. Natl. Acad. Sci. USA* **102**: 12712–12717.

Huelsenbeck, J.P. (1997) Is the Felsenstein zone a fly trap? *Syst. Biol.* **46**: 69–74.

Huelsenbeck, J.P. and Bollback, J.P. (2001) Empirical and hierarchical Bayesian estimation of ancestral states. *Syst. Biol.* **50**: 351–366.

Hunt, D.M., Dulai, K.S., Partridge, J.C., Cottrill, P., and Bowmaker, J.K. (2001) The molecular basis for spectral tuning of rod visual pigments in deep- sea fish. *J. Exp. Biol.* **204**: 3333–3344.

Jeffroy, O., Brinkmann, H., Delsuc, F., and Philippe, H. (2006) Phylogenomics: the beginning of incongruence? *Trends Genet.* **22**: 225–231.

Jermann, T.M., Opitz, J.G., Stackhouse, J., and Benner, S.A. (1995) Reconstructing the evolutionary history of the artiodactyl ribonuclease superfamily. *Nature* **374**: 57–59.

Jones, D.T., Taylor, W.R., and Thornton, J.M. (1992) The rapid generation of mutation data matrices from protein sequences. *Comput. Appl. Biosci.* **8**: 275–282.

Jukes, T.H. and Cantor, C.R. (1969) Evolution of protein molecules. In *Mammalian Protein Metabolism* (Munro, H. N., ed.), pp. 21–132, Academic Press, New York.

Kimura, M. (1980) A simple method for estimating evolutionary rate of base substitutions through comparative studies of nucleotide sequences. *J. Mol. Evol.* **16**: 111–120.

Kishino, H., Miyata, T., and Hasegawa, M. (1990) Maximum-likelihood inference of protein phylogeny and the origin of chloroplasts. *J. Mol. Evol.* **31**: 151–160.

Kolaczkowski, B. and Thornton, J.W. (2004) Performance of maximum parsimony and likelihood phylogenetics when evolution is heterogeneous. *Nature* **431**: 980–984.

Koshi, J.M. and Goldstein, R.A. (1996) Probabilistic reconstruction of ancestral protein sequences. *J. Mol. Evol.* **42**: 313–320.

Lewis, P.O. (1998) Maximum likelihood as an alternative to parsimony for inferring phylogeny using nucleotide sequence data. In *Molecular Systematics of Plants II: DNA Sequencing* (Soltis, P.S., Soltis, D.E., and Doyle, J.J., eds), pp. 132–163. Kluwer, Boston.

Lockhart, P., Novis, P., Milligan, B.G., Riden, J., Rambaut, A., and Larkum, T. (2006) Heterotachy and tree building: a case study with plastids and eubacteria. *Mol. Biol. Evol.* **23**: 40–45.

Lopez, P., Casane, D., and Philippe, H. (2002) Heterotachy, an important process of protein evolution. *Mol. Biol. Evol.* **19**: 1–7.

Malcolm, B.A., Wilson, K.P., Matthews, B.W., Kirsch, J.F., and Wilson, A.C. (1990) Ancestral lysozymes reconstructed, neutrality tested, and thermostability linked to hydrocarbon packing. *Nature* **345**: 86–89.

Matz, M.V., Fradkov, A.F., Labas, Y.A., Savitsky, A.P., Zaraisky, A., Markelov, M., Lukyanov, S.A. (1999) Fluorescent proteins from non-bioluminescent Anthozoa species. *Nature Biotechnology* **17**: 969–973.

Matz, M.V., Lukyanov, K.A., and Lukyanov, S.A. (2002) Family of the green fluorescent protein: journey to the end of the rainbow. *Bioessays* **24**: 953–959.

Menon, S.T., Han, M., and Sakmar, T.P. (2001) Rhodopsin: structural basis of molecular physiology. *Physiol. Rev.* **81**: 1659–1688.

Messier, W. and Stewart, C.B. (1997) Episodic adaptive evolution of primate lysozymes. *Nature* **385**: 151–154.

Misof, B., Anderson, C.L., Buckley, T.R., Erpenbeck, D., Rickert, A., and Misof, K. (2002) An empirical analysis of mt 16S rRNA covarion-like evolution in insects: site-specific rate variation is clustered and frequently detected. *J. Mol. Evol.* **55**: 460–469.

Mizuno, H., Mal, T.K., Tong, K.I., Ando, R., Furuta, T., Ikura, M., and Miyawaki, A. (2003) Photo-induced peptide cleavage in the green-to-red conversion of a fluorescent protein. *Mol. Cell* **12**: 1051–1058.

Müller, J. and Reisz, R.R. (2005) Four well-constrained calibration points from the vertebrate fossil record for molecular clock estimates. *Bioessays* **27**: 1069–1075.

Muse, S.V. and Gaut, B.S. (1994) A likelihood approach for comparing synonymous and nonsynonymous nucleotide substitution rates, with application to the chloroplast genome. *Mol. Biol. Evol.* **11**: 715–724.

Navidi, W.C., Churchill, G.A., and von Haeseler, A. (1991) Methods for inferring phylogenies from nucleic acid sequence data by using maximum likelihood and linear invariants. *Mol. Biol. Evol.* **8**: 128–143.

Nei, M., Zhang, J., and Yokoyama, S. (1997) Color vision of ancestral organisms of higher primates. *Mol. Biol. Evol.* **14**: 611–618.

Nielsen, R. (2005) Molecular signatures of natural selection. *Annu. Rev. Genet.* **39**: 197–218.

Nielsen, R. and Yang, Z. (1998) Likelihood models for detecting positively selected amino acid sites and applications to the HIV-1 envelope gene. *Genetics* **148**: 929–936.

Philippe, H., Delsuc, F., Brinkmann, H., and Lartillot, N. (2005a) Phylogenomics. *Annu. Rev. Ecol. Evol. Syst.* **36**: 541–562.

Philippe, H., Zhou, Y., Brinkmann, H., Rodrigue, N., and Delsuc, F. (2005b) Heterotachy and long-branch attraction in phylogenetics. *BMC Evol. Biol.* **5**: 50.

Phillips, M.J., Delsuc, F., and Penny, D. (2004) Genome-scale phylogeny and the detection of systematic biases. *Mol. Biol. Evol.* **21**: 1455–1458.

Reisz, R.R. and Müller, J. (2004) Molecular timescales and the fossil record: a paleontological perspective. *Trends Genet.* **20**: 237–241.

Roth, M.E., Feng, L., McConnell, K.J., Schaffer, P.J., Guerra, C.E., Affourtit, J.P. *et al.* (2004) Expression profiling using a hexamer-based universal microarray. *Nat. Biotechnol.* **22**: 418–426.

Sainudiin, R., Wong, W.S., Yogeeswaran, K., Nasrallah, J.B., Yang, Z., and Nielsen, R. (2005) Detecting site-specific physicochemical selective pressures: applications to the Class I HLA of the human major histocompatibility complex and the SRK of the plant sporophytic self-incompatibility system. *J. Mol. Evol.* **60**: 315–326.

Shagin, D.A., Barsova, E.V., Yanushevich, Y.G., Fradkov, A.F., Lukyanov, K.A., Labas, Y.A. *et al.* (2004) GFP-like proteins as ubiquitous metazoan superfamily: evolution of functional features and structural complexity. *Mol. Biol. Evol.* **21**: 841–850.

Steel, M. (2005) Should phylogenetic models be trying to "fit an elephant"? *Trends Genet.* **21**: 307–309.

Sun, H.M., Merugu, S., Gu, X., Kang, Y.Y., Dickinson, D.P., Callaerts, P., and Li, W.H. (2002) Identification of essential amino acid changes in paired domain evolution using a novel combination of evolutionary analysis and in vitro and in vivo studies. *Mol. Biol. Evol.* **19**: 1490–1500.

Tavare, L. (1986) Some probabilistic and statistical problems of the analysis of DNA sequences. *Lect. Math. Life Sci.* **17**: 57–86.

Thorne, J.L. (2000) Models of protein sequence evolution and their applications. *Curr. Opin. Genet. Dev.* **10**: 602–605.

Thornton, J.W. (2004) Resurrecting ancient genes: experimental analysis of extinct molecules. *Nat. Rev. Genet.* **5**: 366–375.

Thornton, J.W. and Kolaczkowski, B. (2005) No magic pill for phylogenetic error. *Trends Genet.* **21**: 310–311.

Ugalde, J.A., Chang, B.S.W., and Matz, M.V. (2004) Evolution of coral pigments recreated. *Science* **305**: 1433–1433.

Weinreich, D.M., Delaney, N.F., Depristo, M.A., and Hartl, D.L. (2006) Darwinian evolution can follow only very few mutational paths to fitter proteins. *Science* **312**: 111–114.

Whelan, S., Lio, P., and Goldman, N. (2001) Molecular phylogenetics: state-of-the-art methods for looking into the past. *Trends Genet.* **17**: 262–272.

Wilkie, S.E., Vissers, P., Das, D., Degrip, W.J., Bowmaker, J.K., and Hunt, D.M. (1998) The molecular basis for uv vision in birds—spectral characteristics, cdna sequence and retinal localization of the uv-sensitive visual pigment of the budgerigar (*Melopsittacus Undulatus*). *Biochem. J.* **330**: 541–547.

Williams, P.D., Pollock, D.D., Blackburne, B.P., and Goldstein, R.A. (2006) Assessing the accuracy of ancestral protein reconstruction methods. *PLoS Comput. Biol.* **2**: e69.

Wong, W.S., Yang, Z., Goldman, N., and Nielsen, R. (2004) Accuracy and power of statistical methods for detecting adaptive evolution in protein coding sequences and for identifying positively selected sites. *Genetics* **168**: 1041–1051.

Wong, W.S., Sainudiin, R., and Nielsen, R. (2006) Identification of physicochemical selective pressure on protein encoding nucleotide sequences. *BMC Bioinformatics* **7**: 148.

Yang, Z. (1994) Maximum likelihood phylogenetic estimation from DNA sequences with variable rates over sites: approximate methods. *J. Mol. Evol.* **39**: 306–314.

Yang, Z. (1997) PAML: a program package for phylogenetic analysis by maximum likelihood. *Comput. Appl. Biosci.* **13**: 555–556.

Yang, Z. (1998) Likelihood ratio tests for detecting positive selection and application to primate lysozyme evolution. *Mol. Biol. Evol.* **15**: 568–573.

Yang, Z., Kumar, S., and Nei, M. (1995) A new method of inference of ancestral nucleotide and amino acid sequences. *Genetics* **141**: 1641–1650.

Yang, Z., Nielsen, R., Goldman, N., and Pedersen, A.M. (2000) Codon-substitution models for heterogeneous selection pressure at amino acid sites. *Genetics* **155**: 431–449.

Yang, Z., Wong, W.S., and Nielsen, R. (2005) Bayes empirical bayes inference of amino acid sites under positive selection. *Mol. Biol. Evol.* **22**: 1107–1118.

Zhang, J.Z. and Rosenberg, H.F. (2002) Complementary advantageous substitutions in the evolution of an antiviral RNase of higher primates. *Proc. Natl. Acad. Sci. USA* **99**: 5486–5491.

Zhang, J., Nielsen, R., and Yang, Z. (2005) Evaluation of an improved branch-site likelihood method for detecting positive selection at the molecular level. *Mol. Biol. Evol.* **22**: 2472–2479.

V

Experimental synthesis of ancestral proteins to test biological hypotheses

CHAPTER 16

Using ancestral gene resurrection to unravel the evolution of protein function

Joseph W. Thornton and Jamie T. Bridgham

16.1 Introduction

In the century and a half since Darwin, the central goal of evolutionary biology has been to provide historical explanations for the diversity of species and their myriad adaptations. The recent advent of molecular biology and genomics presents us with a new kind of biodiversity that is equally astonishing: thousands of genes in every genome, each with specific, exquisitely tuned functions. How this functional biodiversity of genes and proteins evolved is arguably the central question in molecular evolution.

Most work to date on the evolution of gene function has relied on statistical methods to infer process from patterns in present-day sequence data (for overviews, see Li, 1997; Page and Holmes, 1998). Many valuable insights have emerged from this approach, but the hypotheses that have been generated remain for the most part empirically untested. Recently, however, advances in phylogenetics and DNA-synthesis techniques have made it possible to infer the sequences of ancestral genes and then synthesize and express them in the laboratory. As a result, hypotheses about the functions of ancient genes—and the mechanistic basis for their evolution—can now be empirically tested using the reductionist power of experimental molecular biology.

In this chapter, we review our use of ancestral gene resurrection to understand how the members of a biologically crucial gene family, the steroid hormone receptors, evolved their diverse and highly specific functions. We also discuss some methodological questions and concerns—particularly related to uncertainty in the reconstruction of ancestral sequences—and point to potential future directions for the budding field of ancestral gene resurrection.

16.2 The evolution of molecular interactions

Virtually everything that living cells do is regulated by specific molecular interactions, such as those between enzymes and substrates, receptors and ligands, and transcription factors and their DNA-binding sites. Genomic diversity is also largely due to the diversity of molecular interactions: members of most gene families have a core conserved function (such as DNA binding or a specific mode of catalysis) but have diversified by changing their specific binding partners. Despite the biological importance of specific molecular interactions, however, there has been very limited work, theoretical or empirical, to understand the general dynamics by which they evolve (see Fryxell, 1996; Aharoni et al., 2005; Haag and Molla, 2005; Bridgham et al., 2006).

Tightly integrated molecular partnerships also exemplify an important and largely unresolved evolutionary issue: the evolution of complexity. The classic model for the evolution of complex systems is that they result from a gradual process of elaboration and optimization under the influence of selection. This model is well supported for some complex structures, such as metazoan eyes: the presence of eyes of intermediate complexity in a variety of taxa indicates that more complex eyes evolved gradually (and repeatedly) from a primitive light-sensing organ (Futuyma, 1998). It is not clear, however, how this model can explain the evolution of tightly integrated molecular systems, in which the function of each part depends on its interaction with the other parts. Simultaneous emergence of more than one element by mutational processes is unlikely, so it is not apparent how selection can drive the evolution of any part or the system as a whole. What, for example, is the selection pressure that drives the evolution of a new hormone if there is not already a receptor to transduce its signal? Conversely, what is the function of a new receptor if there is not already a hormone for it to receive?

Darwin was well aware of this puzzle. He wrote in *The Origin of Species*, "If it could be demonstrated that any complex organ existed, which could not possibly have been formed by numerous, successive, slight modifications, my theory would absolutely break down" (Darwin, 1859). He also recognized that for many present-day complex systems, it would be difficult to reconstruct the stepwise process by which they evolved: "In order to discover the early transitional grades through which the organ has passed, we should have to look to very ancient ancestral forms, long since become extinct." This is a particular problem for studying the evolution of molecular complex systems, the ancestral forms of which, unlike those of morphological features, are not preserved as fossils.

The advent of ancestral gene resurrection provides a way to study the ancestral forms of molecules that would otherwise be scientifically inaccessible. We can now resurrect and characterize the functions of ancient genes, including those that participate in specific interactions. Thus we can begin to unravel the events by which tightly integrated molecular complexes emerged by stepwise Darwinian processes.

16.3 Steroid hormones and their receptors

To understand the evolution of molecular complexity, we study a specific model system: the tight functional interactions between steroid hormones and their intracellular receptors. Steroid hormone receptors are ligand-regulated transcription factors. They are activated by contact with specific steroid hormones, such as testosterone, estradiol, progesterone, cortisol, and aldosterone. These hormones are produced in the gonads or adrenal/interrenal glands through a pathway of enzymatic modifications beginning with cholesterol (Figure 16.1). Humans have six steroid receptors (SRs): two for estrogens (ERα and ERβ) and one each for testosterone and other androgens (AR), progestins (PR), glucocorticoids (GR), and mineralocorticoids (MR). The classic effects of steroid hormones include control of secondary sexual differentiation, and reproductive function in females (estrogens and progestins) and males (androgens), response to stress (glucocorticoids), and maintenance of osmotic homeostasis (aldosterone).

Mechanistically, SRs are molecular mediators (Gronemeyer *et al.*, 2004). In the absence of the hormone, SRs are typically in the cytosol. Steroid hormones are hydrophobic, so they cross the cell membrane by diffusion. Each hormone binds with extraordinary affinity and specificity to a receptor. Hormone-binding triggers a change in the receptor's conformation that allows it to dimerize, translocate to the nucleus, and bind tightly to specific response elements in the nucleus; short DNA sequences in the control region of target genes. The receptor then attracts coactivator proteins that modify chromatin, attract elements of the basal transcription complex, or otherwise increase transcription of the target gene (Figure 16.1). SRs have a conserved modular structure, consisting of a highly conserved DNA-binding domain (DBD), which recognizes and binds to response elements, and a moderately conserved ligand-binding domain (LBD), which binds to the hormone (the ligand) and contains the hormone-activated transcriptional activation function. Receptors also contain a poorly conserved flexible hinge region, which orients the DBD and LBD relative to each other, and a non-conserved N-terminal domain, which contains an autonomous transcriptional function. The DBD and LBD are functionally separable from the rest of the sequence, allowing the construction of chimeric proteins that combine

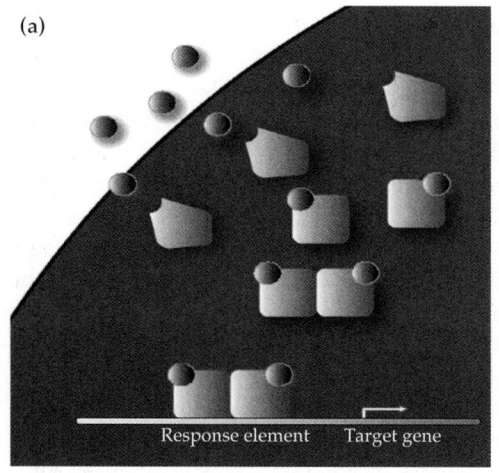 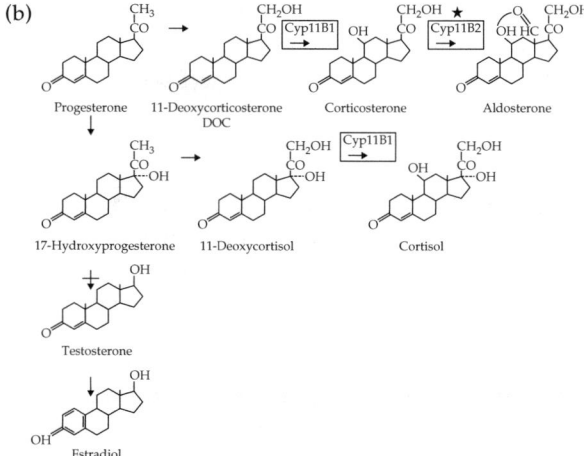

Figure 16.1 Steroid hormone and receptor biology. (a) Simplified mechanism of hormone receptor action. Steroid hormones (dark balls) are hydrophobic, so they cross cell membranes by diffusion. In the cytosol, a hormone molecule binds tightly to a specific receptor, conferring a conformational change that allows the receptor to dimerize, enter the nucleus, and bind to specific response elements in the promoters of target genes. The hormone-bound SR attracts coactivator proteins that increase expression of the target gene. (b) Steroid synthesis pathway. All major steroid hormones are produced in a pathway of enzyme-mediated modifications, beginning with a progesterone precursor. Reactions catalyzed by one of these enzymes, cytochrome P450–11B, are boxed. Only in tetrapods can this protein also catalyze the hydroxylation of corticosterone to produce aldosterone (star).

the functions of one protein's DBD with those of another protein's LBD or activation domain (Green and Chambon, 1987).

The goal of research in our laboratory is to describe the specific mechanisms and dynamics by which new hormone–receptor and receptor–DNA relationships evolved. SRs have several characteristics that make them very suitable for ancestral gene resurrection. First, they form a monophyletic group within a larger superfamily of genes, so phylogenetic methods can be used to reconstruct their proliferation and divergence from a common ancestral protein (Thornton and DeSalle, 2000). Second, there are efficient, well-established molecular assays for determining the intrinsic functions of SRs, which can be used to characterize ancient receptors resurrected in the laboratory. Finally, there is an extensive database on the sequences, structures, and functions of extant receptors, providing a rich context for interpreting reconstructed ancestral sequences.

In this chapter, we review how we have used ancestral gene resurrection to address two evolutionary questions. First, how did the specific interactions of GR and MR with glucocorticoids and aldosterone evolve? Second, how were the PR's and AR's partnerships with progesterone and testosterone established. Our ultimate goal is to determine the evolutionary dynamics and molecular mechanisms by which all of the receptor-specific functions of this important gene family evolved.

16.4 Evolution of corticoid receptor specificity

MR and GR are sister receptors that descend from a gene duplication deep in the vertebrate lineage (Thornton, 2001) and now have distinct signaling functions. GR is specifically activated by the stress hormone cortisol in most vertebrates to regulate metabolism, inflammation, and immunity (Bentley, 1998). MR is activated by aldosterone to control electrolyte homeostasis and other processes (Bentley, 1998; Farman and Rafestin-Oblin, 2001). MR can also be activated by cortisol, although the presence of a cortisol-clearing enzyme in most MR-expressing tissues makes the receptor a largely aldosterone-specific factor (Farman and Rafestin-Oblin, 2001).

Aldosterone has only been detected in tetrapods, so it has long been assumed that the GR, which is insensitive to aldosterone, retains the ancestral functions, with the MR's affinity for aldosterone being derived (Baker, 2001). We sought to test this hypothesis and determine the mechanistic basis for the evolution of GR/MR specificity by resurrecting the ancestral corticoid receptor (AncCR): the ancient protein from which GR and MR descend by gene duplication. Our work on this ancient gene was first presented in Bridgham *et al.* (2006), which provides further details on methods and results.

The first requirement for ancestral gene resurrection is an ample data-set of sequences from extant taxa. The accuracy with which an ancestral sequence is inferred depends strongly on the length of the branches that descend from the ancestral node and, to a lesser extent, on the accuracy of the phylogeny itself. If the node is surrounded by long branches on which the majority of phylogenetic information has been erased by subsequent substitutions, accurately reconstructing the ancestral state becomes very difficult (Zhang and Nei, 1997). A large number of SR sequences from tetrapods and teleosts were publicly available, but few sequences were known from agnathans and elasmobranches—the basal lineages that diverged from other vertebrates just before and after the node represented by AncCR. To improve the robustness of the sequence database, we used degenerate PCR and rapid amplification of cDNA ends (RACE) to isolate corticoid receptors from two jawless fishes—the lamprey *Petromyzon marinus* and the hagfish *Myxine glutinosa*—and an elasmobranch, the skate *Raja erinacea*. We recovered a single unduplicated corticoid receptor from the lamprey and hagfish, and clear orthologs of both the GR and MR from the skate (Figure 16.2a).

The second requirement for ancestral sequence reconstruction is a well-corroborated phylogenetic tree. Based on an alignment of 60 broadly sampled SR protein sequences—including the new sequences from basal vertebrates—we used maximum parsimony, Bayesian Markov chain Monte Carlo (BMCMC), and maximum likelihood to determine the phylogeny of corticoid receptors and their outgroups. For BMCMC, we integrated over numerous protein evolutionary models and found that the JTT + gamma model was supported with 100% posterior probability; we therefore used this model for maximum-likelihood analysis. Maximum likelihood, maximum parsimony, and BMCMC all recovered the same phylogeny, increasing confidence in the result (Kolaczkowski and Thornton, 2004; Thornton and Kolaczkowski, 2005). We found that the node that represents AncCR is extremely well supported, with posterior probability and bootstrap confidence measures equal to 1.0; this is particularly important, because ancestral reconstruction is generally robust to errors in the topology except at the node being reconstructed (Zhang and Nei, 1997). The majority of other nodes on the tree were also inferred with high confidence.

The tree (Figure 16.2a) indicates that the duplication of AncCR, which produced separate GR and MR lineages, occurred after the divergence of jawless vertebrates but before the split of cartilaginous from bony fish. AncCR therefore represents the unduplicated corticoid receptor gene, which existed in the genome of the last common ancestor of agnathans and jawed vertebrates, about 450–470 million years ago.

16.5 Resurrecting the AncCR

Using this phylogeny as a scaffold for phylogenetic inference, we next inferred the protein sequence of the AncCR using the maximum-likelihood-based method of (Yang *et al.*, 1995). The analysis assumed the JTT + gamma model of protein evolution, which was strongly supported by our Bayesian analysis of numerous models. The parameter values of the model, such as branch lengths and the shape parameter for among-site rate variation, were estimated by maximum likelihood. We focused on the LBD sequence, because this is the functional domain that confers ligand specificity. The AncCR LBD protein sequence was inferred with high support: the mean posterior probability was 94% per site, and two-thirds of sites had posterior probabilities of more than 99%. A small number of sites were ambiguously reconstructed, however, with alternative states that had nontrivial probability, an issue to which we will return below (Figure 16.2b).

THE EVOLUTION OF PROTEIN FUNCTION 187

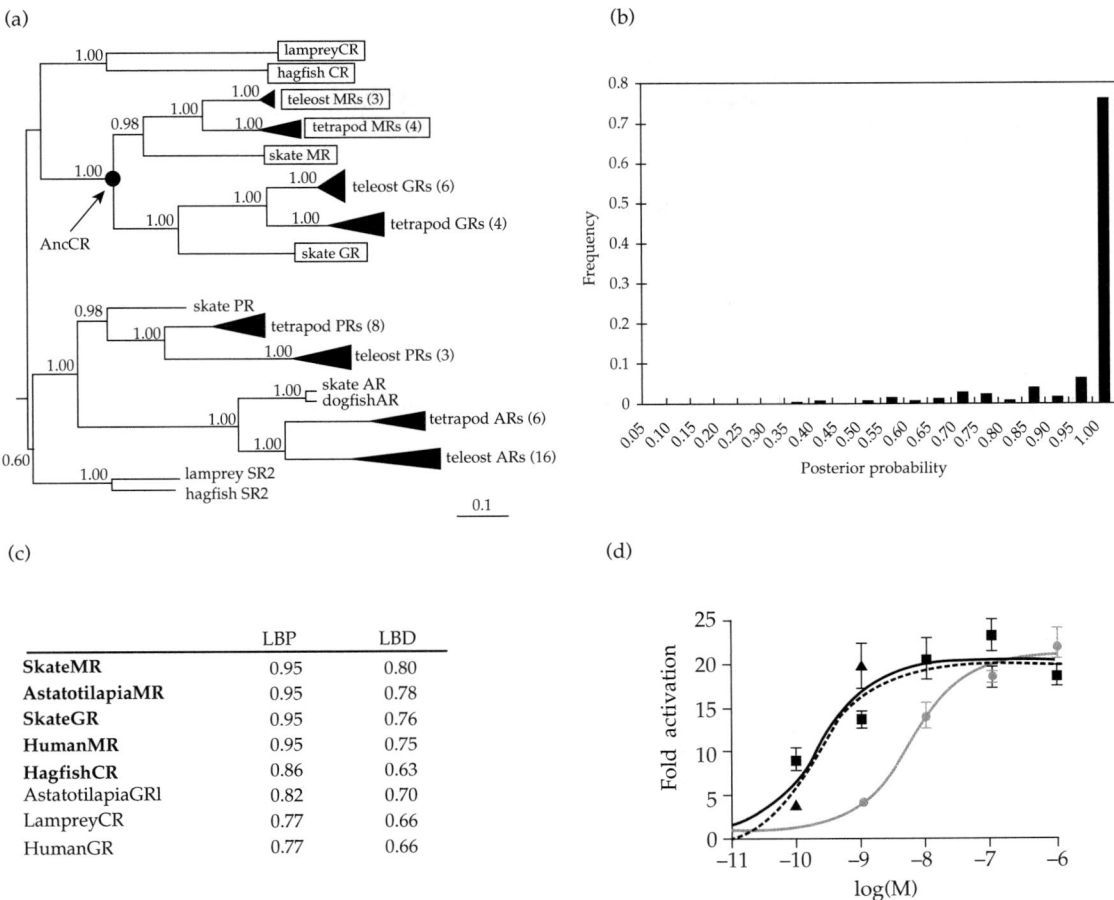

Figure 16.2 Resurrection of the ancestral corticoid receptor (CR). (a) Phylogeny of the steroid receptors. The gene family tree of 59 steroid and related receptor amino acid sequences was inferred using maximum likelihood (ML), Bayesian Markov Chain Monte Carlo (BMCMC), and maximum parsimony. ML branch lengths and BMCMC posterior probabilities for major nodes are shown. Parentheses, number of sequences in each clade. The ancestral corticoid receptor (AncCR) that we reconstructed is marked. Boxes, aldosterone-activated receptors. Reprinted with permission from Bridgham et al. 2006. Evolution of hormone-receptor complexity by molecular exploitation. *Science* **312**: 97–101, © 2006 AAAS. (b) Distribution of the posterior probabilities of the most likely amino acid at each site in the 232-amino acid ligand binding domain of AncCR. (c) Sequence similarity of AncCR to extant corticoid receptors. Receptors that are activated by aldosterone are shown in bold. LBD, entire ligand-binding domain. LBP, ligand-binding pocket, consisting of sites that are known from structural studies to make contact with and coordinate binding of the hormone. (d) The ancestral corticoid receptor is activated by aldosterone. Increase in activation of a luciferase reporter gene by the resurrected AncCR-LBD is shown for increasing doses of aldosterone (black line, squares), cortisol (gray line, circles), and 11-deoxycorticosterone (dashed line, triangles). Fold-activation is relative to activation of the reporter in the absence of hormone. Reprinted with permission from AAASThornton et al. (2003) Resurrecting the ancestral steroid receptor: ancient origin of estrogen signaling. *Science* **301**: 1714–1717 © 2003 AAAS.

Preliminary analysis led us to the hypothesis that the AncCR LBD would have MR-like functions, for two reasons. First, its sequence is most similar to the aldosterone-activated receptors: it differs from them by only one residue in the ligand-binding pocket but is considerably less similar to the aldosterone-insensitive GRs (Figure 16.2c). Second, when we functionally characterized the ligand sensitivity of extant receptors, we found that all the receptors from all the basal vertebrates were activated by very low doses of aldosterone, cortisol, and 11-deoxycorticosterone (DOC; Figure 16.3a and b). They are similar in this respect to MRs of tetrapods and teleosts (Hellal-Levy et al., 1999; Greenwood et al., 2003; Sturm et al., 2005). The only receptors we found to be insensitive to aldosterone were the GRs of tetrapods and teleosts. The most parsimonious scenario

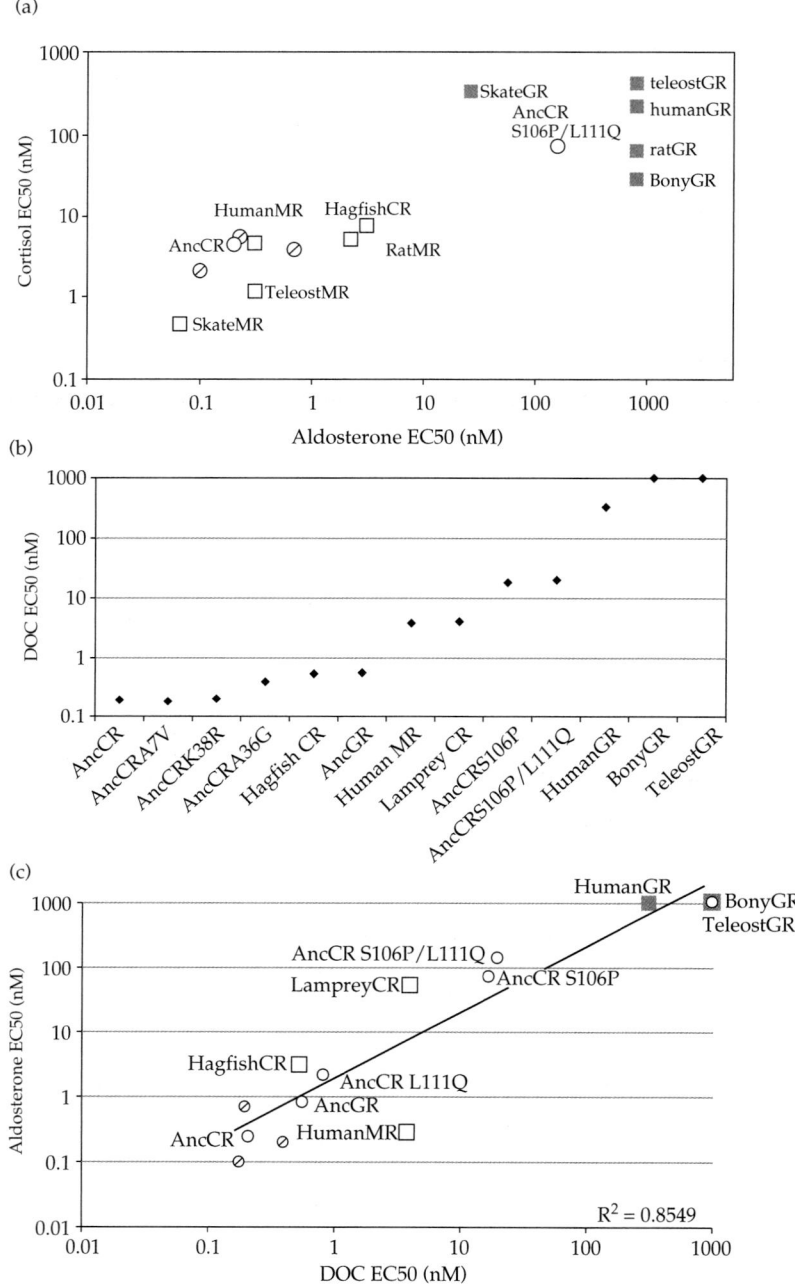

Figure 16.3 Hormone activation of extant and ancestral receptors. (a) Receptor sensitivity to aldosterone and cortisol. Each point represents the sensitivity of one receptor to these two hormones, expressed as the EC50—the concentration (nanomolar, nM) that is required to achieve half-maximal activation of a luciferase reporter gene; a lower EC50 indicates greater sensitivity. White squares are extant MRs and CRs; gray boxes are GRs. White circles are resurrected ancestral receptors; slashed circles are three variant versions that incorporate alternative reconstructions due to phylogenetic uncertainty. The AncCR containing two substitutions that recapitulate the evolution of the GR-like phenotype is also shown. (b) Sensitivity of extant and ancestral receptors to deoxycorticosterone (DOC), the putative ancestral ligand that is present in all vertebrates. (c) Correlation of receptor aldosterone sensitivity with DOC sensitivity. The linear regression shows that variation in a receptor's sensitivity to DOC predicts 85% of variation in aldosterone sensitivity.

a priori is therefore that AncCR was aldosterone-sensitive: the alternative hypothesis would require aldosterone activation to be gained independently in the lineages leading to the agnathan CRs, to the elasmobranch GR, and to the MRs of elasmobranches, teleosts, and tetrapods, all in the absence of the hormone—a most unlikely possibility (Figure 16.2a).

To test this hypothesis, we inferred a cDNA sequence that would code for the AncCR LBD sequence, and optimized it for expression in cultured cells using a standard table of mammalian codon bias. We had this cDNA synthesized commercially, an approach we find practical, accurate, and reasonably priced. We then subcloned the AncCR LBD cDNA into a vector for high-level expression in a fusion protein, and transfected that construct into mammalian Chinese hamster ovary (CHO) K1 cells. CHO-K1 cells are a standard system for characterizing vertebrate receptors for two reasons: they do not express any of their own SRs—so they provide a low-noise cellular background for functional assays—and they contain all the conserved accessory factors for SRs from a wide variety of taxa to function properly.

We used a luciferase reporter-gene assay to determine the AncCR LBD's responsiveness to various corticosteroid hormones. We found that AncCR is a very sensitive and specific aldosterone receptor, activating transcription 20-fold at subnanomolar concentrations. Like the extant CRs and MRs it is also activated by low doses of DOC and, to a lesser extent, cortisol (Figures 16.2d and 16.3). These results corroborated our hypothesis that the ancestral CR had MR like functions and that aldosterone sensitivity is far more ancient than previously assumed.

16.6 Robustness to uncertainty

Ancestral reconstruction using maximum likelihood will converge on the true ancestral states with increasing confidence as the amount of data at each site increases, if the correct tree and evolutionary model are used. As the number of available sequences related to the ancestor increases, uncertainty about the ancestral state declines; the posterior probability of the maximum-likelihood state at every site approaches 1.0, and the probability of error approaches zero.

In reality, however, the number of available sequences that provide phylogenetic information about an ancestor of interest is always finite, so the maximum-likelihood reconstruction of the ancestral sequence will usually be uncertain. That is, at some sites there will be more than one possible state with posterior probability greater than zero. In most cases, the maximum likelihood estimate will be the true ancestral state, but at some sites the ancestral amino acid may be one with a lower likelihood, resulting in erroneous reconstruction.

There are three potential causes of error in ancestral sequence reconstruction. The first is stochastic error. If a 1000-site protein is reconstructed with 0.99 posterior probability at every site, 1% of all sites—or 10 residues overall—in the maximum-likelihood sequence are expected to have the incorrect state. For a site to be reconstructed erroneously, the probability that the state pattern observed in extant sequences would evolve from the true ancestral state must be lower than the probability that the state pattern would evolve from a different (untrue) state. That is, a low-probability set of evolutionary events must have taken place instead of more-likely scenarios; this sort of error will occur, for example, if two sister sequences that descend from their ancestor on short branches share state i, but this is due to two convergent substitutions from ancestral state j rather than the higher-probability scenario of conservation from state i in the ancestor. Low-probability events do occur, albeit with low frequency; over a large number of sites, states with suboptimal likelihoods will be the true state in a few cases. The probability that the maximum-likelihood state is erroneous at a site is, of course, inversely proportional to the posterior probability of that state. Sites that have alternative reconstructions with non-trivial posterior probabilities are therefore the ones at which the true state is most likely to be different from the maximum-likelihood state.

The second and third potential cause of erroneous states arise from the fact that maximum likelihood calculates the probability of ancestral states conditionally on a phylogenetic tree and evolutionary model. If the tree or the model assumed are incorrect, calculated likelihoods will not accurately reflect the actual likelihoods of the possible ancestral states, and the inferred maximum-likelihood sequence may not be the sequence with the highest true likelihood. Because experimental results concerning the functions of a reconstructed ancestral gene are only as good as the inferred sequence on which they rely, it is important to explore whether the maximum-likelihood sequence and its functions are robust to potential errors in sequence reconstruction induced by these factors.

We characterized the robustness of the AncCR sequence to error in several ways. First, we characterized the robustness of AncCR's aldosterone response to stochastic error. We examined all positions that were ambiguously reconstructed on the maximum-likelihood tree, defined as having an alternate state with a posterior probability of more than 0.20. In all cases but one, the alternate state is found in other aldosterone-activated receptors and is therefore not sufficient to abolish aldosterone sensitivity. We introduced the one exception into AncCR using site-directed mutagenesis; it had no effect on ligand activation (Figure 16.3a). We also used structural information to identify sites likely to be of functional importance. We examined sites that make contact with the ligand in the MR crystal structure (Fagart et al., 2005) and found that only one of these was ambiguously reconstructed. The mutagenized AncCR with the alternate state remained activated by very low aldosterone concentrations (Figure 16.3a). These data indicate that AncCR's aldosterone sensitivity is not likely to be due to stochastic errors in the maximum-likelihood reconstruction of the protein sequence.

Second, we examined the effect of uncertainty about the phylogenetic tree on the AncCR sequence and its inferred function. Although the maximum-likelihood tree was generally well supported, there were a few nodes at which alternative topologies were not ruled out. To address the possibility that different trees might change the inferred ancestral sequence and its function, we used BMCMC to collect a large sample of plausible trees (posterior probability > 0.0002), and the reconstructed the AncCR sequence independently on all 467 trees in the 95% credible set. The ancestral sequence on every tree was identical to that on the maximum-likelihood tree at every site but one, which had an alternate state on some trees. When this state was introduced into the reconstructed protein, AncCR became even more sensitive to aldosterone (Figure 16.3a). This result indicates the AncCR's aldosterone sensitivity is not likely to be an artifact of assuming an incorrect tree.

As for model violation, it is possible that the JTT + gamma model we used is erroneous, and violation of this model's assumptions could have produced some inaccurate ancestral states. In our Bayesian analysis, we integrated over a large number of available models and found that the JTT + gamma model was supported with 100% posterior probability; that is, alternative models therefore had vanishingly small probability due to a very poor fit to the data. We therefore did not explore alternative reconstructions using these other models, because the weighted posterior probability of the ancestral sequence states that would be inferred using these models is approximately zero. This does not mean, however, that JTT + gamma represents the true evolutionary model; it implies only that it is the best of the protein evolutionary models currently implemented. The real evolutionary process is almost always more complex than these models, but no methods are available for determining whether the best-fit model is in fact good enough to predict with high accuracy (Thornton and Kolaczkowski, 2005). Thus we cannot rule out the possibility that model violation could have introduced some errors into our reconstruction. We can say only this: based on our Bayesian analysis, the ancestral sequence we inferred—and our experimental results—are robust to ambiguity in the choice of models from among those currently available.

16.7 The evolution of the MR–aldosterone interaction

The aldosterone-sensitivity of AncCR is surprising, because aldosterone is a relatively recent, tetrapod-specific hormone. Aldosterone has been reliably detected in tetrapods and lungfish (Bentley, 1998), but is absent from teleosts (Jiang et al., 1998), elasmobranchs (Simpson and Wright, 1970; Nunez and Trant, 1999), and agnathans (Bridgham et al., 2006). Prior work has shown that aldosterone's emergence is due to evolutionary modification of a key enzyme in the steroidogenic pathway (Figure 16.1b)—cytochrome P450–11B (Cyp11B). The ancestral function of Cyp11B is to catalyze the 11-hydroxylation of DOC in the synthesis of glucocorticoids, a function present in all jawed vertebrates. Only in tetrapods has this enzyme evolved the additional capacity to hydroxylate corticosterone at the 18-position to produce aldosterone (Nonaka et al., 1995; Jiang et al., 1998; Bulow and Bernhardt, 2002). This novel catalytic function appended aldosterone as a new terminal hormone at the end of the more ancient glucocorticoid synthesis pathway (Figure 16.1b).

Our data, together with this knowledge about the evolution of the steroid-synthesis pathway, indicate that the sensitivity of corticoid receptors to aldosterone is more ancient than the hormone itself. The aldosterone-sensitive AncCR existed in an organism that almost certainly did not produce aldosterone, just as present-day teleosts, elasmobranches, and agnathans contain MRs and CRs that are activated by aldosterone, despite the documented absence of the hormone from these taxa. AncCR must have been regulated by a different ligand; one candidate is DOC. DOC is clearly ancient: it is known to be produced by agnathans (Weisbart and Youson, 1977), and tetrapods, teleosts, and elasmobranchs all make it as an intermediate in the synthesis of other corticosteroids (Figure 16.1b). Our experiments show that AncCR is extremely sensitive to DOC, as are both agnathan CRs (Figure 16.3a). DOC is also an effective activator of the human MR (Hellal-Levy et al., 1999) and may be the physiological MR ligand in teleosts (Sturm et al., 2005).

Aldosterone differs structurally from DOC only by the presence of 18-keto and 11-hydroxyl groups; our experiments show that neither of these moieties affect activation of the ancestral or extant CRs and MRs. In all the receptors we examined, both ancestral and extant, there is a very strong correlation between sensitivity to aldosterone and sensitivity to DOC (Figure 16.3c). Whatever the precise identity of the ancestral ligand, AncCR's sensitivity to aldosterone, like that of the CRs and MRs in species that do not produce the hormone, must be due to its similarity to the endogenous steroids that are the receptor's natural ligands.

These results demonstrate how the aldosterone–MR partnership in tetrapods evolved in a stepwise, Darwinian fashion. Our data show that the receptor's affinity for aldosterone preceded the appearance of the hormone; AncCR's sensitivity to aldosterone was a structural byproduct of the receptor's affinity for its natural hormone. In this sense, the receptor was "preadapted" to bind aldosterone when the hormone appeared much later due to modification of the steroidogenic pathway. We call this dynamic molecular exploitation, because it involves a newly evolved molecule—the hormone, in this case—recruiting into a new interaction a more ancient molecule that was previously constrained by selection for an entirely different function.

16.8 The mechanistic basis for GR evolution

One of the most exciting new applications of ancestral gene resurrection is in determining the mechanistic basis for the evolution of gene function. By resurrecting multiple ancestral nodes on a tree it is possible to determine experimentally when a novel function evolved. Candidate amino acid positions can then be identified as those residues that changed state on the same branch on which the new function emerged; the hypothesis is that these substitutions represent the mechanistic basis for the evolution of the new function. These candidate substitutions can then be introduced into the resurrected ancestral gene to determine their actual effect on the function.

We have used this strategy to study the evolution of the GR's derived functions (Bridgham et al., 2006). Our findings concerning the function of the AncCR indicate that the specific MR–aldosterone partnership in tetrapods is due to the loss of the ancestor's aldosterone-sensitivity in the lineage leading to the GRs of bony vertebrates. To determine when this shift occurred, we resurrected two additional ancestral receptors on the tree, both of which existed after the duplication of AncCR: the GR in the ancestor of all jawed vertebrates (AncGR) and the GR in the ancestor of bony vertebrates (BonyGR). Using transcriptional reporter assays, we showed that AncGR is indeed activated by aldosterone, cortisol, and DOC, but the more recent BonyGR displays the full GR-like phenotype, with no aldosterone response and a reduced sensitivity to cortisol (Figure 16.4a). This result indicates that aldosterone sensitivity was lost from the GRs after the elasmobranch divergence but before the tetrapod/teleost split.

To understand the mechanistic basis for this functional shift, we combined ancestral gene resurrection with mutagenesis to identify the specific amino acids responsible for the GR's loss of aldosterone response. We identified candidate substitutions by finding amino acid changes that are phylogenetically and functionally diagnostic, defined as having occurred on the branch where aldosterone sensitivity was lost, with one state conserved in all the aldosterone-activated receptors and a different state in all aldosterone-insensitive GRs. Although it is possible that functionally crucial sites that changed on the key branch may not have been conserved since then in extant sequences, we reasoned that the residues most important to the functions of the GR and MR would probably be constrained by selection, making them reasonable first candidates.

There were four diagnostic changes that met these criteria. To test their functional importance, we introduced each substitution singly and in combination into the AncCR by site-directed mutagenesis. We determined experimentally whether they were capable of producing the GR-like phenotype: loss of aldosterone sensitivity, with moderate cortisol sensitivity maintained. The combination of S106P and L111Q conferred a GR-like function, reducing the receptor's sensitivity to aldosterone by three orders of magnitude while retaining moderate sensitivity to cortisol and DOC (Figures 16.3a and 16.4b). None of the other mutants tested showed this pattern (Bridgham et al., 2006).

Knowing that multiple substitutions were required to yield the GR function, we sought to determine the order in which these substitutions are likely to have occurred. To reconstruct the trajectory of GR evolution, we introduced each replacement in isolation and found that both are required to yield the full GR phenotype. The L111Q mutation alone radically reduced activation by all the ligands tested (Figure 16.4b). S106P strongly impaired both aldosterone and cortisol sensitivity, but this receptor retained significant DOC sensitivity, suggesting a neutral phenotype with regard to the likely ancestral ligand. In the S106P background, L111Q restores the cortisol response to a level characteristic of extant GRs, while further reducing aldosterone response and leaving DOC activation more or less unchanged (Figure 16.4b).

These results indicate that a mutational path beginning with S106P followed by L111Q converts the ancestor to the modern GR phenotype via a functional intermediate step and is therefore the most likely evolutionary scenario (Maynard-Smith, 1970). This result also points to strong intragenic epistasis—that is, the effect of the L111Q substitution depends on the state at site 106—and indicates that that the order of substitutions strongly constrains the potential evolutionary paths that the protein may take through sequence space. The results we observed for these two substitutions suggest that the ancestral sequence evolved along a neutral ridge through sequence space (Gavrilets, 2004), bypassing the nonfunctional valley represented by L111Q alone.

To illuminate how these substitutions altered the evolving GR's response to ligand, we compared the crystal structures of the human GR and MR and found that the two substitutions cooperate to maintain cortisol activation despite the loss of aldosterone sensitivity (Figure 16.4c; see also Fagart et al., 2005; Li et al., 2005). The substitution of proline for serine at the position corresponding to 106 in the AncCR introduces a kink in a loop between two helices; this kink pulls one of the adjacent helices forward, changing the shape of the ligand pocket in a way that partially destabilizes the binding of hormones. As a result, activation by aldosterone is radically reduced, and

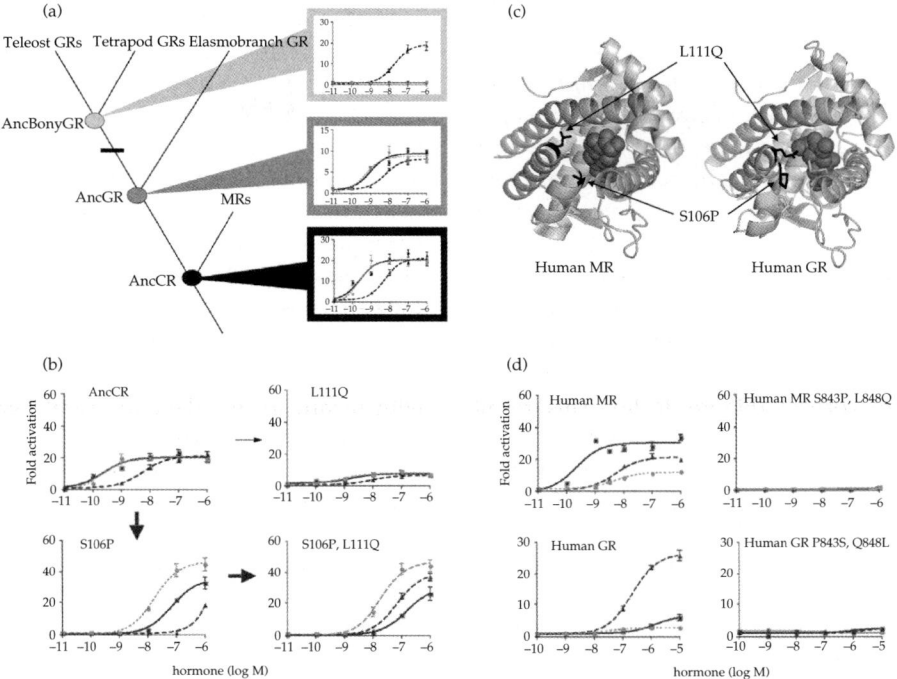

Figure 16.4 Evolution of the GR phenotype. (a) Resurrection of a series of ancestral receptors determines when aldosterone sensitivity was lost from the GR lineage. Ancestral receptors at the three nodes indicated were reconstructed by maximum likelihood, synthesized and functionally characterized in a luciferase reporter assay with increasing doses of aldosterone (solid black line), DOC (gray), and cortisol (dotted black). The GR in the ancestor of all jawed vertebrates retains the aldosterone-sensitive phenotype of AncCR. In contrast, the GR in the ancestor of all bony vertebrates (BonyGR) has the full GR-like phenotype, with response to aldosterone and DOC abolished and moderate sensitivity to cortisol retained. These results indicate that the GR-phenotype emerged on the branch marked with the black bar. (b) Identification of substitutions crucial to the emergence of GR-like function. Four substitutions occurred on the branch where GR function evolved (black bar in panel (a)) and were then conserved, with one state in all aldosterone-insensitive receptors and another in all aldosterone-activated receptors. These were introduced singly and in combination into the ancestral background. One two-fold mutant (S106P/L111Q) had a largely GR-like phenotype, with aldosterone sensitivity reduced by three orders of magnitude and moderate cortisol sensitivity retained. Large arrows show the evolutionary trajectory to the double mutant through a functional intermediate step; small arrow, trajectory involving a nonfunctional intermediate. Reprinted with permission from Bridgham et al. (2006) Evolution of hormone-receptor complexity by molecular exploitation. *Science* **312**: 97–101, © 2006 AAAS. (c) Structural basis for the functional shift caused by the two-fold substitution. S106P and L111Q are plotted on the structures of the human MR and the human GR with their ligands. S106P introduces a kink that excludes aldosterone and moves L111Q into a position where it can form a hydrogen-bond with the unique 17-hydroxyl group of cortisol, stabilizing binding of cortisol in the GR-like structure. (d) The effects of substitutions on function depend on the ancestral sequence background. When introduced into the AncCR sequence, S106P and L111Q together confer the GR-like phenotype, but when the same substitutions are introduced into the extant human MR, they do not have the same effect. The reverse substitutions in the extant human GR background are not capable of reversing the GR-like phenotype to restore the ancestral function.

cortisol—which was a weaker activator to begin with—becomes a completely ineffective ligand. The kink, however, also brings site 111 into a position where it is close to the 17-position on the ligand; the substitution of a polar glutamine for the hydrophobic leucine then forms a hydrogen bond with the 17-hydroxyl, which is only found in cortisol. This bond stabilizes cortisol binding and restores activation by cortisol, compensating for the general reduction in ligand sensitivity induced by S106P. Our findings therefore indicate that the aldosterone specificity of MR evolved by a simple and conserved structural mechanism: two crucial replacements in the GRs that change the general architecture of the binding pocket and then compensate by creating a new stabilizing interaction with one specific ligand. The effect is to wipe out the ancestral sensitivity to aldosterone and retain a moderate response to cortisol.

In summary, our findings demonstrate that the MR–aldosterone partnership evolved in a stepwise fashion consistent with Darwinian theory, but the functions being selected for changed over time. AncCR's sensitivity to aldosterone was present before the hormone, a byproduct of selective constraints on the receptor for activation by the native ligand, just as the agnathan CRs, elasmobranch, and teleost MRs can be activated by aldosterone despite its absence from those organisms. AncCR and its descendant genes were structurally preadapted for activation by aldosterone when that hormone evolved millions of years later. After the duplication that produced GR and MR, only two substitutions in the GR lineage were required to abolish aldosterone sensitivity and yield two receptors with distinct hormone-response profiles. The evolution of an MR that could be independently regulated by aldosterone enabled a more specific endocrine response, because it allowed electrolyte homeostasis to be controlled without also triggering the GR stress response, and vice versa.

16.9 Ancestral gene resurrection for studying structure–function relationships

Many structure–function studies seek to determine the role of individual residues in producing specific protein functions. Candidate residues are often identified by sequence comparisons and then introduced into extant proteins to test the hypothesis that they determine some aspect of function. If intragenic epistasis is important, however, then the effect of a substitution may depend on the sequence at other sites; the only way to reliably determine whether a substitution was crucial for producing a new function is to introduce it into the ancestral background in which it originally occurred.

Understanding the ligand specificity of GR and MR is a ripe goal in molecular endocrinology. These receptors play key roles in numerous diseases, so they are prime drug targets. Their partially overlapping specificities to many synthetic and natural ligands results in unwanted side effects, however. Better insight into the structural basis of the GR's and MR's ligand-binding functions would help guide efforts to design receptor-specific agonists and antagonists.

Having identified S106P and L111Q as phylogenetically and functionally diagnostic substitutions and then verified their ability to produce the GR-like phenotype when introduced into the ancestral sequence, we sought to determine whether these same two substitutions are sufficient to yield a GR-like function when introduced into an extant MR. We constructed a double-mutant humanMR with the homologous S843P and L848Q substitutions. We found that in this background these two substitutions do not produce a cortisol-specific receptor as they do in AncCR; rather, they render the human MR completely unresponsive to cortisol, aldosterone, and DOC (Figure 16.4d). Conversely, we tested whether introducing the ancestral states at these positions into the human GR could restore aldosterone sensitivity as they do in AncCR; however, the P637S/Q642L double substitution in the human GR background also abolished all activation by the receptor (Figure 16.4d).

These data indicate that the functional effect of evolutionarily important residues depends crucially on the amino acids present at other positions in the protein. Our findings show that if we are to understand the mechanistic basis for the evolution of protein functions we must use the ancestral sequence as the substrate for hypothesis testing. Introducing putatively important sequence changes into extant proteins, a common practice, does not reliably affect function in the same way as when the same substitutions occur in the ancestral sequence background.

16.10 Sex-steroid evolution

Our first work in ancestral gene resurrection helped reveal how the evolution of AR's interaction with testosterone and PR's partnership with progesterone evolved. This research involved the resurrection of the common ancestor of the entire SR gene family, or AncSR1 for short. We began by inferring the optimal phylogeny of a large database of SRs and closely related nuclear receptors by both parsimony and BMCMC; both methods found nearly identical trees, and most of the nodes on the tree were strongly supported (Figure 16.5a Thornton, 2001; Thornton et al., 2003). The tree indicated that AncSR1 existed before the divergence of protostomes from deuterostomes, some

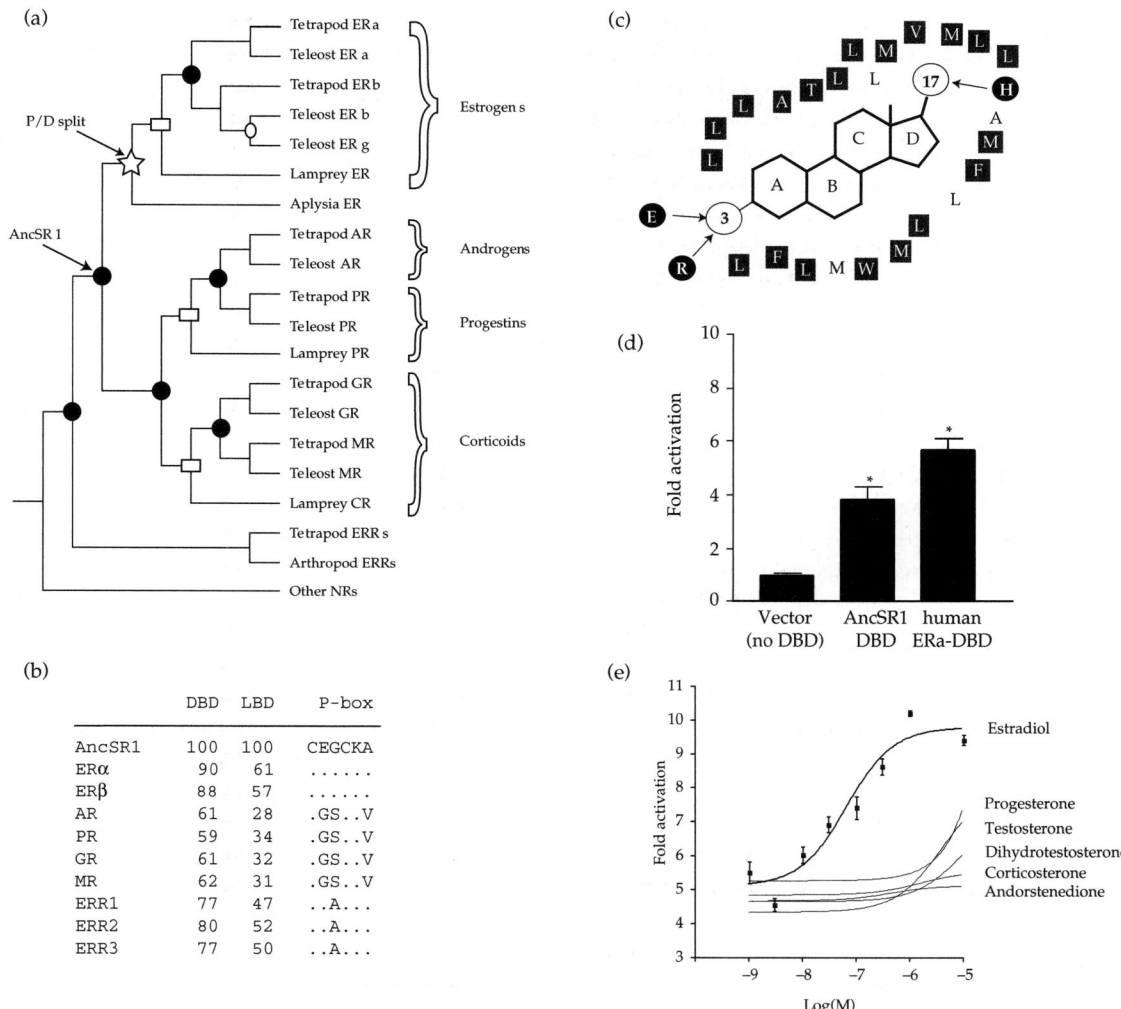

Figure 16.5 The ancestral steroid hormone receptor functioned as an estrogen receptor (see Thornton 2001 and Thornton et al. 2003 for details). (a) Phylogeny of steroid hormone receptors. Dark circles represent gene duplications; star, speciation event that split protostomes from deuterostomes; rectangles, speciation of jawed from jawless vertebrates. AncSR1 is the ancestral gene from which all extant steroid receptors descend. ERRs and other NRs are members of other nuclear receptor families. (b) Sequence similarity of the inferred AncSR1 protein sequence to human steroid and related receptors. Percent amino acid identity of AncSR1 is shown for the DNA-binding domain and the ligand-binding domain. The sequence in the P-box, a region of the DBD that confers specificity for response elements, is also shown. Dots indicate residues identical to AncSR1. (c) The ligand-binding pocket of AncSR1 is almost identical to that of the human estrogen receptor. Residues lining the pocket of AncSR1 are shown in relation to the steroid ligand, based on crystallography of several extant receptors. Gray shaded residues are identical to those in the human ERα. Circled residues interact with steroid moieties at the 3- and 17- positions and confer specificity for the various steroid hormones. (d) AncSR1 activates transcription from estrogen response elements. Activation of an ERE-driven luciferase gene is shown for the DBDs of the resurrected ancestral receptor and the human ERα. Asterisks, significantly different from control, $P<0.01$. Reprinted with permission from Thornton et al. (2003) Resurrecting the ancestral steroid receptor: ancient origin of estrogen signaling. Science 301: 1714–1717, © 2003. (e) AncSR1 activates transcription in response to estrogen but not other hormones. Activation of a luciferase reporter by the ligand-binding domain of the resurrected ancestral receptor is shown over a range of concentrations of several steroid hormones.

600 million years ago, a result consistent with the discovery of ER gene sequences in mollusks (Thornton et al., 2003; Keay et al., 2006). We then inferred the ancestral sequences of the DBDs and LBDs by maximum likelihood, assuming the JTT + gamma model, which was again strongly supported in the Bayesian analysis.

We examined the AncSR1 sequence in light of known structure–function relationships to predict its function, and we hypothesized that it was likely to function like an ER. First, AncSR1 was far more similar to the extant ERs (90% in the DBD and 85% in the residues that line the ligand-binding pocket of the LBD) than it was to the other SRs (at most 62 and 34%, respectively; Figure 16.5b). More specifically, there are six critical residues—three in the LBD and three in the DBD—that are known from crystallographic and mutagenesis studies to discriminate between estrogens and the other steroid hormones and between estrogen-response elements and the elements recognized by the other receptors. Every one of these critical sites in the ancestral protein contained the ER-specific residues, and all were reconstructed with high posterior probability (Figure 16.5b and c).

To test the hypothesis that AncSR1 had ER-like functions, we synthesized cDNA sequences that code for the inferred ancestral protein's functional domains and cloned them into vectors for high-level expression in CHO-K1 cells. Using a luciferase reporter assay, we showed that the AncSR1 DBD activated transcription of genes flanked by estrogen-response elements, to which other SRs do not bind, almost as effectively as modern-day ERs do (Figure 16.5d). This result corroborates the hypothesis that the DNA-binding functions of the ERs represent the conserved ancestral state, and those of the AR, PR, GR, and MR are derived.

The LBD activated transcription in a dose-dependent manner in the presence of low levels of estrogens but was completely insensitive to other steroid hormones (Figure 16.5e). The AncSR1 LBD was less effective than the human ER LBD, activating transcription to a lower level and requiring somewhat higher amounts of estrogens to achieve maximal activation; nevertheless, the specificity of AncSR1 to estrogens was very high. Further, in a ligand-binding assay, the receptor specifically bound radiolabeled estrogens, again with lower affinity than present-day ERs (Thornton et al., 2003).

Together, these data indicated that the ancestor of the entire SR family functioned as a specific ER, activating transcription of genes flanked by estrogen-response elements when (and only when) estrogens are present. The functions of the other members of the SR family—including sensitivity to other hormones, such as testosterone and progesterone—are therefore derived characteristics that emerged after receptor gene duplications and sequence divergence (Thornton et al., 2003). As for the somewhat impaired functions of the AncSR1 LBD compared to the extant ERs, there are two possible interpretations. First, the true ancestral receptor may have been a suboptimal ER, the functions of which were optimized during evolution of the ER lineage. A more likely interpretation in our view is that errors in the inferred sequence of AncSR1 LBD may have impaired the receptor's functions in our hands. Because LBDs are less conserved than DBDs and AncSR1 is such an ancient protein, the AncSR1 LBD was inferred with considerably lower mean posterior probability confidence than either AncSR1 DBD or the AncCR LBD (see Thornton et al., 2003). Accordingly, the expected number of incorrect amino acids in the reconstructed sequence due to stochastic error is higher. Each such mistake has the potential to compromise function, just as nonsynonymous mutations in the sequence would be expected to do. Whatever the explanation for the quantitatively reduced function of the AncSR1, however, it is clear that this ancient protein responded to estrogens, and that sensitivity to the other steroids is a derived function that emerged in lineage leading to the AR, PR, GR, and MR.

In light of the synthesis pathway for producing steroid hormones, our results on the functions of AncSR1 shed light on how the novel interactions of PR and AR with their ligands evolved. Estrogen is the terminal hormone produced in a pathway that uses progesterone and testosterone as intermediates (Figure 16.1b); estrogen synthesis through this pathway appears to be extremely ancient, as all three steroids are present in vertebrates and invertebrates, such as mollusks (D'Aniello et al., 1996; Di Cosmo et al., 2001; Zhu et al., 2003). Our experiments on AncSR1 indicates that that the last hormone to be produced was the first one to serve as an SR ligand. Before progesterone and testosterone served as SR ligands, then, they must have been present as intermediates in the production of estrogen. After the duplication of SR1, duplicated receptors evolved increased affinity for these steroids, turning what had been biochemical stepping stones into *bona fide* hormones.

These results indicate that the interactions of AR and PR with their ligands also evolved by molecular exploitation. In contrast to the MR–aldosterone partnership, the steroids in this case were present before the receptors evolved. For the AR and PR, duplicated receptors diverged in sequence and recruited older ligands, which had previously served an entirely different function, into a novel signaling partnership.

16.11 Future directions in ancestral gene resurrection

We see three major areas for advancement of ancestral gene resurrection. The first is improvements in the methods for phylogenetic inference of ancestral states. Existing evolutionary models do not capture all the dynamics of real evolutionary processes, because they assume a largely homogeneous evolutionary process across sites and lineages. Real sequences are subject to selection pressures that vary considerably among sites and lineages. For example, a site may be subject to strong constraints in one lineage but not in another, a phenomenon called heterotachy (Lopez *et al.*, 2002), which has been shown to undermine the accuracy of likelihood-based phylogenetic methods under some conditions (Kolaczkowski and Thornton, 2004). Several groups, including ours, are working on mixed models to improve performance in the face of heterotachy and other forms of heterogeneity. We need to know whether unincorporated heterogeneity reduces the accuracy of ancestral state inference, and—if it does—determine whether mixed models improve our ability to correctly infer ancestral sequences.

Second, ancestral resurrection can be used even more ambitiously for understanding the evolution of protein function—particularly in determining how specific sequence changes have led to the evolution of new functions. We see three particularly interesting possibilities in this area. The first is to expand on the preliminary work we have reported using site-directed mutagenesis on ancestral sequences to recapitulate evolutionary substitutions in their ancestral background and determine their effect on function. This technique, expanded in a high-throughput framework, could allow the adaptive landscape on which gene sequences have evolved to be characterized, allowing us deep insights into unresolved questions about the evolutionary role of epistasis, the relative importance of substitutions of large and small effect, the prevalence of compensatory and permissive mutations, and the reversibility and contingency of the evolutionary trajectories that actually took place. Characterizing this adaptive landscape will require determining the functions of the substitutions that occurred on critical branches of phylogenies by introducing all the possible combinations of phylogenetically diagnostic residues into ancestral backgrounds.

Another strategy that could yield great insights into the mechanistic basis for the evolution of gene function is to resolve the three-dimensional structures of ancestral proteins—particularly from multiple nodes on a tree—compare them to extant proteins. Together with ancestral mutagenesis experiments to test mechanistic hypotheses, this approach could provide the missing link for understanding *how* specific substitutions generated novel functions, and how protein structures have evolved over time.

A third area for further development is the experimental evolution of ancestral sequences. Microbial experimental evolution systems have proven to be extraordinarily powerful for understanding the nature of the evolutionary process: these systems allow evolution by natural selection to take place in large populations using multiple replicates under controlled laboratory conditions, and the evolutionary intermediates can be sampled at regular intervals, stored in the freezer, and characterized for their sequences, fitness, and functions. We are currently developing a system in which we can subject a recombinant SR to strong selection to evolve affinity for new ligands. The purpose is to understand the mechanisms and evolutionary dynamics by which receptors evolve new specificities. We plan to introduce resurrected ancestral receptors into this system to determine whether the trajectories of sequence evolution taken by receptors in evolving specificity for steroid hormones during real historical evolution represent the only way for that evolutionary problem to be solved. This approach should also let us characterize the nature of the adaptive landscape on which receptors evolve, and the structural constraints that determine the trajectories of sequence evolution.

With these kinds of applications, ancestral gene resurrection will help us gain insight into some of the most difficult, important, and previously intractable problems in evolutionary biology. We have no doubt that others will think of additional ways in which this powerful new tool can advance our understanding of the processes by which genes and their myriad functions have evolved.

References

Aharoni, A., Gaidukov, L., Khersonsky, O., Gould, S.M., Roodveldt, C., and Tawfik, D.S. (2005) The 'evolvability' of promiscuous protein functions. *Nat. Genet.* **37**: 73–76.

Baker, M.E. (2001) Hydroxysteroid dehydrogenases: ancient and modern regulators of adrenal and sex steroid action. *Mol. Cell. Endocrinol.* **175**: 1–4.

Bentley, P.J. (1998) *Comparative Vertebrate Endocrinology.* Cambridge University Press, Cambridge.

Bridgham, J.T., Carroll, S.M., and Thornton, J.W. (2006) Evolution of hormone-receptor complexity by molecular exploitation. *Science* **312**: 97–101.

Bulow, H.E. and Bernhardt, R. (2002) Analyses of the CYP11B gene family in the guinea pig suggest the existence of a primordial CYP11B gene with aldosterone synthase activity. *Eur. J. Biochem.* **269**; 3838–3846.

D'Aniello, A., Di Cosmo, A., Di Cristo, C., Assisi, L., Botte, V., and Di Fiore, M.M. (1996) Occurrence of sex steroid hormones and their binding proteins in *Octopus vulgaris* lam. *Biochem. Biophys. Res. Commun.* **227**: 782–788.

Darwin, C. (1859) *On the Origin of the Species by Means of Natural Selection or the Preservation of Favoured Races in the Struggle for Life.* John Murray, London.

Di Cosmo, A., Di Cristo, C., and Paolucci, M. (2001) Sex steroid hormone fluctuations and morphological changes of the reproductive system of the female of *Octopus vulgaris* throughout the annual cycle. *J. Exp. Zool.* **289**: 33–47.

Fagart, J., Huyet, J., Pinon, G.M., Rochel, M., Mayer, C., and Rafestin-Oblin, M.E. (2005) Crystal structure of a mutant mineralocorticoid receptor responsible for hypertension. *Nat. Struct. Mol. Biol.* **12**: 554–555.

Farman, N. and Rafestin-Oblin, M.E. (2001) Multiple aspects of mineralocorticoid selectivity. *Am. J. Physiol. Renal Physiol.* **280**: F181–F192.

Fryxell, K.J. (1996) The coevolution of gene family trees. *Trends Genet.* **12**: 364–369.

Futuyma, D.J. (1998) *Evolutionary Biology*, 3rd edn. Sinauer Associates, Sunderland, MA.

Gavrilets, S. (2004) *Fitness Landscapes and the Origin of Species.* Princeton University Press, Princeton, NJ.

Green, S. and Chambon, P. (1987) Oestradiol induction of a glucocorticoid-responsive gene by a chimaeric receptor. *Nature* **325**: 75–78.

Greenwood, A.K., Butler, P.C., White, R.B., DeMarco, U., Pearce, D., and Fernald, R.D. (2003) Multiple corticosteroid receptors in a teleost fish: distinct sequences, expression patterns, and transcriptional activities. *Endocrinology* **144**: 4226–4236.

Gronemeyer, H., Gustafsson, J.A., and Laudet, V. (2004) Principles for modulation of the nuclear receptor superfamily. *Nat. Rev. Drug Discov.* **3**: 950–964.

Haag, E.S. and Molla, M.N. (2005) Compensatory evolution of interacting gene products through multifunctional intermediates. *Evolution* **59**: 1620–1632.

Hellal-Levy, C., Couette, B., Fagart, J. Souque, A., Gomez-Sanchez, C., and Rafestin-Oblin, M.E. (1999) Specific hydroxylations determine selective corticosteroid recognition by human glucocorticoid and mineralocorticoid receptors. *FEBS Lett.* **464**: 9–13.

Jiang, J.Q., Young, G., Kobayashi, T., and Nagahama, Y. (1998) Eel (*Anguilla japonica*) testis 11beta-hydroxylase gene is expressed in interrenal tissue and its product lacks aldosterone synthesizing activity. *Mol. Cell. Endocrinol.* **146**: 207–211.

Keay, J., Bridgham, J.T., and Thornton, J.W. (2006) The *Octopus vulgaris* estrogen receptor is a constitutive transcriptional activator: evolutionary and functional implications. Endocrinology **147**: 3861–3869.

Kolaczkowski, B. and Thornton, J.W. (2004) Performance of maximum parsimony and likelihood phylogenetics when evolution is heterogeneous. *Nature* **431**: 980–984.

Li, W.-H. (1997) *Molecular Evolution.* Sinauer Associates, Sunderland, MA.

Li, Y., Suino, K., Daugherty, J., and Xu, H.E. (2005) Structural and biochemical mechanisms for the specificity of hormone binding and coactivator assembly by mineralocorticoid receptor. *Mol. Cell* **19**: 367–380.

Lopez, P., Casane, D., and Philippe, H. (2002) Heterotachy, an important process of protein evolution. *Mol. Biol. Evol.* **19**: 1–7.

Maynard-Smith, J.M. (1970) Natural selection and the concept of a protein space. *Nature* **225**: 563–564.

Nonaka, Y., Takemori, H., Halder, S.K., Sun, T.J., Ohta, M., Hatano, O., Takakusu, A., and Okamoto, M. (1995) Frog cytochrome P-450 (11 beta,aldo), a single enzyme involved in the final steps of glucocorticoid and mineralocorticoid biosynthesis. *Eur. J. Biochem.* **229**: 249–256.

Nunez, S. and Trant, J.M. (1999) Regulation of interrenal gland steroidogenesis in the Atlantic stingray (*Dasyatis sabina*). *J. Exp. Zool.* **284**: 517–525.

Page, R.D.M. and Holmes, E.C. (1998) *Molecular Evolution: a Phylogenetic Approach*. Blackwell Scientific, Oxford.

Simpson, T.H. and Wright, R.S. (1970) Synthesis of corticosteroids by the interrenal gland of selachian elasmobranch fish. *J. Endocrinol.* **46**: 261–268.

Sturm, A., Bury, N., Dengreville, L., Fagart, J., Flouriot, G., Rafestin-Oblin, M.E., and Prunet, P. (2005) 11-deoxycorticosterone is a potent agonist of the rainbow trout (*Oncorhynchus mykiss*) mineralocorticoid receptor. *Endocrinology* **146**: 47–55.

Thornton, J.W. (2001) Evolution of vertebrate steroid receptors from an ancestral estrogen receptor by ligand exploitation and serial genome expansions. *Proc. Natl. Acad. Sci. USA* **98**: 5671–5676.

Thornton, J.W. and DeSalle, R. (2000) A new method to localize and test the significance of incongruence: detecting domain shuffling in the nuclear receptor superfamily. *Syst. Biol.* **49**: 183–201.

Thornton, J.W. and Kolaczkowski, B. (2005) No magic pill for phylogenetic error. *Trends Genet.* **21**: 310–311.

Thornton, J.W., Need, E., and Crews, D. (2003) Resurrecting the ancestral steroid receptor: ancient origin of estrogen signaling. *Science* **301**: 1714–1717.

Weisbart, M. and Youson, J.H. (1977) In vivo formation of steroids from [1,2,6,7-3H]-progesterone by the sea lamprey, *Petromyzon marinus* L. *J. Steroid. Biochem.* **8**: 1249–1252.

Yang, Z., Kumar, S., and Nei, M. (1995) A new method of inference of ancestral nucleotide and amino acid sequences. *Genetics* **141**: 1641–1650.

Zhang, J. and Nei, M. (1997) Accuracies of ancestral amino acid sequences inferred by the parsimony, likelihood, and distance methods. *J. Mol. Evol.* **44** (suppl. 1): S139–S146.

Zhu, W., Mantione, K., Jones, D., Salamon, E., Cho, J.J., Cadet, P., and Stefano, G.B. (2003) The presence of 17-beta estradiol in *Mytilus edulis* gonadal tissues: evidence for estradiol isoforms. *Neuro. Endocrinol. Lett.* **24**: 137–140.

CHAPTER 17

A thermophilic last universal ancestor inferred from its estimated amino acid composition

Dawn J. Brooks and Eric A. Gaucher

17.1 Introduction

The last universal ancestor (LUA) represents a relatively accessible theoretical intermediary between extant cellular organisms and early, pre-cellular "life". Through analysis of modern-day genomes it is possible to infer characteristics of the LUA (Lazcano and Forterre, 1999) and these provide important clues to early evolution. One feature that has attracted significant interest is the temperature of the environment in which the LUA lived (Woese, 1987; Galtier and Lobry, 1997; Galtier et al., 1999; Bocchetta et al., 2000; Brochier and Philippe, 2002; Whitfield, 2004). The earliest evidence, from analysis of a phylogenetic reconstruction of the tree of life, prompted the proposal that the LUA was a (hyper)thermophile; that is, it lived at temperatures above 55°C (Woese, 1987). However, reconstruction of the tree remains controversial, as do the inferences drawn from it (Bocchetta et al., 2000; Brochier and Philippe, 2002). Although it has been asserted that the inferred G+C content of rRNA in the LUA supports a mesophilic lifestyle (Galtier et al., 1999), that claim does not hold up under scrutiny (see the Results and discussion section, below). Characteristics of ancestral proteins have also been brought to bear on this question. Experimentally reconstructed elongation factor EF-Tu proteins of early ancestral bacteria were found to have temperature optima falling between 55 and 65°C, implying that those ancestors were thermophilic in nature (and by the most parsimonious extension, that the LUA was also; Gaucher et al., 2003). In addition, the amino acid composition of the inferred sequences of signal recognition particle and a tRNA synthetase in the LUA was reported to be more similar to that of extant thermophiles than mesophiles (Di Giulio, 2001). Because of the contradictory nature of the evidence regarding the optimal growth temperature (OGT) of the LUA, a fresh approach that provides alternative data should help advance the debate.

There have been several studies reporting a relationship between the OGT and amino acid composition (Kreil and Ouzounis, 2001; Tekaia et al., 2002; Nakashima et al., 2003; Singer and Hickey, 2003). Taking advantage of this relationship, inferred amino acid composition of proteins in the LUA could be used to infer whether it was a thermophile or a mesophile. We described previously an approach analogous to that used by Galtier et al. (1999) to infer the ancestral G+C content of RNA, but addressing proteins rather than RNA (Brooks et al., 2004). Briefly, the expectation-maximization (EM) approach was used to infer ancestral amino acid frequencies, where in each iteration expected counts were derived from posterior distributions at each site (thereby avoiding the bias associated with inference of a single discrete ancestral sequence; Krishnan et al., 2004). Applying this approach to estimate the amino acid composition of 65 proteins in the LUA, we found that composition to be more similar to that of extant thermophiles than mesophiles (Brooks et al., 2004). In the current analysis, we examined whether our previous result is robust with respect to the OGT of the taxa used to infer the amino acid composition of proteins in the LUA. We found that even if only mesophilic species are used to derive the estimated ancestral amino acid composition, that composition is most similar to that of thermophiles, as measured by Euclidean distance. We show that the relative mean Euclidean distance between the amino acid composition in any one species and that of a set of mesophiles or thermophiles can be used unequivocally to classify it. Thus, the inferred amino acid composition in the LUA allows us to classify it as a thermophile.

17.2 Methods

17.2.1 Included taxa and proteins

We sought to include orthologous proteins from as broad a phylogenetic distribution of genomes as possible while representing thermophiles and mesophiles equally in the data-set. Because it was important to utilize orthologs rather than paralogs in the analysis, and because orthologs can be difficult to distinguish from paralogs, we relied upon an established database, the Clusters of Orthologous Groups (COG) database (Tatusov et al., 2001) to aid the selection of proteins. Orthologous groups from complete genomes from 30 major phylogenetic groups were available as of early 2004.

To be relatively confident that a protein family had been present in the LUA, we required two basic criteria be met. First, we required that members of the family be present in the clear majority of taxa (>25). Second, we sought to exclude from the analysis protein families whose presence in the majority of taxa might be due to horizontal transfer between the primary lineages rather than to vertical inheritance. To meet this latter criterion, we required bacterial, archaeal, and eukaryotic family members to form separate phylogenetic clades. In addition, we selected families in which the presence of paralogs would not confound the construction of a phylogenetic tree from the concatenated protein sequences; that is, any paralogs had to be the result of post-speciation duplications, clustering as neighbors on the protein family tree.

Three criteria were used to select genomes for inclusion in the analysis. First, we sought an equal number of thermophiles and mesophiles. Second, inclusion of a genome should not dramatically reduce the number of shared orthologs meeting the criteria listed above. Third, we sought to represent the broadest possible phylogenetic distribution of taxa.

Using the COG database, seven thermophiles—two bacteria, *Aquifex aeolicus* and *Thermotoga maritima*, and five archaea, *Aeropyrum pernix*, *Thermoplasma acidophilum*, *Methanococcus jannaschii*, *Pyrococcus horikoshii*, and *Archaeoglobus fulgidus*—were available for inclusion in our analysis (ignoring closely related taxa such as *P. horikoshii* and *Pyrococcus abyssi*). A phylogenetic tree was built for the 34 concatenated orthologs of the mesophilic taxa, and the seven which represented the greatest total branch lengths, and thus could be assumed to represent the greatest phylogenetic diversity, were selected: one eukaryote, *Saccharomyces cerevisiae*, and six bacteria, *Synechocystis*, *Xylella fastidiosa*, *Helicobacter pylori*, *Treponema pallidum*, *Chlamydia pnemoniae*, and *Bacillus subtilis*. One mesophile, *Halobacterium* sp. NRC-1, was excluded because its inclusion would have reduced the number of sequences meeting our criteria from 34 to 26.

Two additional taxa not available in the COG database were included in the analysis to increase the phylogenetic diversity of the individual mesophilic and thermophilic species sets. *Methanosarcina acetivorans* was included in order to have representation of a mesophilic archaean. Inclusion of *Thermoanaerobacter tengcongensis* allowed for representation of a thermophilic bacterium known not to be located basally in the bacterial lineage (it is a member of *Firmicutes*, and therefore clusters with *B. subtilis*.) To identify the orthologs from *T. tengcongensis* and *M. acetivorans* belonging to each COG protein family, profile hidden Markov models were built using the alignment of proteins collected for the 14 COG taxa. These were then used to search the database of predicted protein sequences for each of the two genomes. Best hits were individually examined to ascertain that their annotation was consistent with the COG protein family. We were unable to identify orthologs for three COG families in both the additional taxa, so that our final set of ortholog families was reduced to 31 (Table 17.1). The OGT of each species analyzed in this study is listed in Table 17.2.

17.2.2 Alignments and phylogenetic trees

The program T-Coffee (Notredame et al., 2000) with default parameter settings was used to build alignments. Columns containing gaps were removed. Concatenation of the ungapped alignments resulted in a single alignment of 4449 residues. A phylogenetic tree was inferred using the neighbor-joining algorithm (Saitou and Nei, 1987)

as implemented in the Phylip software package (Felsenstein, 1993), using its default parameter settings (Figure 17.1a). The topology of the neighbor-joining tree was found to be congruent with a consensus tree for 100 bootstrap replicates (see Figure 17.2), although the bootstrap support was as low as 59 for certain clusters within the bacterial lineage. The Bayesian phylogenetic inference software MrBayes version 3 (Ronquist and Huelsenbeck, 2003) was also used to build a phylogenetic tree (Figure 17.1b). Markov chain Monte Carlo resampling of tree parameters was performed with four chains and allowed to run for 150 000 generations. A mixed model of amino acid substitution was used. Both neighbor-joining and Bayesian trees were midpoint-rooted; however, analyses using alternative rootings of the 16-taxon tree, either at the base of the eukaryotic/archaeal divergence or the base of the bacterial divergence, led to identical conclusions as those using the midpoint-rooted tree. Alignments for the sequences of mesophilic and thermophilic taxa were extracted from the larger alignment of 16 taxa (i.e. preserving the columns thereof). Similarly, trees for the mesophilic and thermophilic taxa were extracted from the 16-taxon tree, using the branch lengths and topology of that tree.

Table 17.1 List of COG database families included in the analysis

COG ID	Protein name
COG0012	Predicted GTPase
COG0024	Methionine aminopeptidase
COG0048	Ribosomal protein S12
COG0051	Ribosomal protein S10
COG0052	Ribosomal protein S2
COG0080	Ribosomal protein L11
COG0081	Ribosomal protein L1
COG0087	Ribosomal protein L3
COG0088	Ribosomal protein L4
COG0090	Ribosomal protein L2
COG0091	Ribosomal protein L22
COG0092	Ribosomal protein S3
COG0093	Ribosomal protein L14
COG0097	Ribosomal protein L6
COG0098	Ribosomal protein S5
COG0100	Ribosomal protein S11
COG0102	Ribosomal protein L13
COG0103	Ribosomal protein S9
COG0180	Tryptophanyl-tRNA synthetase
COG0184	Ribosomal protein S15P/S13E
COG0185	Ribosomal protein S19
COG0186	Ribosomal protein S17
COG0197	Ribosomal protein L16/L10E
COG0199	Ribosomal protein S14
COG0200	Ribosomal protein L15
COG0201	Preprotein translocase subunit SecY
COG0244	Ribosomal protein L10
COG0250	Transcription antiterminator
COG0495	Leucyl-tRNA synthetase
COG0522	Ribosomal protein S4 and related proteins
COG0541	Signal recognition particle GTPase

Table 17.2 Species OGT

Species	OGT (°C)
Saccharomyces cerevisiae	25
Synechocystis	25
Xylella fastidiosa	26
Bacillus subtilis	30
Chlamydia pnemoniae	37
Helicobacter pylori	37
Treponema pallidum	37
Methanosarcina acetivorans	40
Thermoplasma acidophilum	60
Thermoanaerobacter tengcongensis	75
Thermotoga maritima	80
Methanococcus jannaschii	82
Aquifex aeolicus	85
Archaeoglobus fulgidus	85
Aeropyrum pernix	90
Pyrococcus horikoshii	95

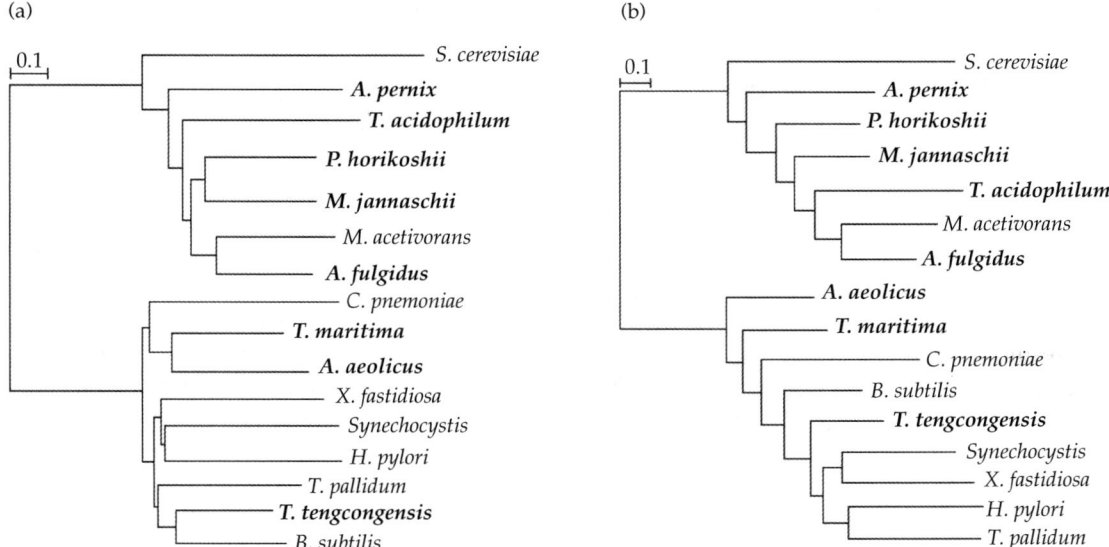

Figure 17.1 The phylogenetic trees used to derive EM estimates of ancestral amino acid composition using all 16 taxa. Mesophiles are in italics and thermophiles are in bold italics. Scale bars indicate 0.1 replacements/site per unit of evolutionary time. (a) Neighbor-joining tree. Trees employed for estimates based solely on mesophilic or thermophilic taxa used the branch lengths and topology of the 16-taxon neighbor-joining tree. (b) Bayesian tree. Genus names are given in the text.

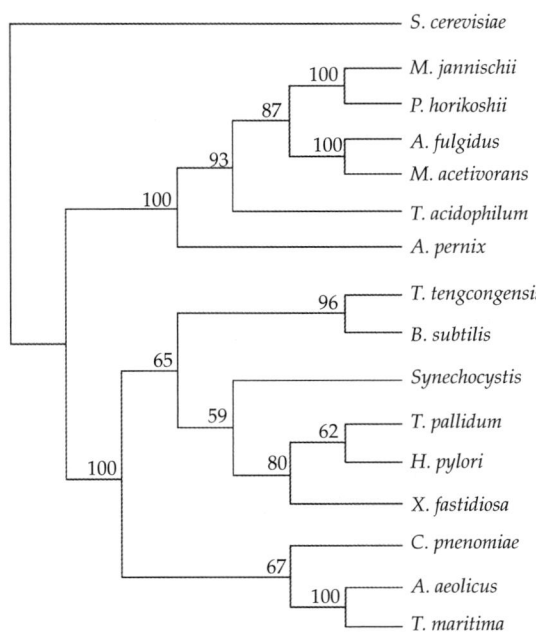

Figure 17.2 Consensus 16-taxon neighbor-joining tree for 100 bootstrap replicates. Genus names are given in the text.

17.2.3 The EM implementation

We used the rate matrix of Jones *et al.* (1992) as the model of evolution. A discrete gamma distribution was used to allow for rate variation between columns of the alignment, the rate categories being estimated using the software package PAML (Yang, 1997). The program implementing the EM method is described elsewhere (Brooks *et al.*, 2004) and available upon request from D.J.B.

17.2.4 Jackknife test of Euclidean distance as a classifier

For each of the 16 taxa in turn, the test species was removed from the set of reference thermophilic and mesophilic species. The Euclidean distance between the amino acid composition of the set of 31 proteins within the test species and each of the reference species was calculated. The mean Euclidean distance between the test and the reference thermophilic species and the test and the reference mesophilic species was determined.

17.3 Results and discussion

Sixteen fully sequenced genomes, including eight thermophiles and eight mesophiles and representing a broad phylogenetic distribution, were included in the analysis. (We did not make the distinction between thermophiles and hyperthermophiles, but instead grouped together all those species with OGT >55°C as thermophiles.) In selecting a set of protein families for analysis, the set was restricted to those in which orthology between members of a family within the 16 taxa was unambiguous and there was an absence of evidence for horizontal transfer between the primary lineages (bacteria, archaea, and eukaryotes). Thirty-one proteins, the majority of them ribosomal, met these criteria. For a list of these proteins see Table 17.1.

Sequences of the 31 families were aligned and used to build a phylogenetic tree for the 16 taxa using either neighbor-joining or Bayesian phylogenetic methods (Figure 17.1). Difficulty resolving the bacterial lineages within the tree is suggested by weak bootstrap support for the branching order of much of the bacterial lineage in the neighbor-joining tree (Figure 17.2) and by the fact that the two taxa, *Bacillus subtilis* and *Thermoanaerobacter tengcongensis*, belonging to the phylum *Firmicutes*, are not correctly clustered in the Bayesian tree. Of particular potential relevance to our investigation, because basal lineages (of moderate or short branch lengths) will have relatively strong influence on the estimated ancestral sequence, is whether the thermophilic species *Aq. aeolicus* and *Th. maritima* truly represent basal branches in the bacterial lineage. (The mesophile *S. cerevisiae* is basal in the eukaryotic/archael lineages in both trees.) Because these two thermophilic species are basal in the Bayesian tree but not in the neighbor-joining tree, we employed an additional phylogenetic tree in which the bacterial taxa are related to each other by a star phylogeny, with branch lengths to the last common ancestor of the bacteria equal to the average distance of the bacterial divergence in the neighbor-joining tree and the eukaryotic/archaeal topology and branch lengths taken from the neighbor-joining tree. If anything, the star phylogeny is biased toward mesophilic species in terms of reconstruction of the ancestral sequence, because all taxa have equal influence on the ancestral sequence and six of the nine bacterial species are mesophiles. Because no outgroup exists for our phylogenies, trees were midpoint-rooted; however, moving the root all the way to the base of the bacterial divergence or to the divergence of the archaea and eukaryotes does not result in qualitatively different results to those reported here.

Five estimates of the amino acid composition in the LUA were derived. For three, we used sequences of all 16 taxa with the three alternative phylogenetic trees described above (neighbor-joining, Bayesian, and neighbor-joining plus star). For the fourth we used only the sequences of the eight mesophilic taxa, and for the fifth we used only sequences of the eight thermophilic taxa. For the estimates using solely mesophilic or thermophilic taxa, phylogenetic trees that had the same topology and branch lengths as the 16-taxon neighbor-joining tree were assumed.

As with any EM approach (Dempster *et al.*, 1977), our method consists of an iteration of expectation and maximization steps. In the expectation step, the posterior probabilities of all 20 amino acids at the root node of a phylogenetic tree are derived for each position of the alignment, assuming their prior probabilities in the ancestral sequence and a model of evolution (Brooks *et al.*, 2004). From these, expected counts of each amino acid in the ancestral sequence can be calculated. In the maximization step, the frequency of each amino acid in the ancestral sequence is estimated as the expected counts of that amino acid in the reconstructed sequence divided by the length of the sequence. These new estimates of ancestral amino acid frequencies are used as the prior probabilities in the next expectation step, and the procedure is iterated to convergence (Brooks *et al.*, 2004). For estimates of amino acid frequencies in the LUA using the different sequence sets and phylogenetic trees see Table 17.3.

To determine whether the amino acid composition inferred in the LUA is more similar to that of extant mesophiles or thermophiles, we calculated the mean Euclidean distance between the estimated amino acid composition in the LUA and the composition observed within the thermophilic and

Table 17.3 Amino acid frequencies estimated in the LUA.

Set All	Amino acid frequency						
	Root NJ	All Bayes	All NJ + star	Thermo NJ	Meso NJ	Mean	SD
Ala	0.0830	0.0831	0.0827	0.0840	0.0857	0.0837	0.0012
Arg	0.0874	0.0897	0.0868	0.0897	0.0844	0.0876	0.0022
Asn	0.0263	0.0239	0.0266	0.0239	0.0329	0.0267	0.0037
Asp	0.0354	0.0350	0.0356	0.0372	0.0369	0.0360	0.0010
Cys	0.0022	0.0027	0.0023	0.0027	0.0018	0.0023	0.0004
Gln	0.0153	0.0150	0.0155	0.0141	0.0184	0.0157	0.0016
Glu	0.0773	0.0770	0.0767	0.079	0.0637	0.0747	0.0062
Gly	0.0865	0.0858	0.0866	0.0846	0.0913	0.0870	0.0026
His	0.0217	0.0213	0.0217	0.0209	0.0222	0.0216	0.0005
Ile	0.1040	0.1031	0.1036	0.1046	0.0994	0.1029	0.0021
Leu	0.0706	0.0696	0.0708	0.0714	0.0722	0.0709	0.0010
Lys	0.1168	0.1182	0.1172	0.1118	0.1056	0.1139	0.0053
Met	0.0199	0.0185	0.0203	0.0194	0.0225	0.0201	0.0015
Phe	0.0261	0.0265	0.0263	0.0243	0.0277	0.0262	0.0012
Pro	0.0422	0.0438	0.0420	0.0445	0.0397	0.0424	0.0019
Ser	0.0232	0.0208	0.0236	0.0200	0.0325	0.0240	0.0050
Thr	0.0363	0.0365	0.0364	0.0369	0.0411	0.0374	0.0021
Trp	0.0016	0.0022	0.0017	0.0045	0.0011	0.0022	0.0013
Tyr	0.0176	0.0188	0.0175	0.0220	0.0140	0.0180	0.0029
Val	0.1063	0.1082	0.1059	0.1042	0.1067	0.1063	0.0014

The three data-sets are indicated as: All, all 16 taxa; Thermo, the eight thermophilic taxa; and Meso, the eight mesophilic taxa. The alternative trees are indicated as NJ, the neighbor-joining tree; Bayes, the Bayesian tree; and NJ + star, the neighbor-joining tree with the bacteria assigned a star phylogeny. The mean and standard deviation (SD) are provided.

mesophilic sequence sets used in the analysis. The mean Euclidean distance between the estimated LUA and the thermophilic amino acid composition is significantly smaller than the mean distance between the LUA and the mesophilic amino acid composition (P value < 0.05) for all data-sets. It is noteworthy that this is the case even for LUA estimates derived using only mesophilic sequences (Figure 17.3).

Using jackknife resampling, we examined whether the relative size of the mean Euclidean distance between the amino acid composition of the set of 31 proteins in one species and that of the mesophilic species and the mean Euclidean distance between the amino acid composition of the protein set in the same species and that of the thermophilic species could be used successfully to classify that species as a mesophile or thermophile; that is, whether a species may be classified according to which set its amino acid composition is more similar to, as measured by Euclidean distance. We found this proposed classifier to have an accuracy of 100% (Figure 17.3). Accordingly, based on the inferred amino acid composition of a set of 31 proteins in the LUA, the LUA can be classified unequivocally as a thermophile, even when proteins of modern-day mesophiles alone are used to derive the estimate.

Our method for estimating amino acid composition of ancestral proteins is closely analogous to that of Galtier et al. (1999), in which EM was used to infer G + C content of ancestral rRNA sequences from extant ones. Those investigators, however, concluded that the inferred composition of rRNA in the LUA is inconsistent with it having been a thermophile. Because their findings are in direct contradiction to ours, we feel it is worthwhile to briefly discuss their data and analysis. It is apparent from the data presented in Galtier et al. (1999) that there is no, or at most a very weak,

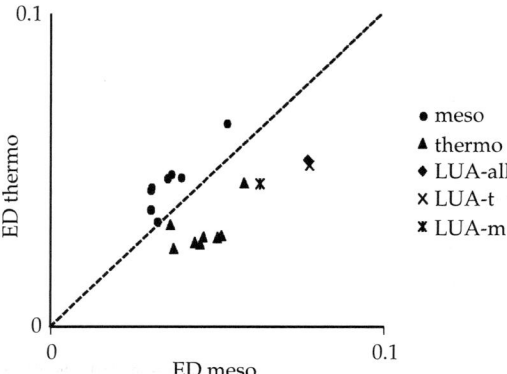

Figure 17.3 Jackknife data examining whether the relative average Euclidean distance (ED) between a species and either a set of mesophiles or a set of thermophiles may be used as a classifier. Each data point represents the mean Euclidean distance between the amino acid composition of the set of 31 proteins in a test species and the reference mesophilic (x axis) and thermophilic (y axis) species. meso, mesophilic test species; thermo, thermophilic test species; LUA-all, LUA-t, and LUA-m, the estimated amino acid compositions in the LUA using, respectively, all 16 taxa, the eight thermophilic taxa, and the eight mesophilic taxa.

correlation between OGT and rRNA G + C content for species with OGT values of less than or equal to 40°C. Consequently, G + C content cannot be a statistically sound means of predicting the OGT of a species. It is apparent from their data, in fact, that the inferred rRNA G + C content in the LUA is compatible with *either* a thermophilic or a mesophilic lifestyle, although not with a hyperthermophilic one (OGT > 80°C).

Although we have focused here on using the estimated amino acid composition in the LUA to make inferences about the OGT of the LUA, this composition may also provide additional clues to the early evolution of life, such as the establishment of the genetic code, a possibility that we have explored elsewhere (Brooks et al., 2002, 2004). Analysis of protein sequences of extant organisms may be a key, yet under-utilized resource for constructing a more complete model of early life on this planet.

17.4 Conclusions

Our results provide strong support for the hypothesis that the LUA was a thermophile, that it lived at temperatures above 55°C. Using Euclidean distance as a measure, the estimated amino acid composition of proteins in the LUA is more similar to that of extant thermophiles than that of mesophiles, even when that estimate is derived using sequences solely from mesophilic species. We show using jackknife sampling that mean Euclidean distance of the protein amino acid composition of a species to the composition in a set of mesophilic or thermophilic species is 100% accurate as a classifier, choosing the set to which it is closest, and thus the LUA may be inferred to be a thermophile.

We note that a majority of data currently supports a proposed hot environment for the LUA. Approaches based on reconstruction of ancestral protein sequences consistently imply a thermophilic LUA (Di Giulio, 2001; Gaucher et al., 2003; Brooks et al., 2004). We are currently extending these previous studies by resurrecting EFs inferred from the ancient amino acid compositions of this study to determine what, if any, effects these prior probabilities have on the reconstructed sequence and subsequent temperature stability of the ancestral proteins. Regardless of this outcome, results from other scientific disciplines support our view. For instance, geologic evidence based on low $\delta^{18}O$ values in 3.5–3.2-billion-year-old cherts from Barberton greenstone belt in South Africa suggest that the ocean temperature was 55–85°C, consistent with the OGT of thermophilic microorganisms (Knauth and Lowe, 2003; Knauth, 2005). Current understanding places the LUA within this Archaean period.

As we noted above, the inferred G + C content of rRNA in the LUA does not exclude the possibility of a thermophilic LUA. Phylogeny-based inferences of ancestral environments, however, have been divided in their conclusions (Bocchetta et al., 2000; Brochier and Philippe, 2002) and seem likely to remain so until agreement can be reached on the appropriate means for phylogeneic reconstruction of such anciently diverging lineages. Nonetheless, we are optimistic that with additional data and analyses, the remaining discrepancies between approaches will be resolved, leading ultimately to a consensus view.

17.5 Acknowledgments

We are grateful to Steve Benner, David Pollock, and Joe Thornton for helpful discussions. This work was supported by grants from the National Science Foundation (D.J.B.) and National Aeronautics and Space Administration's Exobiology and Astrobiology programs (E.A.G.).

References

Bocchetta, M., Gribaldo, S., Sanangelantoni, A., and Cammarano, P. (2000) Phylogenetic depth of the bacterial genera *Aquifex* and *Thermotoga* inferred from analysis of ribosomal protein, elongation factor, and RNA polymerase subunit sequences. *J. Mol. Evol.* **50**: 366–380.

Brochier, C. and Philippe, H. (2002) Phylogeny: a non-hyperthermophilic ancestor for bacteria. *Nature* **417**: 244.

Brooks, D.J., Fresco, J.R., Lesk, A.M., and Singh, M. (2002) Evolution of amino acid frequencies in proteins over deep time: inferred order of introduction of amino acids into the genetic code. *Mol. Biol. Evol.* **19**: 1645–1655.

Brooks, D.J., Fresco, J.R., and Singh, M. (2004) A novel method for estimating ancestral amino acid composition and its application to proteins of the Last Universal Ancestor. *Bioinformatics* **20**: 2251–2257.

Dempster, A.P., Laird, N.M., and Rubin, D.B. (1977) Maximum likelihood from incomplete data via em algorithm. *J. R. Stat. Soc. B Met.* **39**: 1–38.

Di Giulio, M. (2001) The universal ancestor was a thermophile or a hyperthermophile. *Gene* **281**: 11–17.

Felsenstein, J. (1993) *PHYLIP (Phylogeny Inference Package) version 35c.* Department of Genetics, University of Washington, Seattle.

Galtier, N. and Lobry, J.R. (1997) Relationships between genomic G+C content, RNA secondary structures, and optimal growth temperature in prokaryotes. *J. Mol. Evol.* **44**: 632–636.

Galtier, N., Tourasse, N., and Gouy, M. (1999) A non-hyperthermophilic common ancestor to extant life forms. *Science* **283**: 220–221.

Gaucher, E.A., Thomson, J.M., Burgan, M.F., and Benner, S.A. (2003) Inferring the palaeoenvironment of ancient bacteria on the basis of resurrected proteins. *Nature* **425**: 285–288.

Jones, D.T., Taylor, W.R., and Thornton, J.M. (1992) The rapid generation of mutation data matrices from protein sequences. *Comput. Appl. Biosci.* **8**: 275–282.

Knauth, L.P. (2005) Temperature and salinity history of the Precambrian ocean: implications for the course of microbial evolution. *Palaeogeogr. Palaeocl.* **219**: 53–69.

Knauth, L.P. and Lowe, D.R. (2003) High Archean climatic temperature inferred from oxygen isotope geochemistry of cherts in the 3.5 Ga Swaziland Supergroup, South Africa. *Geol. Soc. Am. Bull.* **115**: 566–580.

Kreil, D.P. and Ouzounis, C.A. (2001) Identification of thermophilic species by the amino acid compositions deduced from their genomes. *Nucleic Acids Res.* **29**: 1608–1615.

Krishnan, N.M., Seligmann, H., Stewart, C.B., De Koning, A.P., and Pollock, D.D. (2004) Ancestral sequence reconstruction in primate mitochondrial DNA: compositional bias and effect on functional inference. *Mol. Biol. Evol.* **21**: 1871–1883.

Lazcano, A. and Forterre, P. (1999) The molecular search for the last common ancestor. *J. Mol. Evol.* **49**: 411–412.

Nakashima, H., Fukuchi, S., and Nishikawa, K. (2003) Compositional changes in RNA, DNA and proteins for bacterial adaptation to higher and lower temperatures. *J. Biochem. (Tokyo)* **133**: 507–513.

Notredame, C., Higgins, D.G., and Heringa, J. (2000) T-Coffee: a novel method for fast and accurate multiple sequence alignment. *J. Mol. Biol.* **302**: 205–217.

Ronquist, F. and Huelsenbeck, J.P. (2003) MrBayes 3: Bayesian phylogenetic inference under mixed models. *Bioinformatics* **19**: 1572–1574.

Saitou, N. and Nei, M. (1987) The neighbor-joining method: a new method for reconstructing phylogenetic trees. *Mol. Biol. Evol.* **4**: 406–425.

Singer, G.A. and Hickey, D.A. (2003) Thermophilic prokaryotes have characteristic patterns of codon usage, amino acid composition and nucleotide content. *Gene* **317**: 39–47.

Tatusov, R.L., Natale, D.A., Garkavtsev, I.V., Tatusova, T.A., Shankavaram, U.T., Rao, B.S. *et al.* (2001) The COG database: new developments in phylogenetic classification of proteins from complete genomes. *Nucleic Acids Res.* **29**: 22–28.

Tekaia, F., Yeramian, E., and Dujon, B. (2002) Amino acid composition of genomes, lifestyles of organisms, and evolutionary trends: a global picture with correspondence analysis. *Gene* **297**: 51–60.

Whitfield, J. (2004) Origins of life: born in a watery commune. *Nature* **427**: 674–676.

Woese, C.R. (1987) Bacterial evolution. *Microbiol. Rev.* **51**: 221–271.

Yang, Z. (1997) PAML: a program package for phylogenetic analysis by maximum likelihood. *Comput. Appl. Biosci.* **13**: 555–556.

CHAPTER 18

The resurrection of ribonucleases from mammals: from ecology to medicine

Slim O. Sassi and Steven A. Benner

18.1 Introduction

The family of proteins related to bovine pancreatic ribonuclease A (RNase A) provided the first biomolecular system to be analyzed using experimental paleogenetics. This protein family was chosen in 1979, as it was one of only three families of protein sequences that were sufficiently well represented in the then modest database for which one might consider doing experimental resurrections. Other families that were also considered included cytochrome *c*, which had been developed as a paradigm for molecular evolution by Margoliash (1963, 1964), and hemoglobin, which was studied as a model for biomolecular adaptation (Riggs, 1959; Bonaventura *et al.*, 1974). Cytochromes are substrates for cytochrome oxidases, and it was not considered possible to resurrect both ancestral cytochromes and their ancient oxidases. Hemoglobins are complicated to express, a problem solved only later. This left the RNases.

Fortunately, Jaap Beintema and his colleagues in Groningen had done an excellent job of sequencing (at the level of the protein) ribonucleases from a wide range of ruminants and closely related non-ruminant mammals (Beintema and Gruber, 1967, 1973; Gaastra *et al.*, 1974; Groen *et al.*, 1975; Welling *et al.*, 1975; Emmens *et al.*, 1976; Kuper and Beintema, 1976; Muskiet *et al.*, 1976; Vandenberg *et al.*, 1976; Vandijk *et al.*, 1976; Welling *et al.*, 1976; Gaastra *et al.*, 1978; Beintema *et al.*, 1979, 1984, 1985; Jekel *et al.*, 1979; Lenstra and Beintema, 1979; Beintema and Martena, 1982; Breukelman *et al.*, 2001). Done before the so-called age of the genome, this work exploited classical Edman degradation of peptide fragments derived by selective cleavage of the protein. Such work required substantial amounts of protein, making convenient the large amount of RNase found in the digestive tracts of oxen and their immediate relatives. Beintema had also inferred the sequences of the ancestral proteins throughout the recent history of the digestive enzymes, using parsimony tools that adapted the ideas that Margaret Dayhoff had laid out (Dayhoff *et al.*, 1978).

18.2 Background

Members of the secreted RNase family of proteins are typically composed of a signal peptide of about 25 amino acids and a mature peptide of about 130 amino acids. Most members of the RNase family have three catalytic residues (one lysine and two histidines, at positions 41, 12, and 119 in RNase A, respectively). These come together in the folded enzyme to form an active site. In addition, RNases generally have six or eight cysteines that form three or four disulfide bonds. Except for these conserved residues, the sequences of RNases have diverged substantially in vertebrates, with sequence identities as low as 20% when comparing oxen and frog homologs (for example). RNase was well known as a digestive enzyme. As expected for enzymes found in the digestive tract, RNases are themselves biochemically robust. For example, the first step in the purification of RNase A involved the treatment of an extract from ox pancreas with 0.25 M sulfuric acid. This procedure precipitates most other proteins and removes the glycosyl groups from RNase, but otherwise leaves the protein intact.

RNases proved to present several opportunities for biological interpretation and discovery. As digestive enzymes, pancreatic RNases lie at one interface between their host organisms and their changing environments, and are expected to evolve with the environment. Not all mammals, however, have large amounts of pancreatic RNase. In fact, RNase is abundant in the digestive systems primarily in ruminants (which include the oxen, antelopes, and other bovids, together with the sheep, the deer, the giraffe, okapi, and pronghorn) and certain other special groups of herbivores (Barnard, 1969).

In 1969, Barnard proposed that pancreatic RNase was abundant primarily in ruminants because ruminant digestion had a special need for an enzyme that digested RNA (Barnard, 1969). Ruminant digestive physiology is considerably different from human digestive physiology (for example). The ruminant foregut serves as a vat to hold fermenting microorganisms. The ox delivers fodder to these microorganisms, which produce digestive enzymes (including cellulases) that the ox cannot. The microorganisms digest the grass, converting its carbon into a variety of products, including low-molecular-mass fatty acids. The fatty acids then enter the circulation system of the ruminant, providing energy.

The ox then eats the microorganisms for further nourishment. According to the Barnard hypothesis, this digestive physiology creates a need for especially large amounts of intestinal RNase. The fermenting microorganisms are rich in ribosomes and rRNA, tRNA, and mRNA. Fermenting bacteria therefore deliver large amounts of RNA to the gastric region of the bovine stomach and the small intestine. Barnard estimated that between 10 and 20% of the nitrogen in the diet of a typical bovid enters the lower digestive tract in the form of RNA.

Barnard's hypothesis was certainly consistent with the high level of digestive enzymes in ruminants generally. For example, ruminants have large amounts of lysozyme active against bacterial cell walls in their digestive tracts. Was the Barnard hypothesis merely a just-so story, based on correlations that did not require causality or functional necessity? The first experimental paleogenetics program set out to test this.

The available sequences were adequate to support the inference, with little ambiguity, of the sequence of the RNase represented (approximately) by the fossil ruminant *Pachyportax* (Stackhouse et al., 1990). This was also the case for the more ancient *Eotragus*, which lived in the Miocene. The available RNase sequences also permitted the inference, with only modest ambiguity, of sequences for RNases in the first ruminant, approximated in the fossil record by the genus *Archaeomeryx*. With slightly more ambiguity, the contemporary RNase sequences allowed the inference of the sequences of RNase in the first artiodactyl, the order of mammals having cloven hooves that includes the true ruminants as well as the camels, the pigs, and the hippos. This ancestor is approximately represented in the fossil record by the genus *Diacodexis*. A collaboration between Barbara Durrant at the Center for Reproduction of Endangered Species at the San Diego Zoo and the Benner laboratory yielded several additional sequences that assisted in making these inferences (TrabesingerRuef et al., 1996).

Once the ancestral sequences were reconstructed, the Benner group prepared by total synthesis a gene for RNase that was specially designed to support the resurrection of ancient proteins (Nambiar et al., 1984; Stackhouse et al., 1990). From this gene, approximately two dozen candidate ancestral genes for intermediates in the evolution of artiodactyl ribonucleases were synthesized, cloned, and expressed to resurrect the ancestral proteins for laboratory study (Stackhouse et al., 1990; Jermann et al., 1995).

To assess whether reconstructions yielded proteins that were plausible as intermediates in the evolution of the RNase family, the catalytic activities, substrate specificities, and thermal/proteolytic stabilities of the resurrected ancestral RNases were examined. Most of the resurrected proteins, and all of those corresponding to proteins expected in artiodactyls living after *Archaeomeryx*, behaved as expected for digestive enzymes. This was especially apparent from their kinetic properties (Table 18.1). Modern digestive RNases are catalytically active against small RNA substrates and single-stranded RNA (Blackburn and Moore, 1982). The RNase from *Pachyportax* was also, as were many of the earlier RNases. Thus, if one assumes that these catalytic properties are indicative of a digestive enzyme, these ancestral proteins were digestive enzymes as well.

This was also true quantitatively. Thus, the k_{cat}/K_m values for the putative ancestral RNases with the ribodinucleotide uridylyl $3' \rightarrow 5'$-adenosine (UpA) as a substrate (Ipata and Felicioli, 1968) in many ancient artiodactyls proved not to differ more than 25% from those of contemporary bovine digestive RNase (Table 18.1). With single-stranded poly(U) as substrate, the variance in catalytic activity was even smaller (18%).

Modern digestive RNases, like most digestive enzymes, are stable to thermal denaturation and cleavage by proteases. This suggested another metric for determining whether an ancestral protein acted in the digestive tract. Using a method developed by Lang and Schmid (1986), the sensitivity of the ancestral RNases to proteolysis as a function of temperature was measured (Table 18.2). Again, little change was observed in thermal stability of the ancestral RNases back to the ancestral artiodactyls approximated by *Archaeomeryx* in the fossil record. The midpoints in the activity/temperature curves for these ancient proteins varied by only ±1.1°C when compared with RNase A. This can be compared with typical experimental errors of ±0.5°C.

Had all of the ancestral RNases behaved like modern RNases, the resulting evolutionary narrative would have had little interest. The experiments in paleogenetics became interesting because the behavior of RNases resurrected from organisms more ancient than the last common ancestor of the true ruminants (*Archaeomeryx* and earlier) did not behave like digestive enzymes using these metrics.

These more ancient resurrected ancestral RNases displayed a 5-fold increase in catalytic activity against double-stranded RNA (poly(A)-poly(U)). This is not necessarily a digestive substrate. Further, the ancestral RNases showed an increased ability to bind and melt double-stranded DNA. Bovine digestive RNase A has only low catalytic activity against duplex RNA under physiological conditions, and does not bind and melt duplex DNA; these activities are presumably not needed for a digestive enzyme. At the same time, the catalytic activity of the candidate ancestral sequences against single-stranded RNA and short RNA fragments, the kinds of substrate that are

Table 18.1 Kinetic properties of reconstructed ancestral ribonucleases

RNase	Ancestor of	k_{cat}/K_m (UpA × 10⁶)	Relative to RNase A		
			k_{cat}/K_m (%)	Poly(U)	Poly(A)-poly(U)
RNase A		5.0	100	100	1.0
a	Ox, buffalo, eland	6.1	122	106	1.4
b	Ox, buffalo, eland, nilgai	5.9	118	112	1.0
c	b, Gazelles	4.5	91	97	0.8
d	Bovids	3.9	78	86	0.9
e	Deer	3.6	73	77	1.0
f	Deer, pronghorn, giraffe	3.3	67	103	1.0
g	Pecora	4.6	94	87	1.0
h1	Pecora and seminal RNase	5.5	111	106	5.2
h2	Pecora and seminal RNase	6.5	130	106	5.2
i1	Ruminata	4.5	90	96	5.0
i2	Ruminata	5.2	104	80	4.3
j1	Artiodactyla	3.7	74	73	4.6
j2	Artiodactyla	3.3	66	51	2.7

RNase names refer to nodes in the evolutionary tree shown in Figure 18.1. All assays were performed at 25°C. UpA, uridylyl $3' \rightarrow 5'$-adenosine.

Table 18.2 Thermal transition temperatures for reconstructed ancient ribonucleases

Enzyme	T_m (°C)	ΔT_m (°C)
RNase A[a]	59.3	0.0
RNase A[b]	59.7	+0.4
a	60.6	+1.3
b	61.0	+1.7
c	60.7	+1.4
d	58.4	−0.9
e	61.1	+1.8
f	58.6	−0.7
g	59.1	−0.2
h1	58.9	−0.5
h2	59.3	0.0
i1	58.2	−1.1
i2	58.7	−0.6
j1	56.5	−2.8
j2	57.1	−2.2

Thermal unfolding/proteolytic digestion temperatures (±0.5°C) were determined by incubating the RNase ancestor in 100 mM sodium acetate (pH 5.0) in the presence of trypsin.
[a] Expressed in *Escherichia coli*.
[b] Boehringer Mannheim.

expected in the digestive tract, was substantially lower (by a factor of five) than in the modern proteins. Proposing that these behaviors can be used as metrics, Jermann *et al.* (1995) concluded that RNases in artiodactyls that were ancestral to *Archaeomeryx* were not digestive enzymes.

A similar inference was drawn from stability studies. The more ancient ancestors displayed a modest but significant decrease in thermal-proteolytic stability using the assay of Lang and Schmid. A less stable enzyme, and a lower activity against single-stranded RNA, for example, might imply simply that the incorrect amino acid sequence was inferred for the ancestral protein. The fact that catalytic activity against double-stranded RNA, and the ability to melt duplex RNA, was higher in the ancestors argued against this possibility.

The issue was probed further by considering the ambiguity in the tree. The connectivity of deep branches in the artiodactyl evolutionary tree is not fully clarified by either the sequence data or the fossil record (Graur, 1993). This created a degree of ambiguity in the ancestral sequences. To manage this ambiguity, Jermann *et al.* (1995) synthesized a variety of alternative candidate ancestral RNase sequences. These effectively covered all of the ambiguity in the tree topology, and the resulting ambiguity in the sequences. The survey showed that the measured behavior and the consequent biological interpretation were robust with respect to the ambiguity.

Site 38 proved to be especially interesting. The variant of h1 (Figure 18.1) that restores Asp at position 38 (as in RNase A) has a catalytic activity against duplex RNA similar to that of RNase A (Jermann *et al.*, 1995; Opitz *et al.*, 1998). Conversely, the variant of RNase A that introduces Gly alone at position 38 has catalytic activity against duplex RNA essentially that of ancestor h. These results show that substitution at a single position, 38, accounts for essentially all of the increased catalytic activity against duplex RNA in ancestor h.

The reconstructed amino acids at position 38 are unambiguous before and after the *Archaeomeryx* sequence. Thus, it is highly probable that the changes in catalytic activity against duplex RNA in fact occurred in RNases as the ruminant RNases arose. In one interpretation, catalytic activity against duplex RNA was not necessary in the descendent RNases, and therefore was lost. This implies that the replacement of Gly-38 by Asp in the evolution of ancestor g from ancestor h was neutral. Jermann *et al.* (1995) could not, however, rule out an alternative model, that Asp-38 confers positive selective advantage on RNases found in the ruminants.

18.3 Understanding the origin of ruminant digestion

The experimental paleobiochemical data within the pancreatic RNase family suggested a coherent evolutionary narrative consistent with the Barnard hypothesis. RNases with increased stability, decreased catalytic activity against duplex RNA, decreased ability to bind and melt duplex DNA, and increased activity against single-stranded RNA and small RNA substrates emerged near the time when *Archaeomeryx* lived. The properties that increased are essential for digestive function; the properties that decreased are not. *Archaeomeryx*

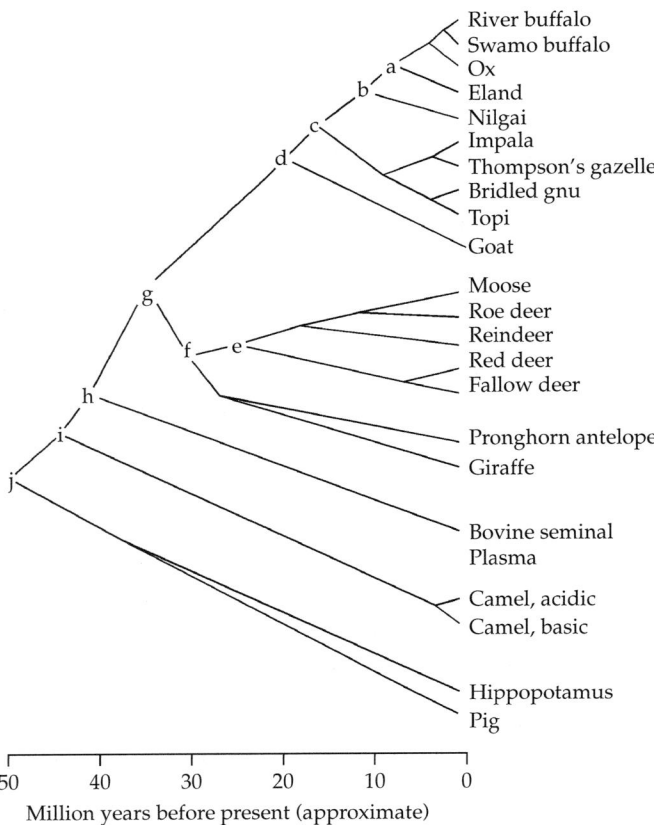

Figure 18.1 The evolutionary tree used in the analysis of ancestral pancreatic RNases. Lower-case letters at the nodes designate putative intermediates in the evolution of the protein family. Upper-case letters (D and G) indicate the residue at position 38 in the contemporary and reconstructed RNases. The time scale is approximate. The tree was adapted from Beintema et al. (1988) with a single alteration to join the pig and the hippopotamus together in a separate subfamily that branches together from the main line of descent. In the Beintema–Fitch tree, the pig and the hippopotamus diverge from the main line at separate points. Reprinted with permission from Benner et al. (2002) Planetary biology: paleontological, geological, and molecular histories of life. Science **296**: 864–868, © 2002 AAAS.

was the first artiodactyl to be a true ruminant. This implies that a digestive RNase emerged when ruminant digestion emerged.

This converts the Barnard hypothesis into a broader and robust narrative. This narrative became still more compelling when the molecular behavior is joined to the historical record as known from the fossil and geological records. These records suggested that the camels, deer, and bovid artiodactyl genera diverged ca.40 million years ago, together with ruminant digestion and the digestive RNases to support it, at the time of global climate change that began at the end of the Eocene, extended through the Oligocene, and reached a climax with the ice ages in the Pliocene and Pleistocene.

This climate change eventually involved the lowering of the mean temperature of Earth by approximately 17°C, and the drying of large parts of the Earth's surface (Janis et al., 1998). This, in turn, was almost certainly causally related to the emergence of grasses as a predominant source of vegetable food in many ecosystems. Tropical rainforests receded, grasslands emerged, and the interactions between herbivores and their foliage changed. Grasses offer poor nutrition compared to many other flora, and ruminant physiology appears to have substantial adaptive value when eating grasses.

This, in turn, may help explain why ruminant artiodactyls were enormously successful in competition with the herbivorous perissodactyls (for example, horses, tapirs, and rhinoceroses) as the global climate change proceeded. Today, nearly 200 species of artiodactyls have displaced the approximately 250 species of perissodactyls that were found in the tropical Eocene. Today, only three species groups of perissodactyl survive. This is the principal reason why resurrection of enzymes from the dawn horse will remain outside of the reach of contemporary paleomolecular biologists, unless ancestral DNA is extracted from the fossil of the dawn horse directly.

18.4 Ribonuclease homologs involved in unexpected biological activities

The paleobiochemical experiments with pancreatic RNases suggested that RNases having digestive function emerged in artiodactyls from a non-digestive precursor about 40 million years ago. This implies, in turn, that non-digestive cousins of digestive RNases might remain in the genomes of modern mammals, where they might continue to play a non-digestive role there.

This suggestion, generated from the first experiments in paleogenetics, emerged at the same time as researchers were independently discovering non-digestive paralogs of digestive RNase A. These were termed RIBAses (ribonucleases with interesting biological activities) by D'Alessio et al. (1991). They include RNase homologs that display immunosuppressive (Soucek et al., 1986), cytostatic (Matousek, 1973), anti-tumor (Ardelt et al., 1991), endothelial-cell-stimulatory (Strydom et al., 1985), and lectin-like activities (Okabe et al., 1991). These proteins all appeared to be extracellular, based on their secretory signal peptides and the presence of disulfide bonds. Their existence suggested to some that perhaps a functional RNA existed outside of cells (Benner, 1988).

These results suggested that the RNase A superfamily was extremely dynamic in vertebrates, with larger than typical amounts of gene duplication, paralog generation, and gene loss. In humans, for example, prior to the completion of the complete genome sequence, eight RNases were already known. These included the poorly named human pancreatic ribonuclease (RNase 1; which does not appear to be a protein specific for the pancreas), the equally poorly named eosinophil-derived neurotoxin (EDN, or RNase 2; which does not appear to have a physiological role as a neurotoxin), the eosinophil-cationic protein (ECP, or RNase 3; aptly named in the sense that the name captured all we knew about the protein), RNase 4, angiogenin (RNase 5), RNase 6 (sometimes known as k6), RNase 7 (Harder and Schroder, 2002; Zhang et al., 2003), and RNase 8 (Zhang et al., 2002).

The analysis *in silico* of the human genome showed that the human RNase 1–8 genes lie on chromosome 14q11.2 as a cluster of approximately 368 kb. In order from the centromere to the telomere, the genes are angiogenin (RNase 5), RNase 4, RNase 6, RNase 1, ECP (RNase 3), an EDN pseudogene, EDN itself (RNase 2), RNase 7, and RNase 8, separated from each other by 6–90-kb intervals. The genome also helped identification of two new human RNase homologs (RNases 9 and 10) in this cluster, preceding angiogenin. In addition, three new open reading frames sharing a number of common features with other RNases were found. Beintema therefore proposed to name these RNases 11, 12, and 13. RNases 11 and 12 are located between RNase 9 and angiogenin. RNase 13 lies on the centromere side of RNase 7, and has a transcriptional direction opposite to that of RNases 7 and 8. The human genome reveals no other open reading frames with significant similarity to these RNase genes. Therefore, it is likely that all human RNase A superfamily members have been identified.

As in humans, rat RNase genes are located on one chromosome (15p14) in a single cluster. The cluster in the rat genome contains the RNase family in the same syntenic order and transcriptional direction as in human, with only a few exceptions. The RNase 1 family (RNase1h, RNase1g, and RNase1y), the eosinophil-associated RNase family (EAR; R15–17, ECP, R-pseudogene, and Ear3), and the angiogenin family (Ang1 and Ang2) have undergone expansion in the rat (Zhao et al., 1998; Singhania et al., 1999; Dubois et al., 2002). Further, orthologs of human RNases 7 and 8 are not present in the rat genome. This permits us to propose a relatively coherent model for the order of gene creation in the time separating primates and rodents, and a listing of the RNase homologs likely to have been present in the last common ancestor of primates and rodents.

The dynamic behavior of this group of genes is shown by the differences separating the rat and mouse groups. In mouse, two RNase gene clusters are found, on mouse chromosome 14qB–qC1 (bcluster AQ) and chromosome 10qB1 (bcluster

BQ). Cluster A is syntenic to the human and rat clusters and is essentially identical to the rat cluster in gene content and order except for substantial expansions of the EAR and angiogenin gene subfamilies. Cluster B emerged in mouse after the mouse/rat divergence, and contains only genes and pseudogenes that belong to the EAR and angiogenin subfamilies. It also includes a large number of pseudogenes.

This level of diversity presents many why-type questions that might be addressed using molecular paleoscience. To date, three of these have been pursued, one in the Rosenberg laboratory, and two in the Benner laboratory.

18.5 Paleogenetics with eosinophil RNase homologs

In an effort to understand more about the function of these abundant RNase paralogs, Zhang and Rosenberg (2002) examined the EDN and ECP in primates. These proteins arose by gene duplication some 30 million years ago in an African primate ancestral to humans and Old World monkeys. Zhang and Rosenberg first asked the basic question: why do eosinophils have two RNase paralogs? Eosinophils are associated with asthma, infective wheezing, and eczema (Onorato *et al.*, 1996); their role in the non-diseased state remains enigmatic. Some textbooks say that eosinophils function to destroy larger parasites and modulate allergic inflammatory responses. Others suggest that eosinophils defend their host from outside agents, with allergic diseases arising as an undesired side effect.

Earlier work by Zhang, Rosenberg and their associates had suggested that ECP and EDN might contribute to organismic defense in other ways. ECP kills bacteria *in vitro*, whereas EDN inactivates retroviruses (Rosenberg and Domachowske, 2001). *In silico* analysis of reconstructed ancestral sequences in primates suggested that the proteins had suffered rapid sequence change near the time of the duplication that generated the paralogs, a change that might account for their differing behaviors *in vitro* (Zhang *et al.*, 1998). This suggests that, in primate evolution, mutations in EDN and ECP may have adapted them for different, specialized roles during the episodes of rapid sequence evolution.

To obtain a more densely articulated tree for the protein family, Zhang and Rosenberg (2002) sequenced additional genes from various primates. They used these sequences to better reconstruct ancestral sequences for ancient EDN/ECPs. They estimated the posterior probabilities of these ancestral sequences using Bayesian inference. Then they resurrected these ancient proteins by cloning and expressing their genes. Guiding the experimental work was the hypothesis that the anti-retroviral activity of EDN might be related to the ability of the protein to cleave RNA. Studies of the ancestral proteins allowed Zhang and Rosenberg to retrace the origins of the anti-retroviral and RNA-cleaving activities of EDN. Both the ribonuclease and antiviral activities of the last common ancestor of ECP and EDN, which lived *ca.* 30 million years ago, were low. Both activities increased in the EDN lineage after its emergence by duplication.

Zhang and Rosenberg showed that replacements at sites 64 and 132 in the sequence were required together to increase the ribonucleolytic activity of the protein; neither alone was sufficient. Zhang and Rosenberg then analyzed the three-dimensional crystal structure of EDN to offer possible explanations for the interconnection between sites suffering replacement and the changes in biomolecular behavior that they created. They concluded that in the EDN/ECP family, either of the two replacements at sites 64 and 132 individually had little impact on behavior. Each does, however, provide the context for the other to have an impact on behavior. This provides one example where a neutral (or, perhaps better, behaviorally inconsequential) replacement might have set the stage for a second adaptive replacement.

This observation influences how protein engineering is done in general. Virtually all analyses of divergent evolution treat protein sequences as if they were linear strings of letters (Benner *et al.*, 1998). With this treatment, each site is modeled to suffer replacement independent of all others, future replacement at a site is viewed as being independent of past replacement, and patterns of replacements are treated as being the same at each site. This has long been known to be an approximation, useful primarily for mathematical analysis (the "spherical cow"). Understanding higher-order features of protein sequence divergence has offered *in silico* approaches to some of the most puzzling conundrums in biological chemistry, including how to predict the folded structure of proteins from sequence data (Benner *et al.*, 1997a), and how to assign function to protein sequences (Benner *et al.*, 1998). The results of Zhang and Rosenberg provide an experimental case where higher-order analysis is necessary to understand a biomolecular phenomenon.

Another interpretive strategy involving resurrected proteins (Benner *et al.*, 1997b) was suggested from the results produced by Zhang and Rosenberg. This strategy identifies physiologically relevant behaviors *in vitro* for a protein where new biological function has emerged, as indicated by an episode of rapid (and therefore presumably adaptive) sequence evolution. The strategy examines the behavior of proteins resurrected from points in history before and after the episode of adaptive evolution. Those behaviors that are rapidly changing during the episode of adaptive sequence evolution, by hypothesis, confer selective value on the protein in its new function, and therefore are relevant to the change in function, either directly or by close coupling to behaviors that are. The properties *in vitro* that are the same at the beginning and end of this episode are not relevant to the change in function. This idea is fully implemented in the example of seminal RNases reviewed next in this chapter.

Whereas the number of amino acids changing is insufficient to make the case statistically compelling, the rate of change in the EDN lineage is strongly suggestive of adaptive evolution (Zhang *et al.*, 1998). The antiviral and ribonucleolytic activities of the proteins before and after the adaptive episode in the EDN lineage are quite different. Benner (2002), interpreting the data of Zhang *et al.* (1998), suggested that these activities are important to the emerging physiological role for EDN. This adds support, perhaps only modest, for the notion that the antiviral activity of EDN became important in Old World primates *ca.* 30 million years ago.

The timing of the emergence of the ECP/EDN pair in Old World primates might also contain information. The duplication occurred near the start of a global climatic deterioration that has continued until the present, with the Ice Ages in the past million years being the culmination (we hope) of this deterioration. These are the same changes as those that presumably drove the selection of ruminant digestion. If EDN, ECP, and eosinophils are part of a defensive system, it is appropriate to ask: what happened during the Oligocene that might have encouraged this type of system to be selected? Why might new defenses against retroviruses be needed at this time? If we are able to address these questions we might better understand how to improve our immune defenses against viral infections, an area of biomedical research that is in need of rapid progress.

18.6 Paleogenetics with ribonuclease homologs in bovine seminal fluid

New biomolecular function is believed to arise, at least in recent times, largely through recruitment of existing proteins with established roles to play new roles following gene duplication (Ohno, 1970; Benner and Ellington, 1990). Under one model, one copy of a gene continues to divergently evolve under constraints dictated by the ancestral function. The duplicate, meanwhile, is unencumbered by a functional role, and is free to search protein structure space. It may eventually come to encode new behaviors required for a new physiological function, and thereby confer selective advantage.

This model contains a well-recognized paradox. Because duplicate genes are not under selective pressure, they should also accumulate mutations that render them incapable of encoding a protein useful for any function. Most duplicates therefore should become pseudogenes (Lynch and Conery, 2000) or inexpressible genetic information (junk DNA; Li *et al.*, 1981) in just a few million years (Jukes and Kimura, 1984; Marshall *et al.*, 1994). This limits the evolutionary value of a functionally unconstrained gene duplicate as a tool for exploring protein structure space in the search of new behaviors that might confer selectable physiological function.

One of the non-digestive RNase subfamilies offered an interesting system to use experimental paleogenetics to study how new function arises in proteins. This focused on the seminal RNase paralogs found in ruminants that arose by duplication of the RNase A gene just as it was becoming a digestive protein. In ox, seminal RNase is 23 amino acids different from pancreatic RNase A. As suggested by its name, the paralog is expressed in the seminal plasma, where it constitutes some 2% of total protein (D'Alessio et al., 1972). Seminal RNase has evolved to become a dimer with composite active sites. It binds tightly to anionic glycolipids (Opitz, 1995), including seminolipid, a fusogenic sulfated galactolipid found in bovine spermatozoa (Vos et al., 1994). Further, seminal RNase has immunosuppressive and cytotoxic activities that pancreatic RNase A lacks (Soucek et al., 1986; Benner and Allemann, 1989).

Laboratory reconstructions of ancient RNases (Jermann et al., 1995) suggested that each of these traits was not present in the most recent common ancestor of seminal and pancreatic RNase, but rather arose in the seminal lineage after the divergence of these two protein families. To learn more about how this remarkable example of evolutionary recruitment occurred, RNase genes were collected from peccary (*Tayassu pecari*), Eld's deer (*Cervus eldi*), domestic sheep (*Ovis aries*), oryx (*Oryx leucoryx*), saiga (*Saiga tatarica*), yellow backed duiker (*Cephalophus sylvicultor*), lesser kudu (*Tragelaphus imberbis*), and Cape buffalo (*Syncerus caffer caffer*). These diverged approximately in that order within the mammal order Artiodactyla (Carroll, 1988). The newly sequenced genes complemented the known genes for various pancreatic RNases (Carsana et al., 1988) and seminal RNases from ox (*Bos taurus*; Preuss et al., 1990), giraffe (*Giraffa camelopardalis*; Breukelman et al., 1993), and hog deer.

Seminal RNase genes are distinguished from their pancreatic cousins by several marker substitutions introduced early after the gene duplication, including Pro-19, Cys-32, and Lys-62. By this standard, the genes from saiga, sheep, duiker, kudu, and the buffaloes were all assigned to the seminal RNase family. No evidence for a seminal-like gene could be found in peccary. Thus, these data are consistent with an analysis of previously published genes that places the gene duplication separating pancreatic and seminal RNases at *ca.* 35 million years before present (Beintema et al., 1988), preceding the divergence of giraffe, sheep, saiga, duiker, kudu, Cape buffalo, and ox, in this order, consistent with mitochondrial sequence data (Allard et al., 1992) and global phylogenetic analyses of Ruminanta (Hassanin and Douzery, 2003; Hernandez Fernandez and Vrba, 2005).

Sequence analysis shows that the seminal RNase genes from giraffe, hog deer, roe deer, and Cape buffalo almost certainly could not produce folded stable protein to serve a physiological function. Deletions or insertions create frame shifts in these genes. Further, the seminal RNase genes from okapi, kudu, and saiga were found to encode substitutions at active-site residues. Thus, these proteins are not likely to have catalytic activity.

To show that these seminal genes were indeed not expressed in semen, seminal plasmas from 15 artiodactyls were examined (ox, forest buffalo (*Syncerus caffer nanus*), Cape buffalo, kudu, sitatunga (*Tragelaphus spekei*), nyala (*Tragelaphus angasi*), eland (*Tragelaphus oryx*), Maxwell's duiker (*Cephalophus monticola maxwelli*), yellow-backed duiker, suni (*Neotragus moschatus*), sable antelope (*Hippotragus niger*), impala (*Aepyceros melampus*), saiga, sheep, and Eld's deer). Catalytically active RNase was not detected in the seminal plasma in significant amounts in any artiodactyl genus diverging before the Cape buffalo, except in *Ovis*. Independent mutagenesis experiments showed that the proteins encoded by these genes, all carrying a Cys at position 32, should form dimers (Trautwein, 1991; Raillard, 1993; Jermann, 1995; Opitz, 1995). By Western blotting, however, only small amounts of a monomeric, presumably pancreatic, RNase were detected in these seminal plasmas. In contrast, the seminal plasmas of forest buffalo, cape buffalo, and ox all contained substantial amounts of Western blot-active RNase (Kleineidam et al., 1999). Only in the seminal plasma of ox, however, is seminal RNase expressed. Even though the gene is intact in water buffalo, no expressed protein could be found in its seminal plasma.

The seminal plasma from the *Ovis* genus (sheep and goat) was a notable exception. Sheep seminal plasma contained significant amounts of RNase protein and the corresponding ribonucleolytic activity. To learn whether RNases in the *Ovis* seminal plasma were derived from a seminal RNase gene, the RNase from goat seminal plasma was isolated, purified, and sequenced by tryptic cleavage and Edman degradation. Both Edman degradation (covering 80% of the sequence) and matrix-assisted laser-desorption ionization (MALDI) mass spectroscopy showed that the sequence of the RNase isolated from goat seminal plasma is identical to the sequence of its pancreatic RNase (Beintema *et al.*, 1988; Jermann, 1995). This shows that the RNase in *Ovis* seminal plasma is not expressed from a seminal RNase gene, but rather from the *Ovis* pancreatic gene. To confirm this conclusion, a fragment of the seminal RNase gene from sheep was sequenced, and shown to be different in structure from the pancreatic gene.

These results could be perceived as inconsistent with a model that the seminal RNase gene family gradually developed a new seminal function by stepwise point mutation and continuous selection under functional constraints in the seminal plasma following gene duplication. Rather, the duplicate RNase gene seems initially to have served no function at all. It therefore suffered damage, only to be repaired much later in evolution, after the divergence of kudu, but before the divergence of Cape buffalo, from the lineage leading to ox. Clades containing the saiga, duiker, and sheep are known in the early Miocene (23.8–16.4 million years ago), whereas clades containing the kudu and cape buffalo are known in the late Miocene (11.2–5.3 million years ago). Despite the incompleteness of the fossil record, we might conclude that the damaged gene was repaired extremely rapidly in only a few million years (TrabesingerRuef *et al.*, 1996). The paleogenetic study, however, will show support for an alternative scenario.

But what was this new function of bovine seminal RNase? What is the molecular basis of the newly acquired function? To address these questions, we set out to reconstruct and resurrect the ancestral seminal proteins. Figure 18.2 shows the nodes where sequences were reconstructed using a likelihood method. These nodes include the evolutionary period where the new biological function might be arising. Three different evolutionary models, one amino acid-based and two codon-based, were used to make the reconstructions. Two outgroups were also considered, those holding the pancreatic RNases and brain RNases, as the data did not unambiguously force the conclusion that one of these two RNAse subfamilies was the closest outgroup. Next, ambiguity at the level of the phylogeny was considered.

To determine the phylogeny four methods were used: a Bayesian analysis as implemented by the program MrBayes, and three different maximum-likelihood models implemented by the PAUP software package. Both outgroups were considered for the different methods. Using the brain sequences as the outgroup, paleontologically unreasonable topologies resulted, as judged by comparison to species trees based on much larger data-sets. With the pancreatic RNases as the outgroup, however, each tool generated the same set of trees, but with slightly different ranking based on slightly different scores. These trees generally agreed with the accepted species trees based on large sets of data (Hassanin and Douzery, 2003; Hernandez Fernandez and Vrba, 2005). The same was observed if *both* pancreatic and brain RNases were used to construct trees. Therefore, the following tree topologies were considered (see Figure 18.2).

1 Topology 1. Preferred when the work began, this topology also received the highest score from a complete Bayesian analysis. Topology 1 groups the okapi with the deer, and models the saiga and duiker as diverging separately from the lineage leading to oxen after the divergence of deer.

2 Topology 2. Preferred today based on a global analysis of all available sequence and paleontological data. Topology 2 places okapi as an outgroup separate from deer, with the giraffe, and diverging before deer diverged from oxen. It also groups saiga and duiker.

3 Topology 3. This topology places okapi in a clade with deer, and places saiga and duiker together.

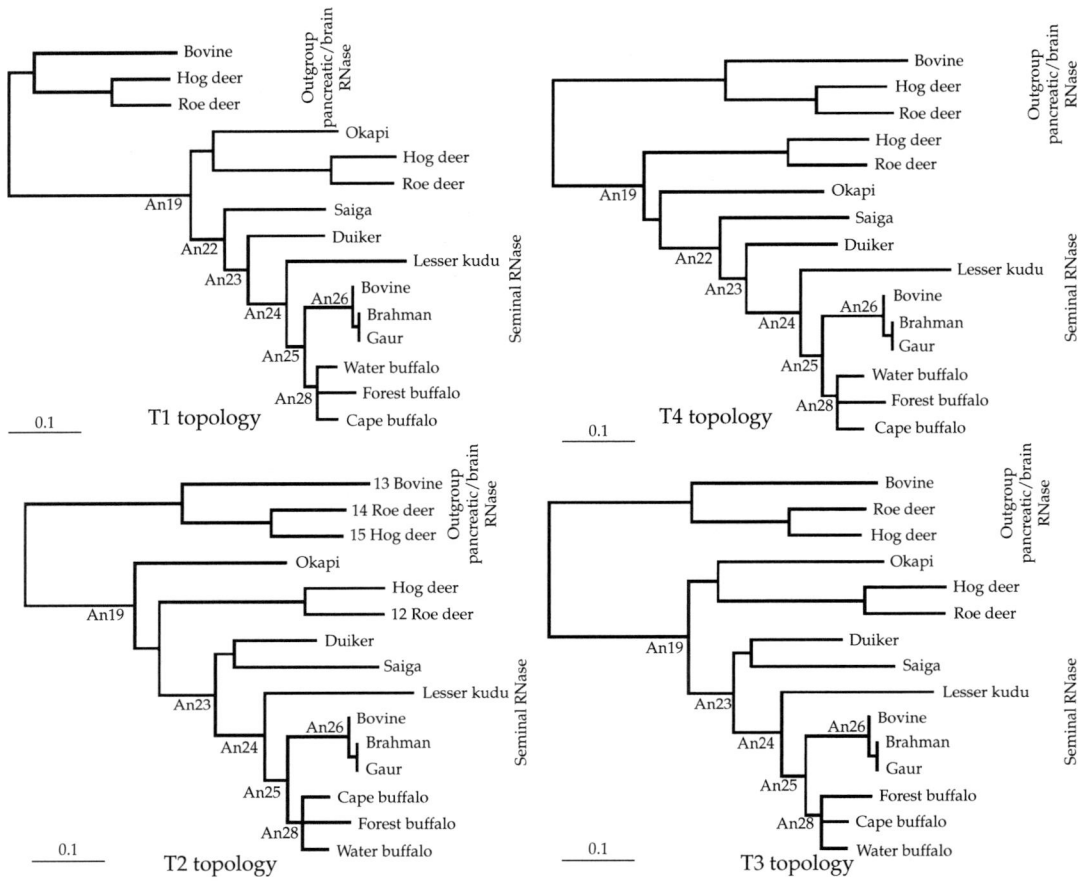

Figure 18.2 Four candidate trees describing the relationship between the artiodactyls providing genes for seminal ribonuclease. Ancestral proteins from the marked nodes (An19–An28) were resurrected in paleogenetic study. See text for discussion.

4 Topology 4. This topology was considered reasonable when this work began, but less so now in light of subsequently emerging data. Here, okapi lies in a clade separate from deer, but diverges from the lineage leading to oxen after deer diverge. Duiker and saiga again were represented as diverging separately from the lineage leading to ox after the divergence of deer.

The closely scoring alternative trees are the consequence of the seminal RNase family having a remarkable amount of homoplasy (parallel and convergent sequence evolution); homoplasy is found at sites 9, 18, 22, 53, 55, 64, 101, and 113. Given this level of homoplasy, no tree can be unambiguously viewed as being correct. Therefore, alternative trees were considered in proposing candidate ancestral sequences. In an effort to manage ambiguities, all possible sequences were resurrected whenever the reconstructions disagreed.

The distribution of ancestral replacements on the three-dimensional structure of seminal RNase followed a specific pattern. All of the active-site residues remained conserved after the gene duplication. Moreover, the RNA-binding site was also conserved. Most of the replacements were concentrated on the surface of the protein and away from the RNA-binding site. This replacement pattern is consistent with an evolutionary path where the enzymatic function of the protein was conserved; it is not consistent with an inference that the ancestral seminal RNase genes were pseudogenes. Furthermore, the lesions causing the pseudogene formation in the different lineages are different. These two observations taken together imply that the ancestral seminal RNases were enzymatically active, and that independent inactivation events converted active genes in the different lineages into pseudogenes in many of the modern artiodactyls. Consistent with this model, the resurrected ancestral seminal RNases were all enzymatically active (hydrolyzing a fluorescently labeled RNA substrate; Kelemen et al., 1999).

What then were the properties of seminal RNases that were the targets for natural selection over the past 30 million years? As noted above, there are many behaviors *in vitro* to choose from. For some (such as anti-proliferative activity against cancer cells in culture), it is difficult to rationalize how such behaviors might be important for a protein that exists in seminal plasma. But the site of expression of a protein is changeable over short periods of evolutionary time, meaning that we cannot be certain where seminal RNase has been expressed over its history.

We hypothesized that since seminal RNase is expressed in the seminal fluid and has immunosuppressive activity, it could have evolved to confer a selective reproductive advantage to bulls when the female reproductive tract mounts an immune response against the invading sperm. Indeed, it has been shown in reproductive biology that in many species sperm encounters a defensive immune response and that in many cases seminal plasma is capable of repressing this response (James and Hargreave, 1984; Schroder et al., 1990; Kelly and Critchley, 1997).

To test whether this is true, Benner et al. (2007) exploited the strategy to identify the physiologically relevant behaviors *in vitro* for a newly emerging function. As noted above, the strategy examines the behavior of proteins resurrected from points in history before and after the presumed episode of adaptive evolution. The behaviors *in vitro* that are rapidly changing during this episode are inferred to be those relevant to adaptive change. The behaviors *in vitro* that are the same at the beginning and end of this episode are not relevant to the change in function.

Episodes of adaptive evolution are frequently inferred from high normalized non-synonymous/synonymous (d_N/d_S) ratios (significantly greater than unity), where amino acid replacements conferred new behaviors that conferred enhanced fitness on a protein subject to new functional demands. Thus, they characterize episodes where the derived sequence, at the end of the episode, has (in some sense) a physiological function different from that of the ancestral sequence at its beginning. Low ratios (<1; although these ratios approach zero in highly conserved proteins) characterize episodes where the ancestral and derived sequences at the beginning and end of the episode have the same physiological function.

The application of this tool in this gene family detected a phase of evolution during the emergence of bovine seminal ribonuclease after the ox diverged from the buffalo. A variety of models within PAML were used to determine d_N/d_S ratios for individual branches in the tree. The Akaike Information Criterion (AIC; Posada and Buckley, 2004) was then used to select the model that best fits the data. Model comparison showed that regardless of the ambiguities in the evolutionary model, the outgroup, or the tree, only the branch leading to the modern seminal RNase in ox, in its three forms (the gaur, Brahman, and ox), underwent adaptive evolution; for no other branch of the tree is this conclusion required. The d_N/d_S ratio was in the range of 1.6–6, depending on the historical model, including the tree topology, choice of outgroup, and choice of codon model.

This strongly suggests that the functional constraints on protein structure, and a correlated change in the physiological function of the protein, occurred in this episode. To identify which *in vitro* behaviors are also changing at this time, the genes encoding ancestral seminal RNase were synthesized by site-directed mutagenesis of a previously prepared RNase synthetic gene. The ancestral RNase candidates were expressed in *Escherichia coli* and purified using newly developed oligonucleotide affinity chromatography.

To address the biomolecular behaviors changing during this adaptive phase, Sassi et al. (unpublished data) examined several biochemical and cell-based biomolecular behaviors. The k_{cat}/K_m ratios characterizing the enzyme's ability to catalyze the hydrolysis of a fluorescently labeled model RNA substrate (carboxyfluoresceinylhexyl-pdAUdAdAp-hexyl-tetramethylrhodamine, IDT) were not significantly different from that of bovine seminal RNase. This implies that this biomolecular behavior is not key to the newly emerging biological function in bovine seminal plasma. All of the candidate ancestral RNases could form dimers under oxidizing conditions, as does seminal RNA, implying that this behavior was not key to the newly emerging function. The rates of folding and other gross physical properties of the ancestors were also not greatly different in the ancestral and modern seminal RNases.

In contrast, the immunosuppressivity of the seminal RNases, measured *in vitro* using a mixed lymphocyte reaction assay exploiting bovine leukocytes isolated from fresh peripheral bovine blood, increased noticeably in the descendents following the branch having adaptive evolution. This suggests that immunosuppression, as measured in this assay *in vitro*, is physiologically relevant for the new function of seminal RNase. This result was paralleled by results in mitogen induction assays which have less physiological relevance. This suggests that the cell-based assays are measuring a property that is important for the new biological function of seminal RNase.

Raines recently suggested that the cell-based activities of RNase might require a swapping of residues 1–20, mentioned above, to form composite active sites (Kim *et al.*, 1995; Lee and Raines, 2005). Accordingly, the extent of the swap was measured using a divinylsulfone crosslinking reagent following the procedure of Ciglic *et al.* (1998). Whereas the extent of swapping may be sensitive to the precise conditions under which the proteins were renatured, the extent of swapping measured *in vitro* also increases during the episode of adaptive evolution. This confirms the hypothesis of Raines, and suggests a structural feature relevant to an adaptive change as well as a biomolecular behavior.

It is important to note that these paleogenetic experiments suggest inferences about the structural changes and behavioral changes that may be important to changing physiological function without recourse to specific studies on the living animals. Further, these inferences are robust with respect to the ambiguities inherent in the reconstruction of historical states from derived sequences.

This study presents an example where the evolutionary history of a gene and the physiological function of the protein were both unknown but the resurrection of the ancestral protein provided evidence for a hypothesis and hints at the evolutionary events shaping this gene's history.

As the number of genomic sequences available increases, the paleogenetic strategy to connect biomolecular behavior *in vitro* to physiological relevance will become easier to apply. Whereas we do not expect it to replace specific studies on the living animals, it should be useful to direct those studies in fruitful directions, and add an interpretive dimension. Paleogenetics is applicable to any biomolecular system where evolutionary reconstructions indicate an episode of adaptive evolution, either through high non-synonymous/synonymous ratios or by high absolute rates of protein sequence divergence deduced using geobiological markers (e.g. fossils) for dating times of divergence.

18.7 Lessons learned

The emergence of systems biology as a paradigm in modern science has brought new attention to a longstanding problem in reductionist biology, which asks whether a particular behavior of an isolated protein, measured *in vitro*, is physiologically relevant in a complex living organism. As Darwinism admits natural selection as the only mechanism for obtaining functional behaviors in biology, this question is equivalent to asking whether the *in vitro* behavior in question, if changed, would lead to a host organism with a diminished ability to survive and reproduce. It has proven remarkably difficult to correlate specific behaviors to fitness although biomolecular behaviors must correlate with fitness in a general sense. A strategy mainly demonstrated in the seminal RNase example uses resurrected ancestral proteins from extinct organisms to help identify biomolecular behaviors *in vitro* that are physiologically relevant to newly emerging biomolecular function.

The ribonuclease family contains the best-developed example of the use of paleomolecular resurrections to understand protein function. It also demonstrates most of the key issues that must be addressed when implementing this paradigm. This includes the management of ambiguities. In all of the cases reviewed here, additional sequences were obtained from additional organisms to increase the articulation of the evolutionary tree, and thereby reduce the ambiguity in the inferred ancestral sequences. When ambiguities remained, multiple candidate ancestral sequences were resurrected to determine that the behavior subject to biological interpretation was robust with respect to the ambiguity.

These examples also show the value of maximum likelihood and empirical Bayes tools in reconstructing ancestral sequences. The simplest parsimony tools, which minimize the number of changes in a tree, are easily deceived by swaps around short branches. Ancestral character states are less likely to be confused by incorrect detailed topology of a tree when they are constructed using maximum-likelihood tools than by maximum-parsimony tools.

More important, however, these examples show the potential of molecular paleoscience as a strategy to sort out the complexities of biological function in complex genome systems. Here, the potential of this strategy has only begun to be explored. In the long term, we expect that paleomolecular resurrections will allow us to understand changing biomolecular function in the context of ecological and planetary systems. Margulis and others have referred to this as planetary biology (Margulis and West, 1993; Margulis and Guerrero, 1995; Benner et al., 2002).

Last, these examples show the value of paleomolecular resurrections in converting just-so stories into serious scientific narratives that connect phenomenology inferred by correlation into a comprehensive historical-molecular hypothesis that incorporates experimental data and suggests new experiments. Thus they offer a key example of how paleobiology might enter the mainstream of molecular biology as the number of genome sequences becomes large, and the frustration with their lack of meaning becomes still more widespread.

References

Allard, M.W., Miyamoto, M.M., Jarecki, L., Kraus, F., and Tennant, M.R. (1992) DNA systematics and evolution of the artiodactyl family Bovidae. *Proc. Natl. Acad. Sci. USA* **89**: 3972–3976.

Ardelt, W., Mikulski, S.M., and Shogen, K. (1991) Amino acid sequence of an anti-tumor protein from *Rana pipiens* oocytes and early embryos. Homology to pancreatic ribonucleases. *J. Biol. Chem.* **266**: 245–251.

Barnard, E.A. (1969) Biological function of pancreatic ribonuclease. *Nature* **221**: 340–344.

Beintema, J.J. and Gruber, M. (1967) Amino acid sequence in rat pancreatic ribonuclease. *Biochim. Biophys. Acta* **147**: 612–614.

Beintema, J.J. and Gruber, M. (1973) Rat pancreatic ribonuclease. 2. Amino acid sequence. *Biochim. Biophys. Acta* **310**: 161–173.

Beintema, J.J. and Martena, B. (1982) Primary structure of porcupine (hystrix-cristata) pancreatic ribonuclease: close relationship between african porcupine (an old world hystricomorph) and new world caviomorphs. *Mammalia* **46**: 253–257.

Beintema, J.J., Gaastra, W., and Munniksma, J. (1979) Primary structure of pronghorn pancreatic ribonuclease: close relationship between giraffe and pronghorn. *J. Mol. Evol.* **13**: 305–316.

Beintema, J.J., Wietzes, P., Weickmann, J.L., and Glitz, D.G. (1984) The amino acid sequence of human pancreatic ribonuclease. *Anal. Biochem.* **136**: 48–64.

Beintema, J.J., Broos, J., Meulenberg, J., and Schuller, C. (1985) The amino acid sequence of snapping turtle (chelydra-serpentina) ribonuclease. *Eur. J. Biochem.* **153**: 305–312.

Beintema, J.J., Schuller, C., Irie, M., and Carsana, A. (1988) Molecular evolution of the ribonuclease superfamily. *Prog. Biophys. Mol. Biol.* **51**: 165–192.

Benner, S.A. (1988) Extracellular 'communicator RNA'. *FEBS Lett.* **233**: 225–228.

Benner, S.A. (2002) The past as the key to the present: resurrection of ancient proteins from eosinophils. *Proc. Natl. Acad. Sci. USA* **99**: 4760–4761.

Benner, S.A. and Allemann, R.K. (1989) The return of pancreatic ribonucleases. *Trends Biochem. Sci.* **14**: 396–397.

Benner, S.A. and Ellington, A.D. (1990) Evolution and structural theory: the frontier between chemistry and biology. In *Bioorganic Chemistry Frontiers* (Dugas, H., ed.), pp. 1–54. Springer Verlag, Berlin.

Benner, S.A., Cannarozzi, G., Gerloff, D., Turcotte, M., and Chelvanayagam, G. (1997a) Bona fide predictions of protein secondary structure using transparent analyses of multiple sequence alignments. *Chem. Rev.* **97**: 2725–2843.

Benner, S.A., Haugg, M., Jermann, T.M., Opitz, J.G., Raillard-Yoon, S.-A., Soucek, J. et al. (1997b) Evolutionary reconstructions in the ribonuclease family. In *Ribonucleases* (D'Alessio, J.R.a.G., ed.), pp. 214–244. Academic Press, New York.

Benner, S.A., Trabesinger, N., and Schreiber, D. (1998) Post-genomic science: converting primary structure into physiological function. *Adv. Enzyme Regul.* **38**: 155–180.

Benner, S.A., Caraco, M.D., Thomson, J.M., and Gaucher, E.A. (2002) Planetary biology: paleontological, geological, and molecular histories of life. *Science* **296**: 864–868.

Benner, S.A., Sassi, S.O., and Gaucher, E.A. (2007) Molecular paleosciences. Systems biology from the past. In *Advances in Enzymology and Related Areas of Molecular Biology: Protein Evolution* (Toone, E., ed.), vol. 75, pp. 1–132. Wiley, Chichester.

Blackburn, P. and Moore, S. (1982) *Pancreatic Ribonucleases, The Enzymes*, pp. 317–433. Academic Press, New York.

Bonaventura, J., Bonaventura, C., and Sullivan, B. (1974) Urea tolerance as a molecular adaptation of elasmobranch hemoglobins. *Science* **186**: 57–59.

Breukelman, H.J., Beintema, J.J., Confalone, E., Costanzo, C., Sasso, M.P., Carsana, A. et al. (1993) Sequences related to the ox pancreatic ribonuclease coding region in the genomic DNA of mammalian species. *J. Mol. Evol.* **37**: 29–35.

Breukelman, H.J., Jekel, P.A., Dubois, J.Y.F., Mulder, P., Warmels, H.W., and Beintema, J.J. (2001) Secretory ribonucleases in the primitive ruminant chevrotain (*Tragulus javanicus*). *Eur. J. Biochem.* **268**: 3890–3897.

Carroll, R.L. (1988) *Vertebrate Paleontology and Evolution*. WH Freeman & Co, New York.

Carsana, A., Confalone, E., Palmieri, M., Libonati, M., and Furia, A. (1988) Structure of the bovine pancreatic ribonuclease gene: the unique intervening sequence in the 5' untranslated region contains a promoter-like element. *Nucleic Acids Res.* **16**: 5491–5502.

Ciglic, M.I., Jackson, P.J., Raillard, S.A., Haugg, M., Jermann, T.M., Opitz, J.G. et al. (1998) Origin of dimeric structure in the ribonuclease superfamily. *Biochemistry* **37**: 4008–4022.

D'Alessio, G., Floridi, A., De Prisco, R., Pignero, A., and Leone, E. (1972) Bull semen ribonucleases. 1. Purification and physico-chemical properties of the major component. *Eur. J. Biochem.* **26**: 153–161.

D'Alessio, G., Di Donato, A., Parente, A., and Piccoli, R. (1991) Seminal RNase: a unique member of the ribonuclease superfamily. *Trends Biochem. Sci.* **16**: 104–106.

Dayhoff, M.O., Schwartz, R.M., and Orcutt, B.C. (1978) A model of evolutionary change in proteins. In *Atlas of Protein Sequence and Structure* (Dayhoff, M.O., ed.), pp. 345–352. National Biomedical Research Foundation, Washington DC.

Dubois, J.Y.F., Jekel, P.A., Mulder, P., Bussink, A.P., Catzeflis, F.M., Carsana, A., and Beintema, J.J. (2002) Pancreatic type ribonuclease 1 gene duplications in rat species. *J. Mol. Evol.* **55**: 522–533.

Emmens, M., Welling, G.W., and Beintema, J.J. (1976) Amino acid sequence of pike-whale (lesser-rorqual) pancreatic ribonuclease. *Biochem. J.* **157**: 317–323.

Gaastra, W., Groen, G., Welling, G.W., and Beintema, J.J. (1974) Primary structure of giraffe pancreatic ribonuclease. *FEBS Lett.* **41**: 227–232.

Gaastra, W., Welling, G.W., and Beintema, J.J. (1978) Amino acid sequence of kangaroo pancreatic ribonuclease. *Eur. J. Biochem.* **86**: 209–217.

Graur, D. (1993) Towards a molecular resolution of the ordinal phylogeny of the eutherian mammals. *FEBS Lett.* **325**: 152–159.

Groen, G., Welling, G.W., and Beintema, J.J. (1975) Amino acid sequence of gnu pancreatic ribonuclease. *FEBS Lett.* **60**: 300–304.

Harder, J. and Schroder, J.M. (2002) RNase 7, a novel innate immune defense antimicrobial protein of healthy human skin. *J. Biol. Chem.* **277**: 46779–46784.

Hassanin, A. and Douzery, E.J. (2003) Molecular and morphological phylogenies of ruminantia and the alternative position of the moschidae. *Syst. Biol.* **52**: 206–228.

Hernandez Fernandez, M. and Vrba, E.S. (2005) A complete estimate of the phylogenetic relationships in Ruminantia: a dated species-level supertree of the extant ruminants. *Biol. Rev. Camb. Philos. Soc.* **80**: 269–302.

Ipata, P.L. and Felicioli, R.A. (1968) A spectrophotometric assay for ribonuclease activity using cytidylyl-(3', 5')-adenosine and uridylyl-(3', 5')-adenosine as substrates. *FEBS Lett.* **1**: 29–31.

James, K. and Hargreave, T.B. (1984) Immunosuppression by seminal plasma and its possible clinical significance. *Immunol. Today* **5**: 357.

Janis, C.M., Effinger, J.E., Harrison, J.A., Honey, J.G., Kron, D.G., Lander, B. et al. (1998) Artiodactyla. In *Evolution of Tertiary Mammals of North America* (Janis, C.M., Scott, K.M., and Jacobs, L., eds), pp. 337–357. Cambridge University Press, Cambridge.

Jekel, P.A., Sips, H.J., Lenstra, J.A., and Beintema, J.J. (1979) Amino acid sequence of hamster pancreatic ribonuclease. *Biochimie* **61**: 827–839.

Jermann, T.M. (1995) *Der Ursprung und die Evolution der Ribonuklease aus dem Pankreas und aus der Samenfluessigkeit*. ETH Dissertation no. 11059, Zurich.

Jermann, T.M., Opitz, J.G., Stackhouse, J., and Benner, S.A. (1995) Reconstructing the evolutionary history of the artiodactyl ribonuclease superfamily. *Nature* **374**: 57–59.

Jukes, T.H. and Kimura, M. (1984) Evolutionary constraints and the neutral theory. *J. Mol. Evol.* **21**: 90–92.

Kelemen, B.R., Klink, T.A., Behlke, M.A., Eubanks, S.R., Leland, P.A., and Raines, R.T. (1999) Hypersensitive substrate for ribonucleases. *Nucleic Acids Res.* **27**: 3696–3701.

Kelly, R.W. and Critchley, H.O. (1997) Immunomodulation by human seminal plasma: a benefit for spermatozoon and pathogen? *Hum. Reprod.* **12**: 2200–2207.

Kim, J.S., Soucek, J., Matousek, J., and Raines, R.T. (1995) Mechanism of ribonuclease cytotoxicity. *J. Biol. Chem.* **270**: 31097–31102.

Kleineidam, R.G., Jekel, P.A., Beintema, J.J., and Situmorang, P. (1999) Seminal type ribonuclease genes in ruminants, sequence conservation without protein expression? *Gene* **231**: 147–153.

Kuper, H. and Beintema, J.J. (1976) Amino acid sequence of topi pancreatic ribonuclease. *Biochim. Biophys. Acta* **446**: 337–344.

Lang, K. and Schmid, F.X. (1986) Use of a trypsin pulse method to study the refolding pathway of ribonuclease. *Eur. J. Biochem.* **159**: 275–281.

Lee, J.E. and Raines, R.T. (2005) Cytotoxicity of bovine seminal ribonuclease: monomer versus dimer. *Biochemistry* **44**: 15760–15767.

Lenstra, J.A. and Beintema, J.J. (1979) Amino acid sequence of mouse pancreatic ribonuclease: extremely rapid evolutionary rates of the myomorph rodent ribonucleases. *Eur. J. Biochem.* **98**: 399–408.

Li, W.H., Gojobori, T., and Nei, M. (1981) Pseudogenes as a paradigm of neutral evolution. *Nature* **292**: 237–239.

Lynch, M. and Conery, J.S. (2000) The evolutionary fate and consequences of duplicate genes. *Science* **290**: 1151–1155.

Margoliash, E. (1963) Primary structure and evolution of cytochrome C. *Proc. Natl. Acad. Sci. USA* **50**: 672–679.

Margoliash, E. (1964) Amino acid sequence of cytochrome C in relation to its function and evolution. *Can. J. Biochem. Physiol.* **42**: 745–753.

Margulis, L. and West, O. (1993) Gaia and the colonization of Mars. *GSA Today* **3**: 277–280, 291.

Margulis, L. and Guerrero, R. (1995) Life as a planetary phenomenon: the colonization of Mars. *Microbiologia* **11**: 173–184.

Marshall, C.R., Raff, E.C., and Raff, R.A. (1994) Dollo's law and the death and resurrection of genes. *Proc. Natl. Acad. Sci. USA* **91**: 12283–12287.

Matousek, J. (1973) The effect of bovine seminal ribonuclease (AS RNase) on cells of Crocker tumour in mice. *Experientia* **29**: 858–859.

Muskiet, F.A.J., Welling, G.W., and Beintema, J.J. (1976) Studies on primary structure of bison pancreatic ribonuclease. *Int. J. Pept. Protein Res.* **8**: 345–348.

Nambiar, K.P., Stackhouse, J., Stauffer, D.M., Kennedy, W.P., Eldredge, J.K., and Benner, S.A. (1984) Total synthesis and cloning of a gene coding for the ribonuclease S protein. *Science* **223**: 1299–1301.

Ohno, S. (1970) *Evolution by Gene Duplication*. Springer Verlag, Berlin.

Okabe, Y., Katayama, N., Iwama, M., Watanabe, H., Ohgi, K., Irie, M. et al. (1991) Comparative base specificity, stability, and lectin activity of 2 lectins from eggs of *Rana catesbeiana* and *R. japonica* and liver ribonuclease from *R. catesbeiana*. *J. Biochem. (Tokyo)* **109**: 786–790.

Onorato, J., Scovena, E., Airaghi, S., Morandi, B., Morelli, M., Pizzi, M., and Principi, N. (1996) Role of serum eosinophil cationic protein (s-ECP), neutrophil myeloperoxidase (s-MPO) and mast cell triptase (s-TRY) in children with allergic, infective asthma and atopic dermatitis. *Riv. Ital. Pediatr.* **22**: 900–911.

Opitz, J.G. (1995) *Maximum parsimony: Ein neuer Ansatz zum besseren Verstaendnis von Protein/Nukleinsaeure-Wechselwirkungen*. ETH Dissertation no. 10952, Zurich.

Opitz, J.G., Ciglic, M.I., Haugg, M., Trautwein-Fritz, K., Raillard, S.A., Jermann, T.M., and Benner, S.A. (1998) Origin of the catalytic activity of bovine seminal ribonuclease against double-stranded RNA. *Biochemistry* **37**: 4023–4033.

Posada, D. and Buckley, T.R. (2004) Model selection and model averaging in phylogenetics: advantages of akaike information criterion and bayesian approaches over likelihood ratio tests. *Syst. Biol.* **53**: 793–808.

Preuss, K.D., Wagner, S., Freudenstein, J., and Scheit, K.H. (1990) Cloning of cDNA encoding the complete precursor for bovine seminal ribonuclease. *Nucleic Acids Res.* **18**: 1057.

Raillard, S.A. (1993) *Veraenderung der Struktur und der biologischen Aktivitaet in RNase A mit Hilfe von gezielter Mutagenese*. ETH Dissertation no. 10022, Zurich.

Riggs, A. (1959) Molecular adaptation in haemoglobins. Nature of The Bohr effect. *Nature* **183**: 1037–1038.

Rosenberg, H.F. and Domachowske, J.B. (2001) Eosinophils, eosinophil ribonucleases, and their role in host defense against respiratory virus pathogens. *J. Leukoc. Biol.* **70**: 691–698.

Schroder, W., Mallmann, P., van der Ven, H., Diedrich, K., and Krebs, D. (1990) Cellular sensitization against spermatic and seminal plasma antigens in women after intrauterine insemination. *Arch. Gynecol. Obstet.* **248**: 67–74.

Singhania, N.A., Dyer, K.D., Zhang, J., Deming, M.S., Bonville, C.A., Domachowske, J.B., and Rosenberg, H.F. (1999) Rapid evolution of the ribonuclease A superfamily: adaptive expansion of independent gene clusters in rats and mice. *J. Mol. Evol.* **49**: 721–728.

Soucek, J., Chudomel, V., Potmesilova, I., and Novak, J.T. (1986) Effect of ribonucleases on cell-mediated lympholysis reaction and on GM-CFC colonies in bone marrow culture. *Nat. Immun. Cell Growth Regul.* **5**: 250–258.

Stackhouse, J., Presnell, S.R., McGeehan, G.M., Nambiar, K.P., and Benner, S.A. (1990) The ribonuclease from an extinct bovid ruminant. *FEBS Lett.* **262**: 104–106.

Strydom, D.J., Fett, J.W., Lobb, R.R., Alderman, E.M., Bethune, J.L., Riordan, J.F., and Vallee, B.L. (1985) Amino acid sequence of human tumor derived angiogenin. *Biochemistry* **24**: 5486–5494.

TrabesingerRuef, N., Jermann, T., Zankel, T., Durrant, B., Frank, G., and Benner, S.A. (1996) Pseudogenes in ribonuclease evolution: a source of new biomacromolecular function? *FEBS Lett.* **382**: 319–322.

Trautwein, K. (1991) *Construction of an Improved Expression System for Bovine Pancreatic Ribonuclease A and Construction and Characterization of RNase A Mutants.* ETH Dissertation no. 9613, Zurich.

Vandenberg, A., Vandenhendetimmer, L., and Beintema, J.J. (1976) Isolation, properties and primary structure of coypu and chinchilla pancreatic ribonuclease. *Biochim. Biophys. Acta* **453**: 400–409.

Vandijk, H., Sloots, B., Vandenberg, A., Gaastra, W., and Beintema, J.J. (1976) Primary structure of muskrat pancreatic ribonuclease. *Int. J. Pept. Protein Res.* **8**: 305–316.

Vos, J.P., Lopes-Cardozo, M., and Gadella, B.M. (1994) Metabolic and functional aspects of sulfogalactolipids. *Biochim. Biophys. Acta* **1211**: 125–149.

Welling, G.W., Groen, G., and Beintema, J.J. (1975) Amino acid sequence of dromedary pancreatic ribonuclease. *Biochem. J.* **147**: 505–511.

Welling, G.W., Mulder, H., and Beintema, J.J. (1976) Allelic polymorphism in arabian camel ribonuclease and amino acid sequence of bactrian camel ribonuclease. *Biochem. Genet.* **14**: 309–317.

Zhang, J.Z. and Rosenberg, H.F. (2002) Complementary advantageous substitutions in the evolution of an antiviral RNase of higher primates. *Proc. Natl. Acad. Sci. USA* **99**: 5486–5491.

Zhang, J., Rosenberg, H.F., and Nei, M. (1998) Positive Darwinian selection after gene duplication in primate ribonuclease genes. *Proc. Natl. Acad. Sci. USA* **95**: 3708–3713.

Zhang, J.Z., Dyer, K.D., and Rosenberg, H.F. (2002) RNase 8, a novel RNase A superfamily ribonuclease expressed uniquely in placenta. *Nucleic Acids Res.* **30**: 1169–1175.

Zhang, J.Z., Dyer, K.D., and Rosenberg, H.F. (2003) Human RNase 7: a new cationic ribonuclease of the RNase A superfamily. *Nucleic Acids Res.* **31**: 602–607.

Zhao, W., Kote-Jarai, Z., van Santen, Y., Hofsteenge, J., and Beintema, J.J. (1998) Ribonucleases from rat and bovine liver: purification, specificity and structural characterization. *Biochim. Biophys. Acta* **1384**: 55–65.

CHAPTER 19

Evolution of specificity and diversity

Denis C. Shields, Catriona R. Johnston, Iain M. Wallace, and Richard J. Edwards

19.1 Introduction

The divergence of proteins following gene duplication has long been recognized as an important process in the evolution of both new and specific protein functions (Serebrovsky, 1938; Ohno, 1967, 1970; Hughes, 1994; Lynch and Conery, 2000). For functional divergence to occur, the duplicated gene has to survive duplication and avoid becoming a pseudogene (gene death). The mechanism by which a gene duplicates survive is still under some debate, but it is thought that maintenance of duplicate pairs can be accomplished by the evolution of novel functions (neofunctionalization; Jacobs, 1996; Chen et al., 1997; Hunt et al., 1998; Cheng and Chen, 1999; Dulai et al., 1999; Kratz et al., 2002; Zhang et al., 2002; Gilad et al., 2003), splitting ancestral functions between duplicate pairs called paralogs (subfunctionalization; Serebrovsky, 1938; Jensen, 1976; Wistow and Piatigorsky, 1987; Hughes, 1994; Force et al., 1999; Stoltzfus, 1999), or some combination of both neo- and subfunctionalization. Although no consensus has yet been reached as to which process plays a more dominant role in the generation and maintenance of duplicates at the genomic or protein level, the distinction is somewhat irrelevant for the bioinformatic prediction of individual specificity-determining sites; that is, those sites that are important for differences in gene function between paralogs. Instead, the evolutionary history and changing selective constraints for individual residues is important for the interpretation of results. Here we focus on the types of substitution that occur at these sites and the phylogenetic signals that they leave.

19.2 Different kinds of changes relating to specificity

Sites of functional change following duplication can be broadly classified into two categories (Table 19.1), which Xun Gu has named type I and type II (Gu, 2001). Type I functional divergence shows a change in selective constraint on a site following duplication, either by relaxation of existing purifying selection or by gaining functional importance at a previously unimportant site. In contrast, sites experiencing type II divergence are important in both duplicates but a different amino acid is favoured in each duplicate. Both type I and type II divergence can occur as the result of either neo- or subfunctionalization. For example, subfunctionalization may occur by partitioning domain functions, with different domains maintained in different paralogs (type I divergence), or by each paralog specializing for a given set of existing substrates (type II divergence). Similarly, new gene function may arise at previously unimportant sites (type I) or by recruiting existing functional sites to the new function (type II), while the paralog fulfills the previous role of the ancestral protein.

19.3 Predictive models and available tools

A simple tool to distinguish residues that are more likely to confer specificity would be very useful for

Table 19.1 Division of methods into two broad groups based on whether they are designed to detect changes in rate of evolution (rate-shifting sites) or alternatively to detect sites conserved in subfamilies, but very different from each other (conservation-shifting sites)

Study	Rate-shifting sites (RSS)	Conservation-shifting sites (CSS)	URL	Estimate of statistical significance?	Explicitly model evolution?	Requires three-dimensional structural data?
Gu and Vander Velden (2002)	Type I	Type II	DIVERGE; http://xgu.zool.iastate.edu/software.html	Y	Y	
Knudsen and Miyamoto (2001)	Likelihood ratio			Y	Y	
Abhiman and Sonnhammer (2005a)		Conservation-shifting sites (CSS)	Results of Pfam surveys at http://FunShift.cgb.ki.se		(Y)	
Caffrey et al. (2000)		Burst after duplication (BAD)			Y	
Edwards and Shields (2005)	BADASP	BADASP	www.bioinformatics.rcsi.ie/~redwards/badasp/		Y	
Lichtarge et al. (2003)		Evolutionary trace	www.cmpharm.ucsf.edu/~marcinj/JEvTrace/			Y
Kalinina et al. (2004)		Mutual information	http://math.genebee.msu.ru/~psn/	Y		
del Sol Mesa et al. (2003)		Ordination	http://industry.ebi.ac.uk/SeqSpace/			

experimental biologists, and would also have the potential to shed light on the evolution of specificity. A number of approaches have been proposed (del Sol Mesa et al., 2003; Edwards and Shields, 2005), ranging from simple rule-based to more complex methods.

19.3.1 Property-change methods

A simple approach first adopted was to compare the amino acid properties of groups of sequences at each residue, and highlight residues that are strongly similar within a sequence grouping but strongly differ between groups (Livingstone and Barton, 1993). This was subsequently extended to compare ancestral predictions of sequences (in the burst-after-duplication (BAD) method). The rationale for this was that the shift in specificity is likely to occur in a defined evolutionary period, and that this will increase the power to define the most important changes (Caffrey et al., 2000). In the course of this ancestral sequence-based analysis, it was found that when trying to predict the set of most highly specific residues simply taking the majority amino acid sequence as the ancestral prediction was as efficient as considering the predicted values across all likelihoods of different ancestral states. The reason for this is that the residues with a high degree of uncertainty around their ancestral states have little power to predict specificity. This method has been implemented within software that takes into account gapped ancestral sequences in a straightforward manner (Edwards and Shields, 2004, 2005). In addition, it considers alternative models, depending on whether the comparison of a subfamily is with the nearest related subfamily, compared with all other subfamilies individually, or compared with the entire evolutionary tree. Routinely considering both statistics is valuable in automated analyses, where alignment quality can decline with evolutionary distance, and where the nearest subfamily and the entire tree may provide varying levels of information depending on the rates of evolution and can help detect by type I and type II divergence of functionally important sites. Different duplication events may also be responsible for different changes in gene function involving different sites, making it useful to be able to focus on the duplication of interest when additional information is available.

19.3.2 Structural modeling

An even simpler approach is the evolutionary trace (or ET) method (Lichtarge et al., 1996), which demands that the residues conferring specificity are completely conserved subsequent to the gene duplication. Amino acid properties are not considered and so a change from leucine to isoleucine, for example, will score as highly as a change from lysine to proline. At the same time, evolutionary trace will eliminate a certain number of specific residues whose properties may be strongly conserved but not identical. Instead, the predictive power of evolutionary trace was increased by taking structural information into account (Lichtarge and Sowa, 2002; Lichtarge et al., 2002, 2003). Structural modeling has been adopted in a number of other models (Johnson and Church, 2000; Armon et al., 2001; Pupko et al., 2002; Doron-Faigenboim et al., 2005; Landau et al., 2005) but has the obvious weakness of limiting analysis to proteins for which known structures are available.

19.3.3 Multivariate statistical modeling

A multivariate statistical approach has been taken by the Valencia group to cluster sequences based on a multiple alignment (Casari et al., 1995; Andrade et al., 1997). Principle-component analysis is used to project the sequences on to a lower-dimensional space which allows viewing of possible sequence clusters. The residues can also be projected onto the same space, so that residues specific to a group can be identified. More recently this method has been automated so that the user does not have to manually identify the sequence groups (del Sol Mesa et al., 2003). This method has been implemented as a software package called SequenceSpace.

19.3.4 Mutual information

More recently, mutual information has been proposed as a good measure to identify residues that confer specificity without needing structural data (Mirny and Gelfand, 2002). Mutual information is frequently used in bioinformatics, for instance to identify residues that covary in RNA (Clarke, 1995). In this case, it is a measure of how often a particular position in a sequence is conserved in one subfamily, and varies in another. The statistical significance ($P(I)$) of the mutual information (I) is also calculated. Positions that have a high I value and a low $P(I)$ value are considered to be specificity-determining positions (or SDPs). This method was validated on the LacI/PurR family of bacterial transcription factors, where 12 specificity-determining positions were identified; three of these residues are known to bind to DNA, whereas eight of them were found to be binding the ligand in the ligand-binding domain. In further work this method was applied to the protein kinase family (Li et al., 2003), of which there are over 500 in the human genome. One of the interesting features of protein kinases is their specificity; despite being highly similar in sequence and structure they are able to identify different substrates. The computational analysis was found to agree with known experimental data, and provided an insight as to how different protein kinases identify substrates. Further improvements were made in 2004 to the algorithm which allowed the method to take into account the non-uniformity of amino acid substitutions via amino acid-substitution matrices (Kalinina et al., 2004a). A method of automatically setting cut-off thresholds was also introduced. Overall this allowed the method to analyse more distantly related sequences in a more automatic fashion. There is a web server available (Kalinina et al., 2004b) that allows the user to upload an alignment and specify the groups.

19.3.5 Detection of changes in rate

Changes in specificity may result in a difference in the rate of evolution between subfamilies (Gaucher et al., 2002a). If the function of a protein is changing some residues will be subject to altered functional constraint. This implies that the evolutionary rate at these sites will vary between the homologous genes of different function. Gu (1999) developed a two-state model to try and detect the residues with differing rates of evolution. In the model, an amino acid can have one of two states: S_0, in which the site has the same mutation rate in both families, and S_1, in which the rate varies. This has been extended to include more complex models (Gu, 2001) and has been implemented in a software package called DIVERGE (Gu and Vander Velden, 2002). The programme calculates the residues that have differing rates of evolution based on a given phylogeny between two subfamilies. It also includes a PDB viewer on which the predicted residues can be displayed on a structure, if available. A Bayesian likelihood ratio method has also been presented to address this (Knudsen and Miyamoto, 2001), which has parallels with another related implementation (Massingham and Goldman, 2005). The method differs from that of Gu in the weighting of amino acid replacements, leading to differences in predictions (Knudsen and Miyamoto, 2001).

19.4 Experimental approaches to testing predictions of specificity

The evolutionary trace method has been used to predict residues conferring specificity, from proteins of known structure (Lichtarge et al., 2002; Madabushi et al., 2004). Prediction of specificity-determining residues in *rab5* and *rab6* using a multivariate approach (Casari et al., 1995) was then tested by creating chimeras (Stenmark et al., 1994), and similarly to assess specificity in small GTP-binding proteins (Bauer et al., 1999). Prediction of specificity determinants by the BAD approach exploiting ancestral sequence information was also tested by creation of chimeras between mitogen-activated protein kinases (Caffrey et al., 2000). However, chimera construction has the disadvantage that frequently chimeras are null for activity, and it is then difficult to determine the effect of these regions on specificity; the disturbance of normal folding in the chimera is not usually in itself a component of specificity, but is an unfortunate byproduct of incompatibility of diverged subregions.

An alternative approach that is relevant when investigating specificity of protein–protein interactions mediated by short linear motifs is to synthesize peptides corresponding to the specificity-determining regions. This approach has been adopted in investigation of chemokine-receptor dimerization (Hernanz-Falcon et al., 2004; de Juan et al., 2005). Rate changes have also identified particular sites of interest, for example in an elongation factor (Gaucher et al., 2002b). We recently used this approach to study platelet signalling pathways in over 20 different transmembrane proteins following BAD (Caffrey et al., 2000; Edwards and Shields, 2005) specificity prediction (Edwards et al., 2007). Pairs of peptides were synthesized from both paralogs, focused on residues predicted to be functionally specific. Although specific differences in activity were detected for some peptide pairs from paralogous proteins, the majority of pairs showed no such difference in the two assays used (data not shown).

Gene knock-in experiments to investigate the role of the ancestral sequences compared with different present-day sequences have the potential to refine the key changes in specificity, and to determine which changes over which evolutionary time span conferred that particular change. A more indirect alternative would be to use reagents that interact with the specificity determinants. Thus, antibodies against the specific regions, or drugs designed to be specific for a particular evolutionary subgroup, will then determine the phenotypic effect of the different proteins. However, they do not demonstrate that the regions targeted confer specificity.

All of these approaches suffer from the lack of good controls (i.e. residues selected for experiments that *have not* used functional specificity predictions). Although the applications of the prediction methods cited have undeniably led to useful biological discoveries, the additional benefits that these methods provide over more simple evolutionary analyses of manually identifying conserved regions from multiple sequence alignments, for example, have not been quantified.

19.5 Surveys of potential changes in specificity

Lysozyme evolution (Stewart et al., 1987) provides a case history of particular amino acid residues undergoing change in response to a well-defined and well-understood shift in external selection pressures. Acquisition of the ability to digest large amounts of plant material in colobine monkeys led to a shift in amino acid sequences at a number of conserved positions over a defined evolutionary period in ancestrally reconstructed sequences (Messier and Stewart, 1997). Some of these amino acid changes were convergent with those seen in the independent ruminant (bovine) lineage, suggesting very strongly that the individual residues have a particular role in conferring a specific phenotype. Additionally, studies of hemoglobin (Braunitzer and Hiebl, 1988) and integrins (Hughes, 1992) are similarly consistent with the theory that adaptation to novel functions gives rise to changes at otherwise constrained (slower-evolving) residues.

The question remains, how often do these very discrete changes occur? The neutral theory of evolution implies that the vast majority of changes will be largely neutral and not dominated by changes conferring specificity. Tests of amino acid change excess over synonymous site excess are an established means of determining a particular class of shifts in selection pressure. However, they are limited to a window of evolution following duplication, which may also be dominated by a high level of relaxed selection (Lynch and Conery, 2000). An alternative approach is to ignore DNA-level change, and simply concentrate on the patterns of change at certain residues compared to other residues.

19.5.1 Surveys of the Pfam alignment database

In a large-scale survey using Pfam (Bateman et al., 2004) domain alignments, we found an excess of changes at conserved amino acid positions following gene duplication (Seoighe et al., 2003). This is consistent with the hypothesis that there is an excess of changes conferring specificity to new diverged functions after a gene is duplicated. This excess was found even over longer evolutionary periods, suggesting it is not simply a byproduct of the rate acceleration seen in amino acid replacement after gene duplication (Lynch and Conery, 2000). However, it does not constitute proof, since relaxed selection pressures cannot be completely discounted as a cause. This analysis was primarily a survey of broad trends, and did not pinpoint particular instances of such evolution.

Such a systematic listing of proteins exhibiting most interesting changes has been recently presented by Abhiman and Sonnhammer (2005a, 2005b). From the Pfam resource a subset of over 4000 alignments were analysed. Abhiman and Sonnhammer (2005b) identified a total of 179 210 subfamily pairs, of which 62 384 were predicted to be functionally shifted in 2881 families (http://FunShift.cgb.ki.se; see Chapter 11 in this volume for more details). This approach uses Enzyme Commission (EC) numbers to group sequences, and will no doubt underestimate functional divergence, some of which occurs within these groups. As this FunShift analysis is to date the most concerted effort to systematically identify functional shifts in protein sequences, it is worth considering in some detail. How robust is the prediction of specificity?

The major issue with the FunShift analysis, and any related analyses, is the question of sample size for making a prediction, and the appropriate cut-off for significance of individual results in the context of a wide survey. In addition, there are a number of important methodological considerations. They classified subfamilies (of at least four sequences) using the Bayesian evolutionary tree estimation (BETE) method of Sjolander (1997), which employs a cost function to maximize homogeneity within subfamilies and minimize the number of subfamilies. They identified two classes of functional shift. Firstly, conservation-shifting sites (CSS), are equivalent to the type II sites defined by Gu (Table 19.1). This method (Abhiman and Sonnhammer, 2005b) is similar to that of Sjolander (1997), and calculates a Z-score based on the sum of the relative entropy between subfamilies in amino acid distribution at each position in an alignment. Whereas their measure of entropy is calculated for individual sites, it is then summed across residues in a protein subfamily comparison (as CSS equal to the percentage of sites with a Z-score exceeding 0.5). Thus, this is not a prediction algorithm for a particular residue, but rather it identifies whether the protein as a whole shows an excess of such sites. It highlights families showing a number of such changes, rather than individual residues This method may not be optimized to consider evolutionary information, as it simply pools all residues in a subfamily, and does not down-weight closely related sequences within a subfamily. Secondly, they investigated rate-shifting sites (RSS), equivalent to the type I sites defined by Gu (Table 19.1). For this, they used the Bayesian likelihood ratio method (Knudsen and Miyamoto, 2001), which models evolutionary information more directly. They calculated likelihoods of a rate shift between subfamilies for each residue independently. This created a rather vast number of all-against-all subfamily comparisons. They then used a previously trained classifier of enzyme subfamilies to combine the information from the percentage of RSS and CSS positions, and by this approach estimated that 35% of subfamilies exhibit functional shifts. Additionally, on their website they mark sites above a certain threshold for CSS or RSS. However, it is not clear whether the residues highlighted are statistically significant in themselves, since a very large number of residues are highlighted. Thus, this resource identifies the top third of subfamilies in which functional shifts are probably enriched, and identifies a large number of residues, which are potentially involved in functional shifts. It does not, however, distinguish clearly the most highly significant residues. These results and related methods are discussed in more detail in Chapter 11.

19.5.2 Domain-specific surveys

A more focused approach to experimental data may well be the way forward. Johnson and Church (2000) investigated ligand-binding specificity in *Escherichia coli* periplasmic binding proteins, where groups of closely and more distantly related proteins were investigated. They found that whole domain identity on its own was a poor predictor of ligand specificity. Phylogenetic analysis of the ligand-binding sites improved the predictions. Finally, matching the sequences to proteins of known structure provided a useful alternative, although it may under-predict similarity of function, since a change in residue at a binding site will not always alter ligand binding. To what extent the similarities of function reflected identity-by-descent, and to what extent they represent convergent evolution, was not clear in their analysis. They concluded that dense sampling of protein structures will be required for their methods to have useful power. Clearly, their approach goes beyond simply predicting specificity, but the findings have implications for specificity predictions.

19.6 Choice of method

19.6.1 Power to detect significant residues for functional shifts

In practice, the evolutionary event of a change in specificity at a single residue has a sample size of one, except in the case of convergent evolution, as highlighted above in the example of lysozyme. Thus, it is secondary information concerning the subsequent rate of evolution that provides any power to distinguish specificity-conferring changes from other changes. In many cases, the current databases have relatively sparse sampling of the downstream evolutionary events that contribute to these comparisons, and this currently limits power for many families. To illustrate this, the cut-off chosen by Abhiman and Sonnhammer for visual representation of a rate change corresponds to a P value of 0.05. However, a strict correction for multiple testing would consider not only all the residues within a protein, but probably even the effect of testing multiple proteins. Thus, if a typical protein is 300 residues long, the true cut-off should be much more extreme. For the bulk of available alignment data, imposing such strict cut-offs for significance will quickly eliminate all suggested residues in large-scale screens.

Statistical power is also reduced by the introduction of error. Alignment and topology error have profound effects on the calculation of such statistics. Gene conversion among related proteins in eukaryotes provides a frequent source of bias (Shields, 2000): whereas it cannot be excluded that gene conversion contributes to the evolution of specificity via the creation of novel hybrid functions, more usually it will simply bias statistical estimation. Alignment error has even more wide-ranging consequences, as it is rare for an alignment of two related subfamilies to be perfect. Typically, a number of sites will be well aligned within each subfamily but misaligned between them, giving spurious signals of specificity. Although tools exist that try to identify such regions of the alignment (R.J. Edwards, unpublished work), it is often necessary to manually inspect alignments to be sure, making it a major problem in large-scale analyses.

Clearly, regardless of their true statistical significance, predictions are useful for guiding experimental evaluations to identify specificity-conferring residues or regions, where the cost of performing all experiments is prohibitive. However, investigators should not be misled by the impressiveness of the statistical methods and claims of previous success into thinking that experiments guided by such considerations have been proven to be more efficient than those where the investigator does not think about them. Although this is probably true, it remains unproven.

19.6.2 Lack of a good training data-set of identified residues conferring specificity

One problem that remains in this field is that there is no good, large data-set of amino acid residues that have been determined experimentally to confer specificity, which can be used for the general training and/or assessment of methods predicting specificity. In light of this, a number of groups have used functional classifications of proteins to find a method that identifies the residues that most closely predict functional differences (Hannenhalli and Russell, 2000; Abhiman and Sonnhammer, 2005b). Although this has some value, it is indirect, and we still do not know whether such methods are really giving us a clear identification of the residues that are experimentally proven to confer specificity. In particular, it is not possible to infer the statistical sensitivity and specificity of these predictions from this indirect method. It is possible, for example, that all the prediction methods discussed here have a very low sensitivity (ability to detect all true occurrences) and specificity (ability to avoid false identification).

19.6.3 Comparisons among methods

In the absence of a clear good training set to develop the best methods to identify specificity-conferring regions, we can at least compare the predictions of different methods. Some papers have compared a few alternatives to justify their own methodology (Casari et al., 1995; Caffrey et al., 2000), but there is an absence in the literature of a systematic comparison. As an illustration of what can happen when methods are compared, we took a multiple sequence alignment of estrogen receptors α and β (generated by HAQESAC; R.J. Edwards, unpublished work) and compared the performance of four different algorithms without using structural information using BADASP (Edwards and Shields, 2005) and SDPpred (Kalinina et al., 2004b; Figure 19.1). The poor overlap in the prediction of the specificity-conferring residues indicates that, although these methods are drawing on a very similar principle (conservation within subfamilies contrasting with differences between subfamilies), the methods are quite sensitive to the particular assumptions of the model; the two property-based methods identify a quite distinct group of residues to the methods that ignore amino acid properties. Furthermore, there is very little difference in this example between the very simple evolutionary

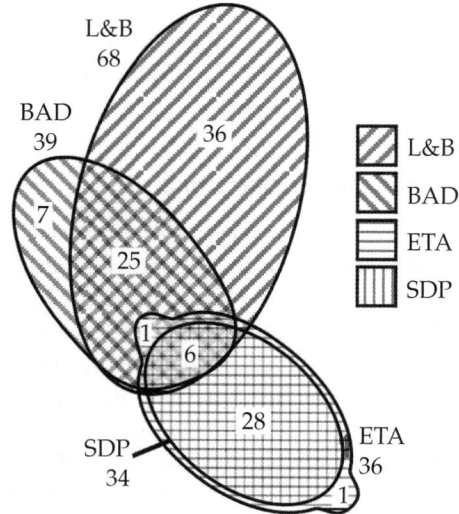

Figure 19.1 Poor concordance of predictions for an example estrogen receptor α and β using four methods. The numbers of sites predicted are shown. BAD, burst after duplication; SDP, SDPpred; L&B, Livingstone and Barton property method; ETA, evolutionary trace analysis (ignoring structural information).

trace method, and the much more complicated SDPpred method, highlighting the possibility that adding to the sophistication and theoretical statistical power of a model does not necessarily improve the quality of the results. The fact that, out of only six residues that are predicted by all four methods, one is in a region of bad alignment also highlights the dangers of biases introduced by alignment errors.

The choice of method for detecting rate shifts has been investigated by Blouin et al. (2005). They recommend that methods such as DIVERGE of Gu and Vander Velden (2002) are appropriate when selecting a limited number of sites, for example mutational studies, whereas the method of Knudsen and Miyamoto (2001) is more appropriate when seeking to identify a large number of sites.

The most sensible approach is to apply the model which best matches our understanding of a given protein's evolution. Although this is slightly more likely to give the best result, we must recognize that most of these predictions are enormously sensitive to such assumptions, and therefore must be interpreted with caution. Essentially, given the low statistical power, and in the absence of any clear evidence regarding their sensitivity and specificity, we cannot make strong statements regarding the relative performance of these methods, until a sizeable and relatively unbiased training data-set is established that can be used to test them.

19.6.4 Practical considerations

In summary, there are numerous practical considerations when trying to predict specificity-conferring residues. The quality and accuracy of multiple sequence alignment and phylogenetic inference, the availability and reliability of three-dimensional structural data, and the process of defining subfamilies and choosing which subfamilies to compare, are all initial considerations. Subsequent to this, there are a variety of methodological considerations.

A number of the rate-altering site-detection methods (Knudsen and Miyamoto, 2001; Gu and Vander Velden, 2002) and the mutual information method, which identifies conservation-shifting sites (Kalinina et al., 2004b), provide estimates of statistical significance. These are important to consider, since many individual sites will not pass a rigorous threshold of significance, suggesting there is insufficient power. However, significance measures cannot be interpreted blindly, since many departures from the assumptions in the models (in particular, alignment error) will radically distort them. Although it has been proposed that ancestral sequence reconstruction improves such predictions (Caffrey et al., 2000), but that a simple majority-rule ancestral sequence is sufficient to predict conservation-shifting residues, the whole area of predicting functional specificity lacks clear strong validation from experimental data of any of its predictions, and many individual predictions lack statistical power to really give confidence to the inferences. In terms of choosing among alternative approaches available as software, many of these complex methods lack simple software implementations and/or rely on additional information, such as structural data, which is not always available. There are available tools for state-of-the-art predictions for divergence of either type I (Gu and Vander Velden, 2002) or type II (Kalinina et al., 2004b) change for single protein families, and for structural data (Pupko); the BADASP method provides a command-line interface to implement the previously published BAD algorithm (Caffrey et al., 2000), plus two variants for identifying type I and type II divergence. The sensitivities of all these methods to the very frequent errors in alignment and inference of topology remain to be investigated, and a reasonable approach would be to compare the outputs of different models and try and interpret the predictions in the light of possible errors in the input data.

19.7 Other considerations of protein evolution

Evolutionary change at a single residue can confer functional specificity in terms of that protein's interaction with some external molecule, such as a substrate at an active site, or a ligand (protein or non-protein) at a binding site. However, such changes may often disturb the balance within the protein. In particular, protein stability may be markedly affected by an internal change within a substrate-binding site. The consequences of such a change are to immediately create a selection pressure for parallel changes elsewhere in the protein, which will restore the stability of the protein (DePristo et al., 2005). Most of the statistical methods proposed here, and the training of methods based on enzyme or other classifications, may be partly combining among their predictions true specificity-determining changes, as well as those subsequent changes that impact on stability. Such stability-related changes may account for a substantial proportion of amino acid changes fixed during evolution (DePristo et al., 2005). In the absence of experimental data-sets that distinguish between specificity-conferring and stability-modifying changes, it is not clear to what extent stability-altering changes tend to be subsequently strongly conserved during evolution, and therefore it is unclear to what extent such changes are indeed mimicking specificity-conferring residues. It is possible that this is less of an issue for surface-accessible residues involved in ligand interactions, which will have fewer destabilizing effects. While stability-modifying changes are interesting in themselves in terms of understanding the total behaviour of a protein, they blur the boundaries in terms of what is considered a specificity-conferring prediction. Thus, both individual results and clusters of changes in evolutionary time or structure space need to be interpreted in the light of alternative explanations. In particular, a single rare amino acid change fixed by drift that reduces protein stability may then be followed by a suite of compensating replacements, and this entire cluster of change may be a consequence of an initial chance event, rather than the result of a selection pressure to make the protein different from the protein that preceded the burst of change.

Acknowledgments

This work was funded by Science Foundation Ireland, the Health Research Board, and the Programme for Research in Third Level Institutions administered by the Higher Education Authority.

References

Abhiman, S. and Sonnhammer, E.L. (2005a) FunShift: a database of function shift analysis on protein subfamilies. *Nucleic Acids Res.* **33**: D197–D200.

Abhiman, S. and Sonnhammer, E.L. (2005b) Large-scale prediction of function shift in protein families with a focus on enzymatic function. *Proteins* **6**: 6.

Andrade, M.A., Casari, G., Sander, C., and Valencia, A. (1997) Classification of protein families and detection of the determinant residues with an improved self-organizing map. *Biol. Cybern.* **76**: 441–450.

Armon, A., Graur, D., and Ben-Tal, N. (2001) ConSurf: an algorithmic tool for the identification of functional regions in proteins by surface mapping of phylogenetic information. *J. Mol. Biol.* **307**: 447–463.

Bateman, A., Coin, L., Durbin, R., Finn, R.D., Hollich, V., Griffiths-Jones, S. et al. (2004) The Pfam protein families database. *Nucleic Acids Res.* **32**: D138–D142.

Bauer, B., Mirey, G., Vetter, I.R., Garcia-Ranea, J.A., Valencia, A., Wittinghofer, A. et al. (1999) Effector recognition by the small GTP-binding proteins Ras and Ral. *J. Biol. Chem.* **274**: 17763–17770.

Blouin, C., Butt, D., and Roger, A.J. (2005) Impact of taxon sampling on the estimation of rates of evolution at sites. *Mol. Biol. Evol.* **22**: 784–791.

Braunitzer, G. and Hiebl, I. (1988) Molecular aspects of high altitude respiration of birds. Hemoglobins of the striped goose (*Anser indicus*), the Andean goose, (*Chloephaga melanoptera*) and vulture (*Gyps rueppellii*). *Naturwissenschaften* **75**: 280–287.

Caffrey, D.R., O'Neill, L.A., and Shields, D.C. (2000) A method to predict residues conferring functional differences between related proteins: application to MAP kinase pathways. *Protein Sci.* **9**: 655–670.

Casari, G., Sander, C., and Valencia, A. (1995) A method to predict functional residues in proteins. *Nat. Struct. Biol.* **2**: 171–178.

Chen, L., DeVries, A.L., and Cheng, C.H. (1997) Evolution of antifreeze glycoprotein gene from a trypsinogen gene in Antarctic notothenioid fish. *Proc. Natl. Acad. Sci. USA* **94**: 3811–3816.

Cheng, C.H. and Chen, L. (1999) Evolution of an antifreeze glycoprotein. *Nature* **401**: 443–444.

Clarke, N.D. (1995) Covariation of residues in the homeodomain sequence family. *Protein Sci.* **4**: 2269–2278.

de Juan, D., Mellado, M., Rodriguez-Frade, J.M., Hernanz-Falcon, P., Serrano, A., Del Sol, A. et al. (2005) A framework for computational and experimental methods: Identifying dimerization residues in CCR chemokine receptors. *Bioinformatics* **21** (suppl. 2): ii13–ii18.

del Sol Mesa, A., Pazos, F., and Valencia, A. (2003) Automatic methods for predicting functionally important residues. *J. Mol. Biol.* **326**: 1289–1302.

DePristo, M.A., Weinreich, D.M., and Hartl, D.L. (2005) Missense meanderings in sequence space: a biophysical view of protein evolution. *Nat. Rev. Genet.* **6**: 678–687.

Doron-Faigenboim, A., Stern, A., Mayrose, I., Bacharach, E., and Pupko, T. (2005) Selecton: a server for detecting evolutionary forces at a single amino-acid site. *Bioinformatics* **21**: 2101–2103.

Dulai, K.S., von Dornum, M., Mollon, J.D., and Hunt, D.M. (1999) The evolution of trichromatic color vision by opsin gene duplication in New World and Old World primates. *Genome Res.* **9**: 629–638.

Edwards, R.J. and Shields, D.C. (2004) GASP: Gapped Ancestral Sequence Prediction for proteins. *BMC Bioinformatics* **5**: 123.

Edwards, R.J. and Shields, D.C. (2005) BADASP: predicting functional specificity in protein families using ancestral sequences. *Bioinformatics* **21**: 4190–4191.

Edwards, R.J., Moran, N., Devocelle, M., Kiernan, A., Meade, G., Signac, W. *et al.* (2007) Bioinformatic discovery of novel bioactive peptides. *Nat. Chem. Biol.*, in press.

Force, A., Lynch, M., Pickett, F.B., Amores, A., Yan, Y.L., and Postlethwait, J. (1999) Preservation of duplicate genes by complementary, degenerative mutations. *Genetics* **151**: 1531–1545.

Gaucher, E.A., Gu, X., Miyamoto, M.M., and Benner, S.A. (2002a) Predicting functional divergence in protein evolution by site-specific rate shifts. *Trends Biochem. Sci.* **27**: 315–321.

Gaucher, E.A., Das, U.K., Miyamoto, M.M., and Benner, S.A. (2002b) The crystal structure of eEF1A refines the functional predictions of an evolutionary analysis of rate changes among elongation factors. *Mol. Biol. Evol.* **19**: 569–573.

Gilad, Y., Bustamante, C.D., Lancet, D., and Paabo, S. (2003) Natural selection on the olfactory receptor gene family in humans and chimpanzees. *Am. J. Hum. Genet.* **73**: 489–501.

Gu, X. (1999) Statistical methods for testing functional divergence after gene duplication. *Mol. Biol. Evol.* **16**: 1664–1674.

Gu, X. (2001) Maximum-likelihood approach for gene family evolution under functional divergence. *Mol. Biol. Evol.* **18**: 453–464.

Gu, X. and Vander Velden, K. (2002) DIVERGE: phylogeny-based analysis for functional-structural divergence of a protein family. *Bioinformatics* **18**: 500–501.

Hannenhalli, S.S. and Russell, R.B. (2000) Analysis and prediction of functional sub-types from protein sequence alignments. *J. Mol. Biol.* **303**: 61–76.

Hernanz-Falcon, P., Rodriguez-Frade, J.M., Serrano, A., Juan, D., del Sol, A., Soriano, S.F. *et al.* (2004) Identification of amino acid residues crucial for chemokine receptor dimerization. *Nat. Immunol.* **5**: 216–223.

Hughes, A.L. (1992) Coevolution of the vertebrate integrin alpha- and beta-chain genes. *Mol. Biol. Evol.* **9**: 216–234.

Hughes, A.L. (1994) The evolution of functionally novel proteins after gene duplication. *Proc. Biol. Sci.* **256**: 119–124.

Hunt, D.M., Dulai, K.S., Cowing, J.A., Julliot, C., Mollon, J.D., Bowmaker, J.K. *et al.* (1998) Molecular evolution of trichromacy in primates. *Vision Res.* **38**: 3299–3306.

Jacobs, G.H. (1996) Primate photopigments and primate color vision. *Proc. Natl. Acad. Sci. USA* **93**: 577–581.

Jensen, R.A. (1976) Enzyme recruitment in evolution of new function. *Annu. Rev. Microbiol.* **30**: 409–425.

Johnson, J.M. and Church, G.M. (2000) Predicting ligand-binding function in families of bacterial receptors. *Proc. Natl. Acad. Sci. USA* **97**: 3965–3970.

Kalinina, O.V., Mironov, A.A., Gelfand, M.S., and Rakhmaninova, A.B. (2004a) Automated selection of positions determining functional specificity of proteins by comparative analysis of orthologous groups in protein families. *Protein Sci.* **13**: 443–456.

Kalinina, O.V., Novichkov, P.S., Mironov, A.A., Gelfand, M.S., and Rakhmaninova, A.B. (2004b) SDPpred: a tool for prediction of amino acid residues that determine differences in functional specificity of homologous proteins. *Nucleic Acids Res.* **32**: W424–W428.

Knudsen, B. and Miyamoto, M.M. (2001) A likelihood ratio test for evolutionary rate shifts and functional divergence among proteins. *Proc. Natl. Acad. Sci. USA* **98**: 14512–14517.

Kratz, E., Dugas, J.C., and Ngai, J. (2002) Odorant receptor gene regulation: implications from genomic organization. *Trends Genet.* **18**: 29–34.

Landau, M., Mayrose, I., Rosenberg, Y., Glaser, F., Martz, E., Pupko, T., and Ben-Tal, N. (2005) ConSurf 2005: the projection of evolutionary conservation scores of residues on protein structures. *Nucleic Acids Res.* **33**: W299–W302.

Li, L., Shakhnovich, E.I., and Mirny, L.A. (2003) Amino acids determining enzyme-substrate specificity in prokaryotic and eukaryotic protein kinases. *Proc. Natl. Acad. Sci. USA* **100**: 4463–4468.

Lichtarge, O. and Sowa, M.E. (2002) Evolutionary predictions of binding surfaces and interactions. *Curr. Opin. Struct. Biol.* **12**: 21–27.

Lichtarge, O., Bourne, H.R., and Cohen, F.E. (1996) An evolutionary trace method defines binding surfaces common to protein families. *J. Mol. Biol.* **257**: 342–358.

Lichtarge, O., Sowa, M.E., and Philippi, A. (2002) Evolutionary traces of functional surfaces along G protein signaling pathway. *Methods Enzymol.* **344**: 536–556.

Lichtarge, O., Yao, H., Kristensen, D.M., Madabushi, S., and Mihalek, I. (2003) Accurate and scalable identification of functional sites by evolutionary tracing. *J. Struct. Funct. Genomics.* **4**: 159–166.

Livingstone, C.D. and Barton, G.J. (1993) Protein sequence alignments: a strategy for the hierarchical analysis of residue conservation. *Comput. Appl. Biosci.* **9**: 745–756.

Lynch, M. and Conery, J.S. (2000) The evolutionary fate and consequences of duplicate genes. *Science* **290**: 1151–1155.

Madabushi, S., Gross, A.K., Philippi, A., Meng, E.C., Wensel, T.G., and Lichtarge, O. (2004) Evolutionary trace of G protein-coupled receptors reveals clusters of residues that determine global and class-specific functions. *J. Biol. Chem.* **279**: 8126–8132.

Massingham, T. and Goldman, N. (2005) Detecting amino acid sites under positive selection and purifying selection. *Genetics* **169**: 1753–1762.

Messier, W. and Stewart, C.B. (1997) Episodic adaptive evolution of primate lysozymes. *Nature* **385**: 151–154.

Mirny, L.A. and Gelfand, M.S. (2002) Using orthologous and paralogous proteins to identify specificity-determining residues in bacterial transcription factors. *J. Mol. Biol.* **321**: 7–20.

Ohno, S. (1967). *Sex Chromosomes and Sex-linked Genes. Monographs on Endocrinology,* pp. 46–73. Springer Verlag, Berlin.

Ohno, S. (1970). *Evolution by Gene Duplication.* Springer Verlag, Berlin.

Pupko, T., Bell, R.E., Mayrose, I., Glaser, F., and Ben-Tal, N. (2002) Rate4Site: an algorithmic tool for the identification of functional regions in proteins by surface mapping of evolutionary determinants within their homologues. *Bioinformatics* **18** (suppl. 1): S71–S77.

Seoighe, C., Johnston, C.R., and Shields, D.C. (2003) Significantly different patterns of amino acid replacement after gene duplication as compared to after speciation. *Mol. Biol. Evol.* **20**: 484–490.

Serebrovsky, A. (1938) Genes scute and achaete in *Drosophila melanogaster* and a hypothesis of gene divergency. *C. R. Acad. Sci. URSS* **19**: 77–81.

Shields, D.C. (2000) Gene conversion among chemokine receptors. *Gene* **246**: 239–345.

Sjolander, K. (1997) Bayesian evolutionary tree estimation. In *Proceedings of the Eleventh International Conference on Mathematical and Computer Modelling and Scientific Computing, Computational Biology Session: Conference Computing in the Genome Era 1997,* Georgetown University Conference Center, Washington DC, 31 March–3 April 1997.

Stenmark, H., Valencia, A., Martinez, O., Ullrich, O., Goud, B., and Zerial, M. (1994) Distinct structural elements of rab5 define its functional specificity. *EMBO J.* **13**: 575–583.

Stewart, C.B., Schilling, J.W., and Wilson, A.C. (1987) Adaptive evolution in the stomach lysozymes of foregut fermenters. *Nature* **330**: 401–404.

Stoltzfus, A. (1999) On the possibility of constructive neutral evolution. *J. Mol. Evol.* **49**: 169–181.

Wistow, G. and Piatigorsky, J. (1987) Recruitment of enzymes as lens structural proteins. *Science* **236**: 1554–1556.

Zhang, J., Zhang, Y.P., and Rosenberg, H.F. (2002) Adaptive evolution of a duplicated pancreatic ribonuclease gene in a leaf-eating monkey. *Nat. Genet.* **30**: 411–415.

Conclusion and a way forward

David A. Liberles

We have seen several controversial issues raised in the course of this book. In concluding, we will hope to give these issues some perspective and present the reader attempting ancestral sequence reconstruction with a potential way forward. Ancestral sequence reconstruction, as discussed in chapters throughout the book, can be used for testing general hypotheses about the environment and lifestyles of extinct species, general hypotheses about the processes driving gene and genome evolution, specific hypotheses about the evolution of gene function in individual gene families, and ultimately the generation of an understanding of the mapping between sequence (and substitution) to molecular function. Understanding the sequence underpinnings to molecular function is a general problem in molecular biology, with direct applications in applied fields like agriculture (broadly defined) and drug design, where modification of protein function through sequence is of inherent interest. Especially in the pharmaceutical industry, the power of molecular evolution and comparative genomics have been underappreciated.

Several chapters have addressed important evolutionary mechanisms of how gene function and resulting evolutionary novelty evolve. Chapter 16 presents a link between sequence evolution and the evolution of specificity, ultimately providing insight into the build-up of complex specific pathways. At a finer level, Chapter 9 examines the process of heterotachy, where individual sights are shifting rates under shifting selective pressures on shifting landscapes. This process may ultimately relate to shifting functions, as discussed in Chapters 11 and 19. Further, the discussion of sequence evolution through structure may also play a crucial role in driving processes like heterotachy (as briefly discussed in Chapter 4), which may over long evolutionary periods lead to divergent sequence and structural alignments by sequences sliding through structures. Chapter 7 examines the direct link between sequence evolution and biophysical properties, which may also directly affect function via structure.

Taking the next step

The original meeting raised a discussion of the relative merits of sequence-based compared with structure-based alignment for ancestral sequence reconstruction. Depending upon the level of divergence of the ancestor, the two alignment methods may not differ so radically. For alignments based upon older divergences, information from both protein-level sequence alignment and structural alignment can be compared, given knowledge of the protein. Common programs for structural alignment include Combinatorial Extension (Shindyalov and Bourne, 1998) and SSAP (Orengo and Taylor, 1996), whereas common programs for sequence-based alignment include Muscle (Edgar, 2004), ClustalW (Thompson et al., 1994), T-Coffee (Notredame et al., 2000), POA (Lee et al., 2002), and various methods encoded in the Darwin package (Gonnet et al., 2000).

Built on top of the multiple sequence alignment, a phylogenetic tree shows the evolutionary relationship of each position in the alignment. Common programs for phylogenetic-tree construction include PAUP, Phylip (Felsenstein, 1989), and MrBayes (Ronquist and Huelsenbeck, 2003). Phylip offers a collection of different methods and MrBayes is considered a state-of-the-art Bayesian method. Chapter 10 shows some of the pitfalls for getting the tree wrong. Therefore, it is desirable to consider alternative tree topologies and their effects on ancestral sequence reconstruction.

External lines of evidence about the relationship of species can also be used to map gene trees to species trees and consider the branching order in light of a minimization of gene-duplication and-loss events (consideration of syntenic information can also be informative in identifying orthologous relationships). Software for mapping gene trees to species trees includes Notung (Chen *et al.*, 2000) and Softparsmap (Berglund-Sonnhammer *et al.*, 2006). One concern is how common gene duplication followed by differential gene loss in different lineages is, making the ultimate development of model-based approaches for mapping gene trees on to species trees desirable (see Arvestad *et al.*, 2004).

For smaller data-sets, integrated methods that combine multiple sequence alignment and phylogenetic tree construction in one step are slow, but may provide a better assessment of homology and the evolutionary history of any given amino acid position. Methods for achieving this in a model-based framework include SATCHMO (Edgar and Sjolander, 2003), BEAST (Lunter *et al.*, 2005), and BAli-Phy (Suchard and Redelings, 2006).

The reader may also consider starting with pre-calculated gene families that already contain multiple sequence alignments and phylogenetic trees for families of interest. Such families can be modified with detailed knowledge and expanded to include new sequences, as available and desired. Some examples of databases of pre-calculated gene families include Pfam (Bateman *et al.*, 2004), the Master Catalog (Benner *et al.*, 2000), the Hovergen/Hobacgen group of gene-family databases (Duret *et al.*, 1994; Perriere *et al.*, 2000; Paulsen and von Haeseler, 2006), and TAED (Roth *et al.*, 2005).

Given a multiple sequence alignment and a phylogeny, PAML (Yang, 1997) and FASTML (Pupko *et al.*, 2002) are the state-of-the-art programs for ancestral sequence reconstruction. In using maximum-likelihood methods, model testing to reconstruct sequences under an appropriate model is highly recommended. The Simmap approach presented in Chapter 6 provides software for substitutional mapping. Alternatives to this include ancestral sequence reconstruction to identify ancestral states explicitly at various nodes and parsimony to map the characters that changed to the branch. The latter approach might be considered if one is interested in coupling computational mapping to experimental reconstruction at the nodes. Relatedly, the DIVERGE software package is presented in Chapter 11, which enables comparison of the sequence divergence properties of two clades of genes that may have diverged in function as well as a prediction of functional divergence.

Chapter 8 presents a case for using Bayesian sampling to reconstruct ancestors, while Chapter 15 presents an experimental strategy for accommodating the recommendations of Chapter 8. Chapter 2 presents a brief rebuttal to the view in Chapter 8, making a case for maximum likelihood. Chapter 14 presents an overview of methods for experimentally reconstructing ancestral sequences. In terms of the use of Bayesian sampling compared with maximum likelihood (or maximum parsimony), this is still an open issue for the researcher to decide upon. At the time of publication, the best advice is to continue to follow the literature surrounding this debate as the community ultimately moves towards a consensus.

In considering all of these methods, be rigorous and open-minded. Good luck with your ancestral sequence reconstructions, as this growing field blossoms.

References

Arvestad, L., Berglund, A.C., Lagergren, J., and Sennblad, B. (2004) Gene tree reconstruction and orthology analysis based on an integrated model for duplication and sequence evolution. RECOMB 2004: 326–335.

Bateman, A., Coin, L., Durbin, R., Finn, R.D., Hollich, V., Griffiths-Jones, S. *et al.* (2004) The Pfam protein families database. *Nucleic Acids Res.* **32**: D138–D141.

Benner, S.A., Chamberlin, S.G., Liberles, D.A., Govindarajan, S., and Knecht, L. (2000) Functional inferences from reconstructed evolutionary biology involving rectified databases: an evolutionarily grounded approach to functional genomics. *Res. Microbiol.* **151**: 97–106.

Berglund-Sonnhammer, A.C., Steffansson, P., Betts, M.J., and Liberles, D.A. (2006) Optimal gene trees from sequences and species trees using a soft interpretation of parsimony. *J. Mol. Evol.* **63**: 240–250.

Chen, K., Durand, D., and Farach-Colton, M. (2000) NOTUNG: a program for dating gene duplications and optimizing gene family trees. *J. Comp. Biol.* **7**: 429–447.

Duret, L., Mouchiroud, D., and Gouy, M. (1994) HOVERGEN: a database of homologous vertebrate genes. *Nucleic Acids Res.* **22**: 2360–2365.

Edgar, R.C. (2004) MUSCLE: multiple sequence alignment with high accuracy and high throughput. *Nucleic Acids Res.* **32**: 1792–1797

Edgar, R.C. and Sjolander, K. (2003) SATCHMO: sequence alignment and tree construction using hidden Markov models. *Bioinformatics* **19**: 1404–1411.

Felsenstein, J. (1989) Phylip–Phylogenetic inference package. *Cladistics* **5**: 164–166.

Gonnet, G.H., Hallett, M.T., Korostensky, C., and Bernardin, L. (2000) Darwin v 2.0: an integrated computer language for the biosciences. *Bioinformatics* **18**: 101–103.

Lee, C., Grasso, C., and Sharlow, M.F. (2002) Multiple sequence alignment using partial order graphs. *Bioinformatics* **18**: 452–464.

Lunter, G., Miklos, I., Drummond, A., Jensen, J.L., and Hein, J. (2005) Bayesian coestimation of phylogeny and sequence alignment. *BMC Bioinformatics* **6**: 83.

Notredame, C., Higgins, D.G., and Heringa, J. (2000) T-coffee: a novel method for fast and accurate multiple sequence alignment. *J. Mol. Biol.* **302**: 205–217.

Orengo, C. and Taylor, W.R. (1996) SSAP: sequential structure alignment program for protein structure comparison. *Methods Enzymol.* **266**: 617–635.

Paulsen, I. and von Haeseler, A. (2006) INVHOGEN: a database of homologous invertebrate genes. *Nucleic Acids Res.* **34**: D349–D353.

Perriere, G., Duret, L., and Gouy, M. (2000) HOBACGEN: database system for comparative genomics in bacteria. *Genome Res.* **10**: 379–385.

Pupko, T., Pe'er, I., Graur, D., Hasegawa, M., and Friedman, N. (2002) A branch and bound algorithm for the inference of ancestral amino acid sequences when the replacement rate varies among sites: application to the evolution of five gene families. *Bioinformatics* **18**: 1116–1123.

Ronquist, F. and Huelsenbeck, J.P. (2003) MrBayes 3: Bayesian phylogenetic inference under mixed models. *Bioinformatics* **19**: 1572–1574.

Roth, C., Betts, M.J., Steffansson, P., Saelensminde, G., and Liberles, D.A. (2005) The Adaptive Evolution Database (TAED): a phylogeny based tool for comparative genomics. *Nucleic Acids Res.* **33**: D495–D497.

Shindyalov, I.N. and Bourne, P.E. (1998) Protein structure alignment by incremental combinatorial extension (CE) of the optimal path. *Protein Eng.* **11**: 739–747.

Suchard, M.A. and Redelings, B.D. (2006) BAli-Phy: simultaneous Bayesian inference of alignment and phylogeny. *Bioinformatics* **22**: 2047–2048.

Thompson, J.D., Higgins, D.G., and Gibson, T.J. (1994) CLUSTAL W: improving the sensitivity of progressive multiple sequence alignment through sequence weighting, position-specific gap penalties, and weight matrix choice. *Nucleic Acids Res.* **22**: 4673–4680.

Yang, Z.H. (1997) PAML–a program package for phylogenetic analysis by maximum likelihood. *Comput Appl Biol Sci* **13**: 555–556.

Index

Note: page numbers in *italics* refer to Figures and Tables.

8-10-4-5 regions 141, 142, 145
Abhiman, S. and Sonnhammer, E.L. 230, 231
Abi-Rached, L. et al. 142
absorption spectra, rhodopsin variants 170, *171*
acceleration transformation (ACCTRAN) assignment 44
acetaldehyde, conversion to ethanol 9, *10*
acetate 9
acetyl-CoA 9
acetyl phosphate 9
ACV1 35
1-acyl-glycerol-3-phosphate O-acyltransferases 142
adaptations, digestive ribonucleases 7
adaptive bursts 91
adaptive evolution 23
  as cause of bias 22
adaptive sequence evolution 216
ADH_A 10–12
Aeropyrum pernix 202, *204*
  optimal growth temperature *203*
Affine gap penalty 53
affinity of steroid hormone receptors 184
agnathans, corticoid receptors 186, *187*, 191
Akaike Information Criterion (AIC) 48
Akistrodon piscivorus, GLP-1 *37*
alanine scanning mutagenesis 34
alcohol dehydrogenase (Adh) 9–13
  expression in *Saccharomyces cerevisiae* 157, *158*
  duplications in yeast genome *16*
  functional assays 159, *160*
  heuristic hierarchical Bayesian approach 22
  interconnecting models 13–17
  molecular properties 161
  site-directed mutagenesis 154
alcohol dehydrogenase class III protein, probabilistic

ancestral reconstruction 65–7
aldosterone 184, *185*, 186
aldosterone-sensitivity, steroid receptors *189*
  ancestral corticoid receptor 188, 190–1
  glucocorticoid receptor 192–4
  mineralocorticoid receptor 191–2, 194
Alfimeprase 35
alignment error 231–2
all-α class proteins 133
all-β class proteins 133
allosteric changes, effects 38
α-helices, context-dependent matrices 49
α + β class proteins 133
α/β class proteins 133
α-proteobacterium, origin of mitochondria 133
Alpha Shift Measure (ASM) 125
α-tubulin studies 108
alternative evolutionary models, use in coral GFP-like protein reconstruction 173–4
ambiguities 8–9
  ADH_A 11
  in ancestral protein reconstruction 166–8
  in EF studies 25
  in ribonuclease reconstructions 212, 222
  see also uncertainties
amino acid composition, relationship to optimal growth temperature 201
amino acid composition estimation, LUA proteins 201–2, 205–7, *206*
  methods 202–4
amino acid matrices 48–9
amino acid models 165
amino acid properties, incorporation into codon-based models 166

amino acid replacements, impact 8–9
amino acids, unnatural 30
aminoacyl-tRNAs 23
among-site rate variation 45
among-site rate variation models 46–8
  model selection 48
among-site substitution variation 49
amphioxus 141, 144
Amphiuma tridactylum, GLP-1 *37*
anaphylaxis 142
ancestral corticoid receptor (AncCR)
  DOC sensitivity 191–2
  pre-adaptation to aldosterone 192
  resurrection 186–8
  robustness to uncertainty 188, 190–1
ancestral gene content reconstruction 139–40
ancestral inference analysis, G_q and G_s proteins 121, 123–4
ancestral proteome reconstruction 128, 136
  comparing proteomes 129–30
  last universal common ancestor 132–3
  maximum parsimony approach 130–1, *132*
  mitochondrial ancestor 133–5
  peroxisomal proteome 135–6
  phylogenomic approaches 131–2
AncSR1 195–7
androgen receptors 184, 197
  phylogeny *196*
angiogenin (RNase 5) 214
angiotensin degradation *160*
Anopheles gambiae, MCH conservation *143*, 144, 145
antelopes, ribonucleases 210
antibody response, biopharmaceuticals 39

anti-coagulation factors, snake venom 37
anti-tumor activities, ribonuclease homologs 214, 220
antiviral activity, EDN 215, 216
aquatic animals, rhodopsin absorption maxima 162
Aquifex aeolicus 202, *204*, 205
  optimal growth temperature 203
Archaeoglobus fulgidus 202, *204*
  optimal growth temperature 203
Archaeomeryx, ribonuclease 210, 211, 212–13
Archea, conserved protein analysis 105
Archosauriformes 168
archosaur rhodopsin protein sequence *171*
archosaur rhodopsin reconstruction 86–90, 168–72
Ardell, David vii
Argiope aurantia, spidroin sequence *82*
arthropods
  Ecdysozoa hypothesis 104
  tissue organization 103
artificial life 30
artiodactyls 213
  candidate trees 218–*19*
  molecular reconstructions 21
  pseudogenes 219–20
  ribonucleases 210, 211, 217
Asp-38 substitution, ribonucleases 212
ASR data 51
assumptions of evolutionary models 50
Astatotilapia, corticoid receptors *187*
atmospheric oxygen levels, role in origin of peroxisomes 135
ATP, use in formation of ethanol 9, 10
ATP exchange, endosymbiotic theory 133, *134*
Australian common brown snake, venom 35, 38

Bacillus subtilis 202, *204*, 205
  optimal growth temperature 203
Bacteria, conserved protein analysis 105
BADASP 124, 232, 233
BAli-Phy 238
Barnard, E.A. 210

base frequencies, nucleotide models 165
Bayesian methods vii, 52, 69–70, 72, 85, 86, 165–6, 222, 238
  see also Markov chain Monte Carlo (MCMC) technique
Bayesian phylogenetic approach 21
Beast 238
Beintema, Japp 7, 209, 214
Benner, S.A. 53, 210
best bidirectional hit (BBH) 130
β-sheets, context-dependent matrices 49
β-thymosin gene, synapomorphy 106
β-tubulin studies 108
BETE method 124
bias 22, 92–3
  in archosaur rhodopsin reconstruction 88–90
  determination of functional bias 91
  in Eukaryota tree reconstruction
    gene sampling and taxon sampling 107–8
    Markov chain Monte Carlo methods 109
    running out of steam 108–9
    towards more frequent amino acids 86, 88–90
Bielawski, J.P. and Yang, Z. 125
bilateria
  genomic cluster reconstruction 141
  MCH conservation 145
  proteome comparisons 140
  rRNA secondary structure analysis 106
bile acid synthesis 135
binary fission, peroxisomes 135
binding partner changes 183
binding pocket architecture
  AncSR1 *196*
  GRs *194*
bioavailability, venoms 35
biofilms, metabolic properties 129
biomedical applications 26
biopharmaceuticals 34–6, 38–9
  exendin-4 36–7
  venoms as source 34–6
biotechnology 26
BLAST search 23, 98, 99, *100*
  Pop1 sequences *100*
blood-clotting, effect of snake venoms 37–8
Blouin, C. *et al.* 233

Bollback, J.P. 72
bone morphology 85
bony vertebrates, GR ancestor (BonyGR) 192
bootstrap values, effect of data concatenation 104
Bos taurus see ox
Bourque, G. and Pevzner, P.A. 140
Bousseau, B. *et al.* 134–5
bovids
  evolutionary tree *213*
  ribonucleases 210, *211*
bovine seminal fluid ribonuclease homologs *213*, 216–18
  ancestral protein reconstruction 218–20
brain ribonucleases 218
brain tumors, drug therapy 35
branch-and-bound algorithm 47–8
Branchiostoma floridae 141, 144
branch lengths, probabilistic models 45, 46
Breedlove, Robert 5
Brown, Michael 6
buffalo
  evolutionary tree *213*
  ribonuclease genes 217
  ribonucleases *211*, 220
Bufo marinus, GLP-1 37
buried structural elements, context-dependent matrices 49
burst-after-duplication (BAD) method 227, 229, *232*

Caenorhabditis clade, intron turnover 105
C. elegans 143
  phylogeny *110*
  Pop1 sequences *99*, *100*
camel, evolutionary tree *213*
cancer-promoting gene expression, alteration by transposons 29, 30
candidate regions, gene cluster reconstructions 141–2
Cape buffalo (*Syncerus caffer caffer*), ribonuclease genes 217
carbohydrate, fruits as source 9, 10, 13
caspase family, functional divergence 118
CASSIOPE 148
catalytic activity, ribonucleases 211
CCR5, hydropathy 73–4, *75*
cell membrane steroid hormone diffusion 184
cellular machines 30

cellular metabolism
  reconstruction 129
cephalochordates, divergence from
  craniates 141, 144
Cephalophus sylvicultor (yellow
  backed duiker), ribonuclease
  genes 217
Cervus eldi (Eld's deer),
  ribonuclease genes 217
Chang, B.S.W. *et al.* 22
chaperone family A, duplications
  in yeast genome 15
character-based phylogenetic
  approaches 21
charge of proteins 83
chemokine receptors
  CCR5, hydropathy 73–4, 75
  dimerization 229
chemostat competition 160
chimera construction 229
chimpanzee, CCR5 73
Chinese hamster ovary (CHO) K1
  cells 188
Chlamydia pneumoniae 202, 204
  optimal growth temperature
  203
chloramphenicol acetyltransferase
  assay 160
chloride-channel agonism, TM-
  601 35
chloroplast genome, amino acid
  matrices 49
choanoflagellates, phylogeny 110
cholesterol, in formation of steroid
  hormones 184
cholesterol synthesis 135
chordates
  MCH conservation 144–5
  tissue organization 103
chromatin modification, steroid
  receptors 184
chromophores 168
  formation pathways 174
  GFP-like proteins 173
Chrysopelea ornata, GLP-1 37
chymase proteases, functional
  assays 160
Ciona species
  C. intestinalis
    β-thymosin gene 106
    MHC-like regions 143, 144,
    145
  C. savignyi 144
    genomic organization 144–5
circulatory system, as target of
  animal venoms 35, 37
citric acid cycle 9

clades, phylogenetic
  reconstructions 103–4
climate change, role in ruminant
  evolution 213
Clustal X / ClustalW 119, 237
clusters of orthologous groups
  (COGs) 130, 202
coagulation factors, effects of snake
  venom 37–8
coarse-grained traits 80–3
codon-based models 49–50,
  165–6
codon bias, incorporation into
  models 49
codon optimization 155–7
codon-substitution matrices 62
Coelomata hypothesis 103–4, 105,
  106, 107, 110
  ancestral protein
  reconstructions 111
coelomates 103
coevolution 91
co-evolving substitutions,
  probability vector 47
cofactor evolution 23
color emission spectra, fluorescent
  proteins 160
combinatorial chemistry 34
Combinatorial Extension 237
complement system 142
complex systems, evolution 183–4
computational efficiency, MCMC
  algorithms 70
concatenation of data 51, 104
cone snail venom 35
conservation-shifting sites
  (CSS) 125, 227, 230, 231, 233
Conservation Shift Measure
  (CSM) 124, 125
conserved gene clusters 139–42
  ancestral region
  reconstruction 146–7
  MCH region 142–5
  significance and
  hypotheses 145–6
conserved protein analysis 105
context-dependent matrices 49
continuity of function 80
convergence in EM methods 205
convergence with positive
  selection 146
convergent translocation 148
coral GFP-like proteins 172
  fluorescence spectra 175
  reconstruction
    evolutionary structure–
    function study 174–5

use of alternative evolutionary
  models 173–4
site variation 167
correlated evolution models,
  MCMC approach 70
corticoid receptors, phylogeny 196
corticosterone 185
cortisol 184, 185, 186
cortisol-sensitivity, steroid
  receptors 189
COS-cells, opsin production 157,
  170
covariation, nucleotide
  residues 74, 76–7
covarion models 50
covarion processes, fitness
  landscapes 34
covariotide evolution 95, 96
CpG hypermutations, Markov
  models 72
craniates, divergence from
  cephalochordates 141, 144
Cretaceous 13, 17
crystal structures
  ancestral corticoid receptor 190
  GR and MR, effect of L111P and
  S106P mutations 194
Cunningham, C.W. 21
cyan GFP-like proteins 173
Cynops pyrrhogaster, GLP-1 37
cytochrome oxidase II,
  hydropathy 73, 74
cytochromes 7, 209
  cytochrome *b*, primates,
    ASR 51–2
  cytochrome p450–11B
    (Cyp11B) 185, 191
cytoplasmic protein domains,
  hydrophilia 73
cytosol, steroid receptors 184
cytotoxic activities, ribonuclease
  homologs 214, 217

Danchin, E.G. *et al.* 145–6, 147
Danio rerio 143
Darwin, Charles 103, 184
Darwin software 59, 60, 65–7, 237
databases, gene families 238
data concatenation, eukaryote tree
  reconstruction 104
data types 51
dawn horse 213–14
Dayhoff matrices 58–9, 61–2, 165
Dayhoff model 118–19
dead genes (pseudogenes) 226
  artiodactyls 219–20
  resurrection 161–2

death stalker scorpion venom 35
deer
  evolutionary tree 213
  ribonuclease genes 217
  ribonucleases 210, 211
degenerate gene synthesis 172–3
  coral GFP-like protein reconstruction 173–7
degenerate oligonucleotides 91–2, 167, 168
delayed transformation (DELTRAN) assignment 44
deleterious variants, incorporation into substitutions 86
deletion events, Eukaryota tree reconstruction 106
deletions, tree-based hidden Markov model (T-HMM) 53
11-deoxcortisol 185
deoxycorticosterone (DOC) 185
  11-hydroxylation 191
  sensitivity of ancestral corticoid receptor 191–2
  sensitivity of steroid receptors 189
deoxynucleotide triphosphates (dNTPs) 154
designability of sequences 81–2
Deuterostomes, conserved gene clusters 145–6
diabetes, exendin-4 35, 36
Diacodexis, ribonuclease 210
digestive proteins 23
  see also lysozymes; ribonucleases (RNases)
dimerization, steroid hormone receptors 184
Dimmic, M.W. et al. 49
dinosaurs
  extinction 13
  replacement by mammals 23
1, 3-diphosphoglycerate 9
directed evolution 26
direct MCMC methods 70, 71–2
disassociation constants, Haldane equation 12
discrete gamma distribution, ASRV models 47, 48, 100
disease-causing genes, single-nucleotide polymorphisms 26
distance-based phylogenetic approaches 21
divergent evolution 7, 8, 35–6
  GLP-1 and exendin-4 36–7
  significance of conserved gene clusters 145–6

DIVERGE software 118, 119, 229, 233, 238
DNA-binding assay 160
DNA-binding domain (DBD), steroid receptors 184, 186
DNA/protein libraries 26, 27, 28
DNA-synthesis techniques 183
Dollo parsimony approach 105
domain-specific surveys 231
Dorit, Robert 3
double-deletion, Adh 1 and Adh 2 11
Drosophila melanogaster
  MCH conservation 143, 144, 145
  phylogeny 110
  Pop1 sequences 99, 100
drug candidates 34–6, 38–9
  exendin-4 36–7
  snake venoms 37–8
duiker, ribonuclease genes 217
duplications see gene duplications
Durrant, Barbara 210
dynamic molecular exploitation 192
dynamic programming 53–4, 61, 62, 63
  multiple sequence alignments 64–5

earliest animal ancestor 103–4
Ecdysozoa hypothesis 104, 105, 106, 107, 110
  ancestral protein reconstructions 111
ecological data 149
ecosystem, impact of ethanol 17
ecosystem changes, adaptive responses 23
Edwards, R.J. and Shields, D.C. 53
EF-1α protein 108
  sequence insertion 106–7
egg positioning, fruit flies 17
eland, evolutionary tree 213
elasmobranches, corticoid receptors 186, 187, 191
Eld's deer (*Cervus eldi*), ribonuclease genes 217
electrolyte homeostasis 186
elongation factors (EFs) 23, 25–6, 30
  EF-Tu proteins 201
  expression 157
  functional assays 160, 161
  gene synthesis 155
  heuristic hierarchical Bayesian approach 22
  thermostability assays 24

empirical amino acid matrices 48–9
empirical codon-substitution matrix 49
Encephalotozoon cuniculi 133
endoplasmic reticulum, peroxisome formation 135
endosymbiotic theory
  mitochondria 133, 134
  peroxisomes 135
endothelial cell-stimulatory activities, ribonuclease homologs 214
England, J.L. and Shakhnovich, E.I. 81
enolase studies 108
enzymatic catalysis 161
enzyme cofactor evolution 23
Enzyme Commission (EC) numbers 124
Eocene, ruminant evolution 213
eosinophil cationic protein (ECP, RNase 3) 214, 215–16
  production in *Escherichia coli* 158–9
eosinophil-derived neurotoxin (EDN, RNase 2) 214, 215–16
  production in *Escherichia coli* 158–9
  site-directed mutagenesis 154
Eotragus, ribonuclease 210
equilibrium, mutational dynamics 82
error, causes of 190
Escherichia spp.
  E. coli
    codon usage 155, 156
    EDN and ACP expression 158–9
    EF thermostability studies 24, 25
    elongation factor expression 157
    temperature optima 23
estradiol 184, 185
estrogen receptors 184, 195–7
  phylogeny 196
estrogen response element activation 160
ethanol
  formation from glucose 9–10
  impact on ecosystem 17
euchordates, MCH conservation 144
euchordates ancestor, genomic cluster reconstruction 141

Euclidean distance, amino acid compositions 202, 204, 205–6, *207*
Eukaryota tree reconstruction 103–4
　methodological biases
　　gene sampling and taxon sampling 107–8
　　long branch attraction 109–11
　　Markov chain Monte Carlo methods 109
　　running out of steam 108–9
　methodological developments
　　conserved protein analysis 105
　　data concatenation 104
　　insertion/deletion/fusion events 106–7
　　intron analysis 105
　　protein domain analysis 105–6
　　rRNA secondary structure analysis 106
eukaryotic ribonucleases
　Pop1 sequences 98–9
　Pop4 sequences *96*, 98
evolutionarily conserved gene clusters 139, 141
evolutionary constraints, protein regions 49
evolutionary rates, among-site variation 45, 46–8
evolutionary structure-function study, coral GFP-like proteins 174–7
evolutionary trace (ET) method 228, *229, 232*
Evolutionary Trace viewer 124
evolutionary trajectories 198
evolution of sequences 95–8
　gene finding with ancestral sequences 98–100
　heterotachy and ancestral state reconstruction 100
　heterotachy and gene finding 98
exendin-4 (exenetide) 35
　relationship to GLP-1 36–7
exoskeletons, Ecdysozoa hypothesis 104
expectation-maximization (EM) methods 201, 204, 205
experimental evolution systems 198
experimental paleogenetics beginnings 6–7
exposed structural elements, context-dependent matrices 49

expression hosts 157–9
extinct species, DNA sequencing 129
eyes, evolution 183–4

Factor IX-/X-binding proteins, snake venom 37
Factor V activators, snake venom 37, 38
Factor X activators, snake venom 37, *38*
FASTML 238
fatty acids
　β-oxidation 135
　production in ruminant gut 210
Felsenstein, J. 44–5, 46, 96, 100
fermentation
　impact on egg positioning, fruit flies 17
　participating organisms 17
fermenting bacteria, role in ruminant digestion 210
fermentive enzymes 23
　see also alcohol dehydrogenase
Fersht, Alan 6
FeS clusters 134
fibrinolysis, effect of snake venoms 38
FIGENIX 148
Firmicutes 202, 205
Fitch, W.M., maximum parsimony algorithm 43–4
Fitch, W.M. and Markowitz, E. 50, 95
fitness landscapes 34, 80
FLAG epitope 159
fluorescence spectra, coral GFP-like proteins *175*
fluorescent proteins 23
　emission wavelengths 162
　functional assays *160*, 161
　reconstruction
　　evolutionary structure–function study 174–7
　　use of alternative evolutionary models 173–4
　synthesis 28
folding stability 80–1
fossil records 4
Fox2 136
fruit flies, egg positioning 17
fruits, as carbohydrate source 9, 10, 13
Fugu rubipres, Factor X 38
functional assays 159–61
　GFP-like proteins 175–7
　rhodopsin variants 170, *171*

functional bias 92–3
　determination 91
functional continuity 80
functional divergence 162
functional divergence analysis 117, 124–5
　DIVERGE2 software 119
　G-protein α subunits 119–24
functional divergence types 117–19, 226
functional inference 129
fungi
　fermentive enzymes 23
　phylogenetic position 103, 107–8
　see also yeast
FunShift database 124, 230
fusion events, Eukaryota tree reconstruction 106–7

G + C content, rRNA 201, 206–7
Gabaldón, T. et al. 135
Gabaldón, T. and Huynen, M. A. 134
Gallus gallus 143
　Factor X *38*
　GLP-1 37
Galtier, N., covarion model 50
Galtier, N. et al. 201, 206
Galtier, N. and Gouy, M. 50–1
gamma + invariant model, ASRV 48
gamma distribution, ASRV models 47, 48
gaps 53, 63–4, 203
gastric emptying, GLP-1 36
Gaucher, Eric viii
gazelle, evolutionary tree *213*
GC content, covarion model study 50
GenBank 73
gene cluster conservation, significance 145–6
gene cluster reconstructions 139, 141
　choice of candidate regions 141–2
　MCH region 142–5
gene content reconstruction 139–40
gene conversion 231
gene death *see* pseudogenes
gene duplications 216, 226
　yeast genome 14, *15–16*, 17
　Adh paralogs 12–13
gene duplication studies
　detection of changes in evolutionary rate 228–9
　mutual information 228

property-change methods 227–8
  structural modeling 228
gene-family databases 238
gene finding
  with ancestral sequences 98–100
  and heterotachy 98
gene knock-in experiments 229
general time reversible (GTR)
    amino acid model 165
gene regulation, evolution ix–x
gene sampling, biases 107–8
gene synthesis 154–5
gene therapy, use of
    transposons 29
Gene-Trace 131
genome-based approaches 166
genomes, evidence of ecosystem
    innovation 17
genome selection, LUA amino acid
    composition estimation 202
genomic organization, *Ciona*
    species 144–5
genomic organization
    reconstruction 140–1
geological data 149
geological evidence, temperature
    conditions 26
GFP (green fluorescent protein)-
    like proteins 172
  fluorescence spectra *175*
  reconstruction
    evolutionary structure–
      function study 174–5
    use of alternative evolutionary
      models 173–4
Giardia lamblia 99, *100*
gila monster (*Heloderma suspectum*)
  exendin-4 35, 36, *37*
  GLP-1 *37*
giraffe (*Giraffa camelopardalis*)
  evolutionary tree *213*
  ribonuclease genes 217
  ribonucleases 210
globular proteins
  first appearance 133
  marginal folding stability 81
glucagon-like peptide-1 (GLP-1) 35
  relationship to exendin-4 36–7
glucocorticoid receptors (GRs)
    184
  evolution of specificity 186–8
  mechanistic basis for
    evolution 192–4
  phylogeny *196*
glucose, conversion to ethanol
    9–10
glutathione transferase proteins 97

glyceraldehyde-3-phosphate 9
glyceraldehyde-3-phosphate
  dehydrogenase family,
  duplications in yeast
  genome 16
glycosomes 135
gnathostomata, emergence 141,
  144
gnu, evolutionary tree *213*
goat
  evolutionary tree *213*
  seminal ribonuclease
    expression 218
golden lion tamarind, CCR5 73
golden-rumped tamarind,
  CCR5 73
Goldman, N. and Yang, Z. 49
Goldstein, Richard vii, 21, 22
gorilla, CCR5 73
Gouret, P. 148
G-protein α subunits, functional
    divergence analysis 119–20
  G_s and G_q proteins 120–4
G-protein-coupled receptor
    genes 157
G-proteins, elongation factors 23
G-protein transducin 168
Grantham, R., physiochemical
    distance matrix 49
grasses, diversification 21
grasslands, emergence 213
green monkey, CCR5 73
Gribaldo S. *et al.* 118
GTR substitution model 76
Gu, X. 117, 118, 119, 226, 233
Gu, X. and Vander Velden, K. 233
GZ-Gamma method, ASM 125

hagfish (*Myxine glutinosa*), corticoid
    receptors 186, *187*
Haldane equation 12, 13
Halobacterium sp. NRC-1 202
HAQESAC 232
Haussler, David viii
Helicobacter pylori 202, *204*
  optimal growth temperature *203*
Heloderma suspectum (gila monster)
  exendin-4 35, 36, *37*
  GLP-1 *37*
Hemissenda crassicornis, β-thymosin
    gene 106
hemoglobins 7, 209
  Pauling and Zuckerkandl's
    work 20–1
  specificity changes 230
hemostasis, snake venom 35, 37–8
herbivores

effects of climate changes 213
  ribonucleases 210
heterogenous evolutionary
    processes 198
  as cause of bias 22
heterologous protein
    expression *156*, 157–9, *160*
heterotachy 22, 95–8, 198, 237
  and ancestral state
    reconstruction 100
  and gene finding 98
heuristic science 5, 6–7, 13, 21–2
hidden Markov models
    (HMMs) 98
hierarchical Bayesian methods
    21–2
hippopotamus, evolutionary
    tree *213*
historical linguistics 5
HIV (human immunodeficiency
    virus) 73
HKY85 substitution model 52
HMMER 98, 100
homeobox domain 142
homogeneity assumption 50
homologous genes 30
homologous proteins,
    identification 129–30
homology vii
homoplasy 7
  ADH_A 11
Homo sapiens
  CCR5 73
  Factor X *38*
  GLP-1 *37*
  phylogeny *110*
  Pop1 sequences 99, *100*
Hoplobatrachus rugulosus,
  GLP-1 *37*
Hoplocephalus stephensii, venom *38*
horizontal gene transfer (HGT),
    enolase protein 108
horizontal gene transfer
    penalties 131
hormones, evolution 184
horses 213–14
Hovergen/Hobacgen 238
Huelsenbeck, J.P. 21
Huelsenbeck, J.P. and Bollback, J.P.
    52–3
Hughes, A.L. and Friedman, R. 140
human pancreatic ribonuclease
    (RNase 1) 214
human RNases 214
Hydra sp.
  H. magnipapillata 111
  phylogeny *110*

hydrogen producing hypothesis, endosymbiotic theory 133, *134*
hydropathy, transmembrane regions 73–4, *75*
hydrophilic regions 73
hydrophobicity 73, 83
  amino acids 49
17-hydroxyprogesterone *185*
hypersurfaces 8
hyperthermophiles 23
hypothesis testing 20, 85

ice ages, ruminant evolution 213
immune responses 142
  to biopharmaceuticals 39
immune RNases, functional assays *160*
immunoproteasome 142
immunosuppression
  ribonuclease homologs 214, 217, 220–1
  ribonucleases 161
impala, evolutionary tree *213*
ingroups, secondary loss coefficient calculation 106
inosine-5'-monophosphate dehydrogenase family, duplications in yeast genome *15*
insertion events
  Eukaryota tree reconstruction 106–7
  tree-based hidden Markov model (T-HMM) 53
in silico sampling 92
insulin biosynthesis, GLP-1 36
integrins, specificity changes 230
internal nodes, probabilistic models 45
intron analysis, eukaryote tree reconstruction 104, *105*
intuition 5
invariable sites 95
invariant sites in ASRV models 48
invertebrates, fluorescent proteins 23
isocitrate dehydrogenase
  functional assays *160*, 161
  site-directed mutagenesis 154
isopropylmalate dehydrogenase, functional assays *160*, 161

jackknife test of Euclidean distance 204, 206, *207*
Jaillon, O. *et al.* 140

Jak family, functional divergence 118
Jermann, T.M. *et al.* 212
Johnson, J.M. and Church, G.M. 231
joint ASR 46
joint probabilities 46
joint reconstruction 46
Jones, D.T. 165
Jones, D.T. *et al.* 49
JTT + gamma model, in ancestral corticoid receptor resurrection 187, 191
JTT matrix 49
JTT model 119
Jukes–Cantor nucleotide 165

k6 (RNase 6) 214
karyotype reconstructions 141
Kasahara, M. *et al.* 142
k_{cat}/K_m values, ancestral ribonucleases *211*
kinetic data, candidate ADH_A *12*
kinetoplasmida 135
kingdoms, eukaryotic 103, 107–8
Kluyveromyces, divergence from *Saccharomyces* 13–14
Knudsen, B. and Miyamoto, M.M. 233
KOGs 107
Koonin, E.V. *et al.* 139–40
Koshi, J.M. and Goldstein, R.A. 45, 46
Kreitman, Martin 3
Krishnan, N.M. *et al.* 51
kudu (*Tragelaphus imberbis*) 217
Kunin, V. and Ozounis, C.A. 131
Kurland, C.G. and Andersson, S.G. 133, *134*
Kyte–Doolittle hydropathy plots 73, *74, 75*

L1 retroposons, functional assays *160*
lactate dehydrogenase 10
lamprey (*Petromyzon marinus*), corticoid receptors 186, *187*
landscapes 30
last universal ancestor (LUA) 128, 139–40, 201–2
  amino acid composition estimation 205–7
  methods 202–4
  ancestral proteome reconstruction 132–3
Late Pleistocene species, DNA analysis 129

Latrodectus spp., spidroin sequence traits 82
leaf glyoxysomes 135
lectin-like activities, ribonuclease homologs 214
lemurs, CCR5 73
libraries 26, *27*, 28
  GFP-like proteins 175, *176*
ligand-binding domain (LBD), steroid receptors 184, 186, 197
ligand-binding sites, phylogenetic analysis 231
ligation, degenerate gene synthesis 173
ligation sites, separation 167
light stimulation, protein activation 161
likelihood phylogenetic approach 21
likelihood ratio tests (LRT) 48, 166
  RSS identification 125
likelihood values / scores 148, 165
linear representations 7–8
linguistics, historical 5
lipid biosynthesis enzymes 135
log-normal distribution, ASRV 47
long branch attraction (LBA) 106, 107, 109–11
long oligonucleotides, use in gene synthesis *154*, 155
loop region mutations, rRNA 106
loops, context-dependent matrices 49
loss threshold 131
luciferase reporter assay, AncSR1 *196*, 197
lysozyme evolution 229–30
lysozymes
  functional assays *160*
  molecular reconstructions 21, 23

major ampullate spidroin 1 (MaSp1) architecture *81*–3
major histocompatibility complex (MHC) region 139, 141, 142, *143*
  ancestral region reconstruction 146–7
  conservation 142, 144–5
majority-rule ancestral sequences 233
make–accumulate–consume pathway, ethanol *9*, 13
mammalian ancestor, genomic organization 140
mammal ribonucleases 7

see also ribonucleases (RNases)
mammoth, DNA analysis 129
marginal fitness 80–1
marginal probabilities 46
marginal reconstructions 46, 47
Margoliash, E. 209
Margulis, L. 133, *134*
Markov chain Monte Carlo (MCMC) technique 52, 70, 165
   in ancestral corticoid receptor resurrection 186–7, 191
   direct methods 71–2
   in Eukaryota tree reconstruction 109
   incorporating phylogenetic uncertainty 71
   sampling mutational paths 70–1
Markovian model 59
   computation of probabilistic ancestral sequences 59–61
   scoring 61–3
Markovian processes 8, 47
marmoset, CCR5 73
Martin, W. and Müller, M. 133, *134*
MasterCatalog 238
Matassi, Georgio vii, viii
mathematical modeling 5–6, 7–9
   interconnecting models 13–17
Mathew's correlation coefficient (MCC) 77
Matz, Mikhail viii
maximum *a posteriori* probability (MAP) estimation 69
maximum-likelihood reconstructions vii, 21, 69–70, 165–6, 222, 238
   of ancestral corticoid receptor 186–7, 188
   causes of error 190
   effect of heterotachy 100
   Eukaryota tree reconstruction 110
maximum likelihood tree
   ADH$_A$ *10*, 11
   bacterial sequences 23, *24*
maximum-parsimony approach vii, 43–4, 86
   ancestral corticoid receptor resurrection 186–7
   ancestral proteome reconstruction 130–1, *132*
   effect of heterotachy 100
   shortcomings 45
Maynard-Smith, J. 80
Mayrose, I. *et al.* 48

mechanistic basis for evolution, glucocorticoid receptor (GR) 192–4
mesophiles 23
metabolic pathways, mapping onto proteome 129, 133
metagenomics projects 129
metalloproteinases, as therapeutic targets 35
Metazoa
   protein domain analysis 106
   steroid hormone receptors 23
Methanococcus jannaschii 202, *204*
   optimal growth temperature 203
methanol oxidation 135
Methanosarcina acetivorans 202, *204*
   optimal growth temperature 203
method of moments, ASM 125
Mexican bearded lizard 36
Michaelis constant (K_m) for ethanol
   Adh 1 and 2 10
   candidate ADH$_A$s 12
microbodies 135
microorganisms
   ethanol toxicity 10
   role in ruminant digestion 210
midpoint-rooted trees 203
Miller, Jeffrey 6
mineralocorticoid receptor (MR) 184
   evolution of aldosterone-sensitivity 191–2, 194
   evolution of specificity 186–8
   phylogeny *196*
   S843P and L848Q substitutions 195
minimal genomes 30
Miocene, bovine evolution 218
mitochondrial ancestor, proteome reconstruction 133–5
mitochondrial DNA, primates, ASR 51
mitochondrial evolution 128
mitochondrial genome, amino acid matrices 49
mitochondrial tRNAs, nucleotide covariation 74, 76–7
mitogen-activated protein kinases, chimera construction 229
model selection
   ASRV models 48
   JTT + gamma model 187, 191
model variability 85–6, 91
molecular cladistic approach, Eukaryota tree reconstruction 105–6

molecular interactions, evolution 183–4
mollusks, steroid hormones 197
molting clade 104
monkey kidney cells, opsin production 157, 170
monoclonal antibody, use in elongation factor gene expression 157
monophyletic protein sets 131
Monosiga brevicollis, phylogeny *110*
Montastraea cavernosa, GRP-like proteins *175*
moose, evolutionary tree *213*
morphology, signs of evolution 69
most probabilistic ancestral sequences (MPASs) 22
most probable ancestor (MPA) 86, 91, 92–3
   archosaur rhodopsin 87, 90
motif search methods 58
mouse (*Mus musculus*) 143
   CCR5 73
   Factor X *38*
   Pop1 sequences *99*, *100*
   RNases 214–15
mouse model, GLP-1 36
mouse tumor suppression/promotion 30
MrBayes 71, 76, 237
multiple sequence alignments (MSAs) 59–60, 64–5
murine stem-cell-virus (MSCV) 30
Muscle 237
muscles
   lactate production 10
   as target of animal venoms 35
Muse, S.V. and Gaut, B.S. 49
mutational mapping 73
   nucleotide covariation in mitochondrial tRNAs 74, 76–7
   SIMMAP software 78
   snake venom proteins 38
   transmembrane regions 73–4, *75*
mutational paths 70
   statistical inference 72
mutational path sampling 70–1
mutation matrices 59
mutual historic information content (MHIC) statistic 76, 77
mutual information 228
Mycoplasma, synthetic genome construction 30
Myxine glutinosa (hagfish), corticoid receptors 186, *187*

Nambiar, Krishnan 3
natural history paradigm 21, 26
natural selection 3, 4
nearly-neutral population genetics models, deleterious variants 86
neighbor-joining phylogenetic approach 21
neighbor-joining tree, G_q and G_s proteins *120*
Neisseria menigitidis, EF thermostability 25
nematodes 110
   Coelomata hypothesis 103
   Ecdysozoa hypothesis 104
   protein domain analysis 106
neofunctionalization 161, 226
neomycin resistance gene transposition 160
Nephila spp., MaSp1 architecture *81–3*
nerves, as target of animal venoms 35, 37
neutral drift 9
neutralist/selectionist dispute 3–4
neutral replacements, significance 215
neutral theory of evolution 230
nicotinic acetylcholine receptor antagonism, ACV1 *35*
Nielsen, R. 72
Nielsen, R. and Huelsenbeck, J. P. 72
Nielsen, R. and Yang, Z. 49
nilgai, evolutionary tree *213*
nocturnal visual capacity, ancestral archosaurs 168, 170, 172
non-classical carbocation problem 4
non-homogenous models 50–1
non-oxidative decarboxylase 9
non-stationary base composition, nucleotide models 165
NONSTAT model 51
nonsynonymous/synonymous ratios 22
Notechis scutatus, venom *38*
Notung 238
N-terminal domain, steroid receptors 184
nuclear translocation, steroid hormone receptors 184
nucleotide covariation 74, 76–7
nucleotide frequencies, NONSTAT model 51
nucleotide models 165
nutrient extraction, artiodactyls 21

okapi, ribonucleases 210
Oligocene
   eosinophil RNase homologs, selection 216
   ruminant evolution 213
oligonucleotides
   short fragments 91–2
   variable frequencies 92
oligonucleotide synthesis, introduction of site variation 166, 167
Olsen, G.J. 47
Omland, K.E. 21
oncogene expression, alteration by transposons *29*, 30
opisthokonts 107, 110
opsin apoprotein production 157
opsins 168
   functional assays *160*
optimal growth temperatures (OGTs) 203
   correlation with rRNA G+C content 206–7
   last universal ancestor 201–2
optimization bias 86
organelle evolution 128
ortholog evolution *96*, 97–8
orthologs, Clusters of Orthologous Groups (COG) database 202
orthology assignment 130
ortholome 149
oryx (*Oryx leucoryx*), ribonuclease genes 217
osmotic homeostasis 184
osmotic stress, effect on fermenting organisms 17
outgroup, secondary loss coefficient calculation 106
Ovis genus
   seminal ribonuclease expression 218
   see also sheep (*Ovis aries*)
ox (*Bos taurus*),
   evolutionary tree *213*
   ribonuclease genes 217
   ribonucleases 210, *211* 217, 220
oxidative decarboxylase 9
oxygen scavenger hypothesis, endosymbiotic theory 133, *134*
Oxyuranus spp., venom *38*
Pachyportax, ribonuclease 210, *211*

Pagel, M. 21, 53
pain relief drugs *35*
palaeogenomics 129

paleoenvironment, inference from thermostability studies 23, 25
paleogenetics 20, 222
   bovine seminal plasma ribonuclease homologs 216–21
   eosinophil ribonuclease homologs 215–16
PAM (point accepted mutations) 59
PAML software 87, 100, 165, 169, 204, 238
pancreatic ribonucleases 210
   extraction 209–10
parallel/convergent evolution, as cause of MPAS bias 22
paralogous sequences, heterotachy *97*, 98
paralogs 130, 226
   Saccharomyces cerevisiae 14
paralogy, hidden 108
parameters, ASRV models 48
parsimony methods 21, 85
   effect of heterotachy 100
Pauling, L. and Zuckerkandl, E. 3, 20–1, 26
PAUP 11, 237
Pax paralogs, functional assays 159, *160*
PBX2 142
PCR (polymerase chain reaction) amplification, gene synthesis 155, 173
PDB viewer, DIVERGE 229
peccary (*Tayassu pecari*), ribonuclease genes 217
pecora, ribonucleases *211*
periossodactyls 213
periplasmic protein domains, hydrophilia 73
peroxide metabolism enzymes 135
peroxisome biogenesis proteins (PEX) 135
peroxisome evolution 128
   ancestral proteome reconstruction 135–6
Petromyzon marinus (lamprey), corticoid receptors 186, *187*
Pfam 230–1, 238
Pfam enzyme families, functional divergence 124
Pfu polymerase 155
pGEM-T PCR cloning kit 172–3
pH, effect on fermenting organisms 17

pharmaceutical candidate
    development 34–6, 38–9
    exendin-4 36–7
    snake venoms 37–8
phenotype of proteins, effect of
    structural variations 166
phenotypic inferences 128
Phillipe, H. *et al.* 95
Phillips, M.J. *et al.* 104, 109
phosphatase/thiamine transport
    family A, duplications in
    yeast genome 15
phospholipase A_2, snake venom 37
phosphorylation, degenerate gene
    synthesis 172–3
photoreceptors 168
photosensitivity 168
Phylip software 44–5, 203, 237
phylogenetic approaches 21–2
phylogenetic tree
    construction 237–8
    eukaryotes 103–4
    LUA amino acid composition
        estimation 203–4, 205
phylogenetic uncertainty,
    incorporation into MCMC
    procedure 71
phylogenomic approaches,
    ancestral proteome
    reconstruction 131–2
physiochemical distance matrix 49
physiochemical information,
    incorporation into
    models 54
pig, evolutionary tree *213*
pigments, visual 23
Pipa pipa, GLP-1 37
plant sequences, homogeneity 67
plasminogen activator, snake
    venom 38
platelet signaling pathways 229
platyhelminths, tissue
    organization 103
pleiotropic constraint 34
Pleistocene, ruminant
    evolution 213
Pliocene, ruminant evolution 213
POA 237
Poisson-like errors 82
Poisson model 165
Pollock, David vii, 22
Polyandrocarpa misakiensis 143
polymerases
    use in gene synthesis 155, 173
    use in site-directed
        mutagenesis 154
polytomy 28

Pop1 sequences 98–9
    BLAST search results *100*
Pop4 protein alignment *96*, 98
position independence, Markov
    model 60–1
positively selected genes, as drug
    candidates 34
positive selection, as driver of
    convergence 146
posterior distribution sampling 86
    archosaur rhodopsin
        reconstruction 86–90
    experimental considerations
        91–2
    proposed strategy 92–3
    theoretical considerations 91
posterior predictive distribution,
    measurement of
    uncertainty 72
posterior probabilities 165
    maximum likelihood tree
        construction 11
post-translational protein
    modifications 128
preadaptation to aldosterone,
    ancestral corticoid
    receptor 192
prey, delivery of venoms 35
Prialt (ziconotide) *35*
primate cytochrome *b* ASR 51–2
primate mitochondrial DNA
    ASR 51
primates
    chemokine receptor CCR5,
        hydropathy 73–4, *75*
    cytochrome oxidase II,
        hydropathy 73, *74*
    digestive function 23
primer design
    gene synthesis 154–5
    site-directed mutagenesis 154
primer-extension reactions 154
principle-component analysis
    228
prior distributions, HKY85
    substitution model 52
probabilistic ancestral sequences
    (PASs) 58, *59*–61
    gaps 63–4
    pairwise alignments 63
    rate variation 65
    reconstruction example 65–7
    scoring 62–3
probabilistic models 43, 44–5
    amino acid matrices 48–50
    among-site rate variation (ASRV)
        models 46–8

ancestral sequence
    computation 45–6
covarion models 50
gaps and unknown
    characters 53–4
model selection 48
non-homogenous models 50–1
structural and physiochemical
    information 54
uncertainties 52–3
use of additional
    information 51–2
probability density functions,
    ASRV 47
probability vectors 47
profile search methods 58
progesterone 184, *185*
progestin receptors 184, 197
    phylogeny *196*
prokaryotic origin,
    peroxisomes 135
promoters, inactivated 161–2
pronghorn
    evolutionary tree *213*
    ribonucleases 210
property-change methods 227–8
proportional models approach 52
proteases 23
    functional studies 159
protein activation 161
protein C activators, snake
    venom 37
protein-DNA interactions 161
protein domain analysis, Eukaryota
    tree reconstruction 105–6
protein engineering 6
protein function evolution 198
protein-protein interactions 97, 161
protein regions, evolutionary
    constraints 49
proteins
    laboratory synthesis 166–8
    three-dimensional structure 198
protein selection, LUA amino acid
    composition estimation 202
protein stability, effect of specificity
    changes 234
proteobacterium, origin of
    mitochondria 133
proteolytic stabilities,
    ribonucleases 211–12
proteome comparisons,
    bilateria 140
proteomes 128–9
    see also ancestral proteome
        reconstruction
ProTest 48

prothrombin activators, snake venom 37, 38
prothrombinase, as therapeutic target 35
Protostomes
    conserved gene clusters 145–6
    orthologs 141
Pseudechis porphyriacus, venom 38
pseudocoelomates 103, 107
pseudogenes 226
    artiodactyls 219–20
    resurrection 161–2
Pseudonaja textilis, venom 35, 38
psi-BLAST 130
Pupko, T. et al. 45, 46, 47
purifying selection 118
Pyrococcus abyssi 202
Pyrococcus horikoshii 202, 204
    optimal growth temperature *203*
pyruvate 9–10
pyruvate decarboxylase family A, duplications in yeast genome 15
Pyxicephalus adspersus, GLP-1 37

Q8010 *35*, 38
Qian, B. and Goldstein, R.A. 53
quadruplicated regions 141–2
QuikChange protocol 154, 170

rab5, *rab6*, specificity-determining residues 229
radical replacements, probability vector 47
Raines, R.T. 221
rainforests, recession 213
Raja erinacea (skate) corticoid receptors 186, *187*
Rana spp., GLP-1 37
rare variants 87, 88–90, 91
Rascol, V.L. 148
rat, RNases 214
rate changes, detection 228–9
rate heterogeneity 46–8, 65
rate matrices (Q) 47, 48–50, 165
rate-shifting sites (RSS) 125, *227*, 231, 233
Rate Shift Measure (RSM) 124, 125
Rattus norvegicus 143
    peroxisomes 135
receptors, evolution 184
recombinant genomes 30
reconstructability of proteins viii
reconstructing evolutionary adaptive paths (REAP) approach 27, 28
red GFP-like proteins 173

reductionist approaches 31
reindeer, evolutionary tree *213*
reproduction, hormone control 184
response elements to steroid receptors 184
resurrection studies 153, 161–2
    functional assays 159–61
    heterologous expression 157–9
    sequence construction
        codon optimization 155–7
        gene synthesis 154–5
        site-directed mutagenesis 153–4
retinoic acid receptors (RARs) 142
retinoid X receptor (RXR) 142
retroposon activation 159
retroviruses, activity of EDN 215, 216
reversibility assumption 50
rhinoceroses 213
rhodopsin phylogeny *170*
rhodopsin protein sequence, ancestral archosaur *171*
rhodopsin reconstruction 86–90, 168–72
rhodopsins 168
    absorption maxima 162
    functional assays *160*
    gene synthesis 155
    production in mammalian cells 157–8
ribonucleases (RNases) 7, 21, 23, 161, 209, 222
    ancestral sequence reconstruction 210
    evolutionary tree *213*
    bovine seminal fluid homologs 216–21
    catalytic activity 211–12
    eosinophil homologs 215–16
    functional studies *160*
    function in ruminants 210
    heuristic hierarchical Bayesian approach 22
    Pop1 sequences 98–9
    Pop4 96, 98
    RIBAses (ribonucleases with interesting biological activities) 214–15
    site-directed mutagenesis 154
    thermal and proteolytic stabilities 211, 212
ribonuclease S gene synthesis 154
ribonucleic acid (RNA), presence in ruminant gut 210
ribosomes, action of EFs 23
RNA hydrolysis 21, *160*

rodent ancestor, genomic organization 140
rod photoreceptors 168
Ronquist, F. 21
root, Fitch's algorithm 44
Roy, S.W. and Gilbert, W. 105
rRNA, G+C content 201, 206–7
rRNA secondary structure analysis, eukaryote tree reconstruction 104, 106
ruminant digestion 23
    origins 212–14
ruminants
    evolutionary tree *213*
    ribonucleases 7, 210, *211*

Saccharomyces spp.
    divergence from *Kluyveromyces* 13–14
S. cerevisiae 202, 204, 205
    alcohol dehydrogenase gene expression 157
    formation of ethanol 9–10
    human domestication 13
    optimal growth temperature *203*
    paralogs 14
    peroxisome-less mutants 135
    phylogeny *110*
saiga (*Saiga tatarica*), ribonuclease genes 217
salmonid fish, Sleeping Beauty (SB) element 28, *29*
sampling effects, reduction by data concatenation 104
Sankoff, D., maximum parsimony algorithm 44
SATCHMO 238
scaffolds in protein frameworks 35–6
Schistosoma mansoni, phylogeny *110*
Schizosaccharomyces spp.
    phylogeny *110*
    S. pombe 100
Schneider, A. et al. 49
Schultz, T.R. 21
Schultz, T.R. and Churchill, G.A. 52
scoring sequences 61
    known sequences 61–2
    probabilistic against known sequence 62–3
    two probabilistic sequences 63
SDPpred 124, *232*, 233
SDS/PAGE 157
secondary loss coefficient 106

secondary sexual differentiation, hormone control 184
selectionists, neutralist/selectionist dispute 3–4
seminal fluid ribonuclease homologs 213, 216–18
  ancestral protein reconstruction 218–20
seminal plasma, ribonuclease expression 217–18
sensory protein studies 168
separate model approach 51–2
sequence-based alignment 237
sequence evolution 95–8
  gene finding with ancestral sequences 98–100
  heterotachy and ancestral state reconstruction 100
  heterotachy and gene finding 98
sequence evolutionary models 43
sequence-function mapping 34
sequence space sampling 34
sex-steroid evolution 195–7
sheep (*Ovis aries*)
  ribonuclease genes 217
  ribonucleases 210
  seminal ribonuclease expression 218
short oligonucleotides, use in gene synthesis 154–5
signaling cascades 129
signature genes 141
silk protein, MaSp1 architecture 81–3
SIMMAP 73, 78, 238
single-gene ortholog identification 108
single-nucleotide mutation matrices 62
single-nucleotide polymorphisms 26
site-directed mutagenesis 91, 153–4, 164, 168, 198
  in ancestral corticoid receptor resurrection 190
  use in archosaur rhodopsin reconstruction 170
site-specific rate shift (type I functional divergence) 118
  G_q and G_s proteins 120–1, 122
site-stripping 109
site variation in protein synthesis 166–7
SIV (simian immunodeficiency virus) 73
skate (*Raja erinacea*) corticoid receptors 186, 187

Skovgaard, M. *et al.* 26
Sleeping Beauty (SB) element 28, 29, 30
snail venom, Ziconotide 35
snake venoms, pharmaceutical uses 35, 37–8
Softparsmap 238
Southern copperhead viper, venom 35
specificity changes 229–30
  consequences 233–4
specificity prediction
  choice of method 231–3
  experimental approaches 229
  models and tools 226–8
specificity of steroid hormone receptors 184
  evolution 186–8
spermine transporter family, duplications in yeast genome 16
spider monkey, CCR5 73
spidroin, MaSp1 architecture 81–3
spidroin sequence traits, phylogenetic variation 82
splice site arrangements, transposons 30
splice variants 128
squirrel monkey, CCR5 73
SSAP 237
stability
  of venoms 35
  see also proteolytic stabilities; thermostability
stability changes 234
stabilizing interaction, GRs 194
Stackhouse, Joseph 3
star phylogeny 205
statistical mechanical models, fitness 80–1
stem region mutations, rRNA 106
steroid hormone receptors 23, 184, 185, 186
  aldosterone-sensitivity 189
  ancestral corticoid receptor resurrection 186–8
  robustness to uncertainty 188, 190–1
  AncSR1 195–7
  evolution of specificity 186–8
  functional assays 159, 160
  mechanistic basis for GR evolution 192–4
  phylogeny 196
  structure–function relationships 195
steroid hormones 184, 185

steroid-synthesis pathway 185
stochastic error 190
stress response 184
stroke treatment 35
structural information, incorporation into models 54
structural modeling, gene duplication studies 228
structure-based alignment 237
structure–function relationships 195
Structure Theory 4, 5, 25
subfunctionalization 161, 226
substitution matrix 45
substrate recognition 161
sugars, conversion to ethanol 9
sugar transporter family A, duplications in yeast genome 15
sugar transporter family B, duplications in yeast genome 16
Sullivan, J. *et al*, α factor calculation 125
SwissProt database 65, 66
SWS1 visual pigment, functional assays 160
synapomorphy, β-thymosin gene 106
Syncerus caffer caffer (Cape buffalo), ribonuclease genes 217
Synechocystis 202, 204
  optimal growth temperature 203
synthesis 6
synthetic biology 26, 30
  REAP approach 28

T4 ligase 172
T4 polynucleotide kinase 172
TAED 238
tapirs 213
Taq (*Thermus aquaticus*) polymerase 155
  codon optimization 156
Taverna, D.M. and Goldstein, R. A. 81
taxon sampling, as source of bias 107–8, 110, 111
Tayassu pecari (peccary), ribonuclease genes 217
TC1/mariner family 28, 29
Tc1/mariner transposons, functional assays 160
T-Coffee software 203, 237
teleosts, aldosterone-sensitive steroid receptors 191

Telford, M.J. 103
temperature conditions of early life 23, 25–6
terrestrial animals, rhodopsin absorption maxima 162
tertiary structure, Markov models 72
tertiary structure-based assays 161
testosterone 184, *185*
Tetranodon nigroviridis 140
thermal environment, last universal ancestor 201–2, 206, 207
Thermoanaerobacter tengcongensis 202, *204*, 205
　EF thermostability 23, 25
　optimal growth temperature *203*
thermodynamic properties of reconstructed sequences 54
thermophiles 23, 26
Thermoplasma acidophilum 202, *204*
　optimal growth temperature *203*
thermostability 23, 161, 162
　EFs 23
　ribonucleases 211, *212*
thermostability assays, EF proteins *24*, 25
Thermotoga spp.
　temperature optima 23
　T. maritima 202, *204*, 205
　　EF thermostability 23, *24*, 25
　　optimal growth temperature *203*
Thermus spp.
　temperature optima 23
　T. aquaticus, EF thermostability 25
thiamine, role in formation of ethanol 9
three-dimensional protein structure 198
　functional divergence 119
thrombin inhibitors, snake venom 37
thrombin-like enzymes, snake venom 37, 38
time-reversible models 46
TM-601 35
topi, evolutionary tree *213*
topology error 231
Tragelaphus imberbis (lesser kudu) 217
training data-sets, specificity studies 232
transcription, action of steroid receptors 184
transcriptional regulation x

transducin activation, rhodopsin variants 170, *171*
transfection 157
transition state stabilization 161
transition/traversion bias, incorporation into models 49
translation, action of EFs 23
translocations, MCH region 142
transmembrane protein regions
　evolutionary constraints 49
　hydropathy 73–4, *75*
transposable elements (transposons) 28–30
transposase gene replacement 30
transposon activation 159
transposons, inactivated 161
tree-based hidden Markov model (T-HMM) 53
tree constructions, bovine seminal fluid RNase reconstruction 218–19
Treponema pallidum 202, *204*
　optimal growth temperature *203*
TREx (transition redundant exchange) clock 9, 14
Triassic, archosaur origins 168, *169*
tRNAs, nucleotide covariation 74, 76–7
Tropheryma whipplei, amino acid metabolism 129
Tropidechis carinatus 38
tumor suppression/promotion, mouse 30
two-states character models 53
TYB11 vector 157
type I functional divergence 118
　G_q and G_s proteins 120–1
type II functional divergence 118–19
　G_q and G_s proteins 121, *122*, *123*, *124*

uncertainties
　in ancestral corticoid receptor reconstruction 188, 190–1
　in ancestral protein reconstruction 166–8
　archosaur rhodopsin reconstruction 87–8
　incorporation into models 52–3, 71
　measurement by posterior predictive distribution 72
　see also ambiguities
under-sampling, rare variants 88–90

unicellular choanoflagellates, protein domain analysis 106
uniflagellate reproductive state, fungi 107
unknown characters 53
unnatural amino acids 30
Upper Permian, ancestral archosaurs 168, *169*
Urbilateria 145
　ancestral region reconstruction 146–7
　genomic cluster reconstruction 141
　proteome 140
　proto-MCH reconstruction *143*
Ureuchordata
　genomic cluster reconstruction 141, 147
　proto-MCH region *143*, 144, 145
Ursus spelaeus, DNA analysis 129
Uzzell, T. and Corbin, K.W. 47

variability, methods of induction 91–2
venoms
　exendin-4 36
　pharmaceutical benefits 35–6
　snake venom and hemostasis 37–8
Venter, J.C. *et al.* 25
vertebrate ancestor, genomic organization 140
vertebrates
　rhodopsin divergence 87
　rhodopsin phylogeny *170*
viruses, CCR5 as target 73
visual pigment paralogs, functional studies 159–60
visual pigments 23
　archosaur rhodopsin reconstruction 168–72
voltage-gated calcium channel antagonism, ziconotide 35

water buffalo, seminal ribonuclease expression 217
Watson–Crick nucleotide pairs 76
Western-blot hybridization *156*, 157
wobble nucleotide pairs 76
Woese, C.R. 31

Xenopus laevis 143
　GLP-1 37
Xenopus tropicalis

Factor X *38*
GLP-1 *37*
Xylella fastidiosa 202, *204*
  optimal growth temperature *203*

Yang, Z. *et al.* 49, 119
Yang, Z.H. 21, 45, 47, 52
Yang, Z. and Kumar, S., α factor calculation 125
Yap, V.B. and Speed, T. 51

yeast 202, *204*, 205
  alcohol dehydrogenase gene expression 157
  formation of ethanol 9–10
  human domestication 13
  optimal growth temperature *203*
  paralogs *14*
  peroxisome-less mutants 135
  phylogeny *110*

yellow backed duiker (*Cephalophus sylvicultor*), ribonuclease genes 217

Zhang, J. and Nei, M. 119
Zhang, J.Z. and Rosenberg, H.F. 215–16
ziconotide (Prialt) 35
Zuckerkandl, Emile viii
  see also Pauling, L. and Zuckerkandl, E.